INDUSTR APPLICATIONS of MOLECULAR SIMULATIONS

INDUSTRIAL APPLICATIONS of MOLECULAR SIMULATIONS

Edited by
Marc Meunier

CRC Press
Taylor & Francis Group
Boca Raton London New York

CRC Press is an imprint of the
Taylor & Francis Group, an **informa** business

CRC Press
Taylor & Francis Group
6000 Broken Sound Parkway NW, Suite 300
Boca Raton, FL 33487-2742

First issued in paperback 2019

© 2012 by Taylor & Francis Group, LLC
CRC Press is an imprint of Taylor & Francis Group, an Informa business

No claim to original U.S. Government works

ISBN-13: 978-1-4398-6101-1 (hbk)
ISBN-13: 978-0-367-38211-7 (pbk)

Library of Congress Cataloging-in-Publication Data

Industrial applications of molecular simulations / edited by Marc Meunier.
     p. cm.
   Includes bibliographical references and index.
   ISBN 978-1-4398-6101-1 (alk. paper)
   1. Molecules--Models--Industrial applications. 2. Chemistry, Physical and theoretical. I. Meunier, Marc.

QD480.I53 2012
541'.220113--dc23   2011029284

Visit the Taylor & Francis Web site at
http://www.taylorandfrancis.com

and the CRC Press Web site at
http://www.crcpress.com

Contents

Foreword

About a quarter of a century after the first commercial modeling software packages became available to industrial researchers, this collection of recent papers provides a snapshot of applications and serves as a very useful status update. Most of the work has been carried out with the Materials Studio® package, a leading multiscale modeling environment and successor to packages such as Insight and Cerius that were launched in the 1980s. The papers in this volume are a testimony to the success of the application-focused developments that were started back then. About half of the papers are based on classical simulations and demonstrate that properties such as density, solubility, and even glass transition and diffusivity can be determined with a high degree of accuracy. This fact is attributable in no small part to the concerted efforts in application-focused force field developments that began with the first BIOSYM Consortium 25 years ago. Although quantum mechanics took hold in industrial applications a little later, mostly due to the computational cost, density functional and semiempirical methods are now well established and are used in the other half of the papers in this issue.

The field of quantum and molecular simulations has experienced strong growth ever since the time of the early software packages. A recent study showed a strong increase in the number of people publishing papers based on *ab initio* methods, from about 3000 in 1991 to about 20,000 in 2009, with particularly strong growth in East Asia.

Despite such strong evidence for a dynamically evolving field, there are still sometimes doubts about whether much has really changed since those early days and whether molecular simulations deliver value in industrial applications. Looking to the future, the question remains how these methods can be further integrated into the R&D value chain and the gap to engineering and manufacturing bridged. The papers in this volume provide ample evidence of the value and insights into the changes that have happened, and give us pointers to fruitful future developments.

First of all, this collection impressively demonstrates that industrial applications of molecular simulation are well established in a wide range of fields, from catalysis to drug delivery. Although similar research applications may have been studied for some time, closer inspection shows that the scope, complexity, and accuracy have reached a qualitatively and quantitatively different level. Simulations procedures for a range of properties have become more established. Also, computational resources no longer confine researchers to looking at just one or two small systems. It has become the norm to scan across a significant number of complex systems and determine a wider range of properties. As a result, simulations can be used more routinely and with greater confidence, hence delivering much more significant value as screening tools and guidance for experiments. Examples in this volume include the work by Derecskei and Derecskei-Kovacs studying a range of ionic liquids and their compatibility with cellulose derivatives and other compounds, and the work by Maus et al. on drug dosage form development, also based on compatibility, but in this case between drugs and excipients. Similarly, zeolite design for various catalytic applications is advanced by screening a range of substitution ratios in the work by Chatterjee and Chatterjee, and reaction mechanisms are clarified by Andersen et al. studying all of 32 crystallographically unique Brønsted acid sites of a particular zeolite.

Another important aspect is that a number of papers test and establish procedures for particular property prediction or screening tasks. A good example is the paper by Konkel and Myerson that establishes a protocol for preformulation screening of suspension stabilizers. Likewise, the paper by Rai and Pradip establishes a method for the design, development, and selection of surfactants and dispersants for applications such as mineral processing, paints, and coatings. The work by Grima et al. includes a combinatorial scan of polymer networks and establishes design criteria for auxetic materials. Last but not least, new method development and establishing their accuracy remains a key task, as shown by the contributions from Fitzgerald on Fukui functions, as well as Todorova and Delley, who provide examples testing the accuracy of surface wetting calculations with the $DMol^3$-COSMO method.

The above considerations point to potential future developments, particularly which approach is likely to lead to further integration of molecular simulation with engineering and manufacturing. New insights and discovery at the molecular scale will be of increasing importance for industrial applications, as the requirements on the property profile of materials are ever more demanding. However, it will be equally important to establish routine procedures with known accuracy levels and the ability to scan phase space more widely. This opens up the possibility of integrating protocols and data obtained from molecular modeling into the workflows of engineers.

In fact, such workflow-based integration is arguably more important than multiscale modeling. This scale-bridging approach aims to connect to engineering relevant scales starting from the quantum or atomistic level (see Figure 1a). Although establishing proven multiscale methods are important for some phenomena, their significance may be overemphasized, as was also concluded by the report on integrated computational materials engineering. Also in this collection, just two of the papers use simulations at more than one level. Rather, the focus of industrial application is to ascertain selection criteria at the molecular scale that can be determined routinely, efficiently, with sufficient accuracy, and preferably with known error margins. This leads to the concept of workflow integration rather than multiscale integration, as shown in Figure 1b. Simulations at different scales play a role, as appropriate for the task in hand. However, they do not provide the integration pathway. Rather, it remains the task of the scientist and the engineer to define criteria that are both specific and selective to the problem at hand and play to the strengths of the modeling approach.

I am confident that the role of molecular simulations as a key method in research across academia and industry will grow further due to the need to meet the challenges of the 21st century considering health, food, and water as well as energy supplies. Its role in engineering and manufacturing design is set for a significant change, strongly increasing its impact. However, it will require a concerted effort by scientists, engineers, and computer scientists to bring this about. One thing is clear, though, molecular simulation remains a dynamic field.

Gerhard Goldbeck

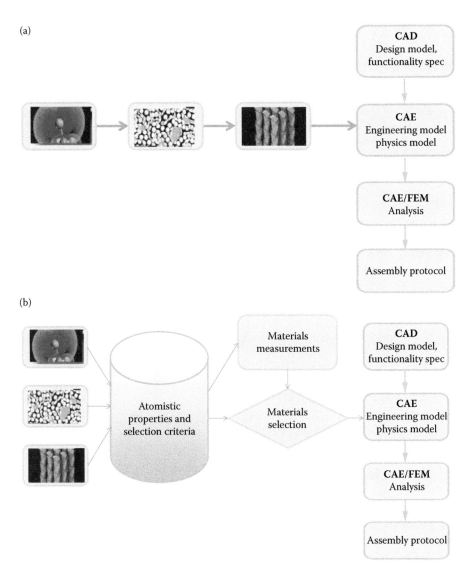

FIGURE 1 (a) Multiscale integration of molecular simulations into engineering. Information is passed from quantum to molecular through mesoscale to engineering simulations. (b) Workflow integration has the definition of key selection criteria for materials at the core. Information from molecular simulation at different levels flows into the decision process.

Preface

Over the last few years, I had the pleasure to edit a couple of special issues of the journal *Molecular Simulation* (Taylor & Francis) on the topic of Industrial Applications of Molecular Simulation (in 2006[1] and in 2008[2] with a focus on Materials Studio).[3] Many authors, from academia, from government research centers or from the industry, have contributed to the success of these special issues. After having received a subsequent number of requests for the manuscripts, the publisher (T&F) invited me to convert both journal publications into a book as it is a better format for distribution. I have then reinvited a short list of authors* who have either resubmitted the original copy of their work or a revised version with new materials or color figures. I, too, have revised my editorials, which I include in this preface.

This book is a collection of scientific papers demonstrating the capability of molecular modeling methods in solving industrial research issues. A wide variety of modeling techniques are presented, ranging from those based on quantum mechanics or so-called *ab initio* methods to those using classical potentials and molecular dynamics. A large range of materials are also studied in this issue, from soft materials such as polymeric blends that are very widely employed in the chemical industry, to hard or inorganic materials such as glasses and alumina as well as organic drug crystals used in the pharmaceutical industry.

In the pharmaceutical research area, modeling and simulation techniques have been extensively used over the last 30 years or so. Some large pharmaceutical companies even have dedicated research centers for drug discovery. In the chemicals, materials, and manufacturing industries, however, the use of computational methods in general (and not only of modeling and simulations) is more recent; companies are equipped with intelligent databases and software packages based on statistical methods that can relatively quickly (and more or less accurately) estimate materials' properties. However, the use of "hard-core" molecular modeling techniques (e.g., *ab initio* or molecular dynamics) in materials science is only growing slowly despite the recent steep increase in computer performance (required by such methods) and the low prices of high-end supercomputers that have allowed the spread of such tools. One exception is academic research centers; they have significantly benefited from these factors and are currently witnessing an increase in the use of modeling tools. Furthermore, most recent commercial modeling software packages contain a wide range of techniques, the so-called "integrated solutions," so that today's modeler can easily access multiple techniques without having to swap from one vendor program and/or computer type to another. This in turn facilitates the development of the so-called "multiscale modeling approach"[4] that is so important for tackling problems as complex as those encountered in materials science.

---

* Due to space constraints, we could not include all the papers from both special issues in this book. I apologize to the authors who did not make the cut. Papers in this book are listed in alphabetical order of the corresponding author.

Thanks to all the advances in molecular modeling techniques, real materials in different conditions and formulations as well as important properties (e.g., polymer membranes and permeation as reported by Schepers and Hofmann[5] and Pelzer and Hofmann[6]) that can be compared to experimental data can nowadays be estimated relatively quickly, allowing the design of new materials (i.e., the so-called "computer-assisted molecular design"). Macromolecular systems are routinely examined using methods based on classical potentials as demonstrated by the work of Derecskei and Derecskei-Kovacs[7,8] on cellulose derivatives. Such large (amorphous) models can be studied with molecular modeling methods due to mathematical tricks such as periodic boundary conditions; in addition, solvent conditions, even for polymers and surfaces, can be included in the models thanks to a conductor-like screening model.[9] Molecular modeling tools can also be used to study complex inorganic materials using empirical pair potentials such as glass substrates as shown by Vyas et al.[10] or to estimate lattice energies and other thermodynamic properties of aluminosilicates as exposed by Vinograd et al.[11] Surface chemistry is a typical topic of study of methods based on quantum mechanics (e.g., Hartree–Fock, DFT). The modeling tools allow the computation of adsorption energy to study, for example, metal deposition on an alumina surface shown by Chatterjee[12] or urea interaction with inorganic surfaces.[13] Simplified *ab initio* methods, such as the so-called semiempirical methods (e.g., MNDO, AM1), exist, which are computationally less demanding without a great loss of accuracy. Such methods were used by Plummer and Cowley[14] to study the decomposition mechanism of hydrogen sulfide, which is relevant to the petrochemical industry. Last but not least, methods based on structure–activity relationships where empirical correlations are drawn between the chemical structure of the models and their activity that are so commonly used in the pharmaceutical area are now finding their way into the "chemical" world. An example of the use of such a tool for the computation of thermodynamics data of oxygen-containing heterocycles is described by Adams et al.[15]

The rise of research funding in the nanotechnology area (and, in particular, in the nanobiology field) is encouraging the development of modeling techniques as these methods are well suited for studying problems (or materials) at the nanoscale level. Additionally, the use of "materials" techniques in the (bio)-pharmaceutical sector to study, for example, drug preformulation and drug delivery systems is also contributing to the spread of such tools in the (materials) industrial arena.

The second special issue of *Molecular Simulation* was a collection of over 50 scientific papers in which the modeling platform Materials Studio (MS) was used to build molecular models, run classical simulations, or perform quantum mechanical calculations.[2,3] The MS package contains many different modeling tools, and this is reflected in the diversity of the articles presented here.

The first section of the issue contains papers related to the use of *ab initio* calculations, using the density functional theory (DFT) method, as well as calculations performed at the semiempirical level of theory as shown by Elliott and Shibuta.[16] The DFT, as implemented in the DMol^3 code[17] seems to be a very popular package among MS users as seen in the number of publications contained in the present edition of the journal (see, e.g., articles on reaction mechanisms by Andersen et al.[18] or Jordaan and Vosloo[19] or on remarkable Fukui functions by Fitzgerald[20]) and in the literature in general.[21] The second section of the issue contains papers related to classical simulations where molecular dynamics is used to determine properties of materials, for example, mechanical properties of the carbon nanotube[22] or of organic networked polymers,[23] diffusion of small gas molecules in amorphous polymers as in the work by Gestoso and Meunier[24], and to predict materials (e.g., drug–polymers) compatibility reported by Maus et al.[25] The third section contains papers using the Monte Carlo (GCMC) technique, where adsorption isotherms in nanoporous materials are computed and the effect of the structure studied. The fourth section contains papers related to the area of crystallography, where analytical instrument simulation tools were used to predict crystal structure[26] or suspension stabilization, as reported by Konkel and Myerson.[27]

MS was first commercially released in June 2000, replacing both Cerius[2] (of MSI) and InsightII (of Biosym) modeling platforms. A decade later, MS contains over 25 modules (each based on techniques such as molecular dynamics, Monte Carlo, and DFT), allowing the development of multiscale modeling approaches required to tackle complex issues encountered in the chemical or pharmaceutical industries. Current model building capabilities are also remarkable in letting users easily construct anything from multiwalled carbon nanotubes, dendrimers, and heterogeneous systems such as layers of molecules and solvents on metal surfaces. Recent additions have been made in the growing area of mesoscale simulation. Such tools can be readily used for the study of complex fluids or to predict the phase diagram of polymer blends as well as their dynamical properties. MS also includes many codes taken from external collaborators or in collaboration with academics, for example, CASTEP[28] or GULP.[29] Some of the recent developments originated from a consortium in nanotechnology that has been running for six years and had members of major chemical companies or renowned academic institutions. New tools, including a linear scaling DFT application (ONETEP),[30] a hybrid QM/MM application (QMera), and a nanoparticle builder came out of the first phase of this consortium. Future developments will see the incorporation of a density functional tight-binding method[31] and a kinetic Monte Carlo tool into MS.

In addition, an extensive scripting application program interface has been included recently to allow users to develop their own scripts inside MS. Users are able to automate tasks and perform customized analysis.

In conclusion, I hope that the current list of selected papers for this book edition will show the readers how molecular modeling tools can be used to tackle research problems relevant to chemical or pharmaceutical industries.

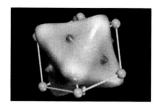

ACKNOWLEDGMENTS

I hope that *Industrial Applications of Molecular Simulations* will contribute to further the dissemination of molecular modeling techniques in the ever-challenging area of materials science in the industrial world. I would like to personally thank all the authors for their contributions and Prof. N. Quirke for his trust in inviting me twice to be guest editor and for his continuous professional support and friendship.

Marc Meunier
Cambridge, U.K.

REFERENCES

1. M. Meunier. Special issue of *Molecular Simulation* on industrial applications of molecular simulation. *Mol. Simul.* **32**, 2006, 2.
2. M. Meunier. Special issue of *Molecular Simulation* on Materials Studio. *Mol. Simul.* **34**, 2008, 10–15.
3. Accelrys Software Inc., San Diego, 2011.
4. S. McGrother, G. Goldbeck-Wood and Y. M. Lam. Integration of modelling at various length and time scales. *Lect. Notes Phys.* **642**, 2004, 223–233.
5. C. Schepers and D. Hofmann. Molecular simulation study on sorption and diffusion processes in polymeric pervaporation membrane materials. *Mol. Simul.* **32**, 2006, 2.
6. S. Pelzer and D. Hofmann. Structure–property relations between silicon containing polyimides and their carbon containing counterparts. *Mol. Simul.* **34**, 2008, 10–15.
7. B. Derecskei and A. Derecskei-Kovacs. Molecular dynamics studies of the compatibility of some cellulose derivatives with selected ionic liquids. *Mol. Simul.* **32**, 2006, 2.
8. B. Derecskei and A. Derecskei-Kovacs. Molecular modelling simulations to predict density and solubility parameters of ionic liquids. *Mol. Simul.* **34**, 2008, 10–15.
9. B. Delley. The conductor-like screening model for polymers and surfaces. *Mol. Simul.* **32**, 2006, 2.
10. S. Vyas, J. E. Dickinson and E. Armstrong-Poston. Towards an understanding of the behaviour of silanes on glass: An atomistic simulation study of glass surfaces. *Mol. Simul.* **32**, 2006, 2.
11. V. L. Vinograd, B. Winkler, A. Putnis, H. Kroll, V. Milman, J. D. Gale and O. B. Fabrichnaya. Thermodynamics of pyrope–majorite, $Mg_3Al_2Si_3O_{12}$–$Mg_4Si_4O_{12}$, solid solution from atomistic model calculations. *Mol. Simul.* **32**, 2006, 2.
12. A. Chatterjee. Comparative behaviour of Pd and Ag deposition phenomenon on clean α-Al_2O_3 (001) surface—A first principle study. *Mol. Simul.* **32**, 2006, 2.
13. A. Singh and B. Ganguly. DFT study of urea interaction with potassium chloride surfaces. *Mol. Simul.* **34**, 2008, 10–15.
14. M. A. Plummer and S. W. Cowley. Chemical mechanisms in hydrogen sulphide decomposition to hydrogen and sulphur. *Mol. Simul.* **32**, 2006, 2.
15. N. Adams, J. Claus, M. Meunier and U. S. Schubert. Predicting thermochemical parameters of oxygen-containing heterocycles using simple QSPR models. *Mol. Simul.* **32**, 2006, 2.
16. J. A. Elliott and Y. Shibuta. A semi-empirical molecular orbital study of freestanding and fullerene-encapsulated Mo nanoclusters. *Mol. Simul.* **34**, 2008, 10–15.
17. B. Delley. An all electron numerical method for solving the local density functional for polytomic molecules *J. Chem. Phys.* **92**, 1990, 508.

18. A. Andersen, N. Govind and L. Subramanian. Theoretical study of the mechanism behind the *para*-selective nitration of toluene in zeolite H-beta. *Mol. Simul.* **34**, 2008, 10–15.
19. M. Jordaan and H. C. M. Vosloo. A DFT computational study of phosphine ligand dissociation versus hemilability in a Grubbs-type precatalyst containing a bidentate ligand during alkene metathesis. *Mol. Simul.* **34**, 2008, 10–15.
20. G. Fitzgerald. On the use of fractional charges for computing Fukui functions. *Mol. Simul.* **34**, 2008, 10–15.
21. http://accelrys.com/products/materials-studio/publication-references/dmol3-references/index.html.
22. Q. Wang. Molecular simulations of in-plane stiffness and shear modulus of double-walled carbon nanotubes. *Mol. Simul.* **34**, 2008, 10–15.
23. J. N. Grima, D. Attard, R. N. Cassar, L. Farrugia, L. Trapani and R. Gatt. On the mechanical properties and auxetic potential of various organic networked polymers. *Mol. Simul.* **34**, 2008, 10–15.
24. P. Gestoso and M. Meunier. Barrier properties of small gas molecules in amorphous *cis*-1,4-polybutadiene estimated by simulation. *Mol. Simul.* **34**, 2008, 10–15.
25. M. Maus, K. G. Wagner, A. Kornherr and G. Zifferer. Molecular dynamics simulations for drug dosage form development: Thermal and solubility characteristics for hot-melt extrusion. *Mol. Simul.* **34**, 2008, 10–15.
26. J. Thun, M. Schoeffel and J. Breu. Crystal structure prediction could have helped the experimentalists with polymorphism in benzamide. *Mol. Simul.* **34**, 2008, 10–15.
27. J. T. Konkel and A. S. Myerson. Empirical molecular modelling of suspension stabilisation with Polysorbate 80. *Mol. Simul.* **34**, 2008, 10–15.
28. M. D. Segall et al. First pinciple simulations: Ideas, illustrations and CASTEP code. *J. Phys.: Condens. Matter* **14**, 2002, 2717–2744.
29. J. D Gale. GULP: A computer program for the symmetry-adapted simulation of solids. *J. Chem. Soc., Faraday Trans.* **93**, 1997, 629–637.
30. C.-K. Skylaris, P. D. Haynes, A. A. Mostofi and M. C. Payne. Introducing ONETEP: Linear-scaling density functional simulations on parallel computers. *J. Phys. Chem.* **122**, 2005, 084119.
31. http://www.dftb.org/.

Editor

Marc Meunier is a fellow at Accelrys, the world leader in scientific informatics software based in Cambridge, U.K. and an adjunct senior research fellow at the Complex and Adaptive Systems Laboratory (CASL), University College Dublin, Ireland. His research interests are in space charges, simulation of polymeric materials, pharmaceutical materials science, and more recently on the growing field of materials informatics. He is an editorial board member of the journal *Molecular Simulation*, and his publications appear in the journals *Chemical Physics*, *Applied Physics*, and *Polymer*.

Contributors

Amity Andersen
William R. Wiley Environmental
   Molecular Sciences Laboratory
Pacific Northwest National Laboratory
Richland, Washington

Daphne Attard
Metamaterials Unit, Faculty of Science
University of Malta
Msida, Malta

Mark Bankhead
National Nuclear Laboratory
Warrington, United Kingdom

Reuben Cauchi
Metamaterials Unit, Faculty of Science
University of Malta
Msida, Malta

Abhijit Chatterjee
Accelrys
Tokyo, Japan

Maya Chatterjee
Research Center for Compact Chemical
   Process
Sendai, Japan

Bernard Delley
Paul Scherrer Institute Switzerland
Villigen, Switzerland

Bela Derecskei
Lyondell Chemicals
Research Center Baltimore
Glen Burnie, Maryland

Agnes Derecskei-Kovacs
Lyondell Chemicals
Research Center Baltimore
Glen Burnie, Maryland

James A. Elliott
Department of Materials Science and
   Metallurgy
University of Cambridge
Cambridge, United Kingdom

James F. Ely
Chemical Engineering Department
Colorado School of Mines
Golden, Colorado

George Fitzgerald
Accelrys
San Diego, California

Bishwajit Ganguly
Analytical Science Discipline
Central Salt and Marine Chemicals
   Research Institute Bhavnagar
Gujarat, India

Ruben Gatt
Metamaterials Unit, Faculty of Science
University of Malta
Msida, Malta

Patricia Gestoso
Accelrys
Cambridge, United Kingdom

William A. Goddard III
Beckman Institute
California Institute of Technology
Pasadena, California

Niranjan Govind
William R. Wiley Environmental
    Molecular Sciences Laboratory
Pacific Northwest National Laboratory
Richland, Washington

Joseph N. Grima
Metamaterials Unit, Faculty of Science
University of Malta
Msida, Malta

Roger Guilard
Institute of Molecular Chemistry
University of Burgundy
Dijon, France

Sergey Gusarov
National Institute for Nanotechnology
National Research Council of Canada
Edmonton, Alberta, Canada

Dieter Hofmann
Center for Biomaterial Development
Institute of Polymer Research
Teltow, Germany

Margaritha Jordaan
Catalysis and Synthesis Research Group
North-West University
Potchefstroom, South Africa

Jamie T. Konkel
Global Research and Development
Baxter Healthcare
Round Lake, Illinois

Andreas Kornherr
Mondi Uncoated Fine Paper
Ulmerfeld-Hausmening, Austria

Andriy Kovalenko
National Institute for Nanotechnology
National Research Council of Canada
Edmonton, Alberta, Canada
and
Department of Mechanical Engineering
University of Alberta
Edmonton, Alberta, Canada

Bin Liu
Chemical Engineering Department
Colorado School of Mines
Golden, Colorado

Mark T. Lusk
Physics Department
Colorado School of Mines
Golden, Colorado

Martin Maus
Boehringer Ingelheim Pharma GmbH
    & Co. KG
Biberach, Germany

Marc Meunier
Accelrys
Cambridge, United Kingdom

Allan S. Myerson
Department of Chemical and Biological
    Engineering
Illinois Institute of Technology
Chicago, Illinois

Scott L. Owens
National Nuclear Laboratory
Warrington, United Kingdom

Silke Pelzer
Center for Biomaterial Development
Institute of Polymer Research
Teltow, Germany

Pradip
Tata Research Development &
    Design Centre
Pune, India

Pluton Pullumbi
Claude-Delorme Research Center
Air Liquide
Jouy-en-Josas, France

Beena Rai
Tata Research Development &
    Design Centre
Pune, India

Sabyasachi Sen
Department of Chemical Engineering &
    Materials Science
University of California, Davis
Davis, California

Yasushi Shibuta
Department of Materials Engineering
The University of Tokyo
Tokyo, Japan

Ajeet Singh
Analytical Science Discipline
Central Salt and Marine Chemicals
    Research Institute Bhavnagar
Gujarat, India

Olivier Siri
Interdisciplinary Nanoscience Center
    of Marseille
Campus de Luminy
Marseille, France

Kevin J. Smith
Department of Chemical and Biological
    Engineering
University of British Columbia
Vancouver, British Columbia, Canada

Stanislav R. Stoyanov
National Institute for Nanotechnology
National Research Council of Canada
Edmonton, Alberta, Canada

Lalitha Subramanian
Accelrys
San Diego, California

Alain Tabard
Institute of Molecular Chemistry
University of Burgundy
Dijon, France

Theodora Todorova
Paul Scherrer Institute Switzerland
Villigen, Switzerland

Adri C.T. van Duin
Beckman Institute
California Institute of Technology
Pasadena, California

H.C. Manie Vosloo
Catalysis and Synthesis Research
    Group
North-West University
Potchefstroom, South Africa

Shyam Vyas
National Nuclear Laboratory
Warrington, United Kingdom

Karl G. Wagner
Department of Pharmaceutical
    Technology
University of Tuebingen
Tuebingen, Germany
and
Boehringer Ingelheim Pharma GmbH
    & Co. KG
Biberach, Germany

Abraham Q. Wang
Department of Mechanical and
    Manufacturing Engineering
University of Manitoba
Winnipeg, Manitoba, Canada

Sharif F. Zaman
Department of Chemical and Biological
    Engineering
University of British Columbia
Vancouver, British Columbia, Canada

Victor Zammit
Metamaterials Unit, Faculty of Science
University of Malta
Msida, Malta

Gerhard Zifferer
Department of Physical Chemistry
University of Vienna
Vienna, Austria

1 Theoretical Study of the Mechanism behind the *Para*-Selective Nitration of Toluene in Zeolite H-Beta*

Amity Andersen, Niranjan Govind, and Lalitha Subramanian

CONTENTS

1.1 INTRODUCTION

Nitration of aromatic compounds such as toluene is important for the synthesis of small-molecule precursors for further use in fine chemicals such as dyes, pharmaceuticals, perfumes, plastics, and explosives.[1–3] Typically, it is the *para* nitration product of monosubstituted benzene compounds that is the most desirable.[4] The traditional aromatic nitration method that is still widely used today is the homogeneous

* A version of published article: *Molecular Simulation* **34** (10–15), 2008, 1025–1039.

solution-phase catalysis of the aromatic compound by nitric acid species in a mixture of substrate, nitric acid, and sulfuric acid.[5] However, the nitration of aromatics such as toluene is not very selective (toluene nitration of the *ortho, meta,* and *para* is about 40%, 3%, and 57%, respectively)[5] and often leads to overnitration and oxidized by-products and the creation of environmentally unfriendly acid waste that is expensive to treat. More recently, a number of promising catalysts have been proposed as "green" alternatives to the traditional method of nitration. These include lanthanide triflate,[6] Nafion-H and other polysulfonic acid resins,[7] bismuth subnitrate and thionyl chloride,[8] montmorillonite clay-supported copper(II) nitrate,[9] and various zeolites (e.g., mordenite,[10] ZSM-11,[11] ZSM-5,[12] zeolite beta,[4] and faujasite[13]).

Of these catalysts, the solid acid zeolites are among the most promising. The advantages of zeolites include easy removal of substrate and product, regioselectivity, operation under relatively mild conditions, and recyclability. The most promising zeolite for the *para*-selective nitration of toluene is H-beta zeolite. The suitability of H-beta as a selective toluene nitration catalyst was shown in the pioneering synthesis work of Smith et al.[4] In their synthesis method, H-beta zeolite was first combined with concentrated nitric acid (90%) followed by acetic anhydride (Ac_2O) to effectively create acetyl nitrate and acetic acid. The acetic anhydride was also added in excess to undergo hydrolysis and effectively remove water molecules that may otherwise poison the acid sites of the H-beta. After adding acetic anhydride, toluene was then added to undergo nitration. The nitration of toluene was quantitative (>99%), and the resulting nitration yielded up to 81% *para*-nitrotoluene. Decreasing the temperature from 30°C to 50°C resulted in a slight increase in the *para* product. Having a stoichiometric ratio of Ac_2O to HNO_3 gave optimal *para* selectivity. Increasing the amount of beta zeolite with respect to the Ac_2O and HNO_3 reagents increased the *para* product. Vacuum distillation proved to be the best means of restoring H-beta to a near-pristine state after one round of toluene nitration. The recycled H-beta could undergo a few more rounds of toluene nitration before diminished nitration ability was seen. High *para* selectivity seemed to be unique to beta. Experiments with H-mordenite, H-ZSM-5, and H-Y did not yield as high a *para* selectivity as observed for H-beta.[4]

Following the work of Smith et al., Prins and coworkers applied liquid- and solid-state magic angle spinning nuclear magnetic resonance (MAS NMR) to Ac_2O/HNO_3 liquid and in beta to try to elucidate the mechanism for beta zeolite's unique *para* selectivity.[14–16] Prins and coworkers suggested that the source of beta zeolite's high selectivity toward *para* selectivity was due to steric hindrance imposed by the acetyl group of acetyl nitrate bound to the octahedral (i.e., Lewis acid) aluminum sites of the zeolite cage via the nitrate group (this mechanism is depicted in Figure 1.1). However, this explanation of beta's selectivity is tentative because they observed octahedral aluminum ^{27}Al MAS NMR nitrate peak disappearance upon the addition of acetic anhydride to the nitric acid-treated beta zeolite. A broad octahedral aluminum peak persisted, but Prins and coworkers did not have a definitive evidence that the octahedral peaks were due to acetyl nitrate binding or to other species binding such as acetic anhydride, acetic acid, or residual nitrate. Some tetrahedral aluminum recovery was observed. Partial Al–O–Si hydrolysis has been proposed for the mechanism responsible for the appearance of octahedral aluminum in the steam

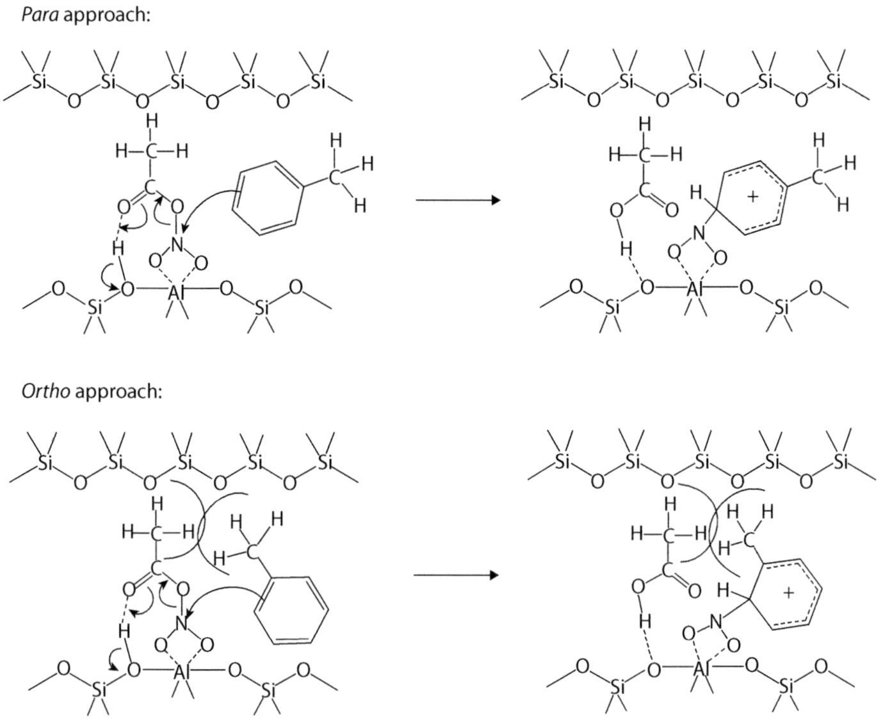

Para approach:

Ortho approach:

FIGURE 1.1 Scheme to explain favorable *para* nitration according to the work of Prins and coworkers.[14,15] Note that steric hindrance from the acetyl and toluene methyl group interaction is proposed to prevent toluene nitration in the *ortho* approach case.

treatment of beta and other zeolites and aluminosilicates.[17–26] Because acetic acid is able to donate another proton to an Al–O–Si linkage (as with nitric acid[18]), it seems that the acetate ion conjugate base would be a more appropriate species for octahedral coordination of aluminum compared to the acetyl nitrate molecule. Therefore, the origin of beta's unique *para* selectivity still has yet to be explained satisfactorily.

There is also a diffusional component to beta's *para* selectivity. Beta has a three-dimensional network of large 12-T pore channels (shown in the two polymorph structures of Figures 1.2 and 1.3), which are the most likely diffusion paths for large molecular species such as toluene and nitrotoluene. Such a network of large channels allow toluene reagent and nitrotoluene product to readily diffuse in and out of the beta zeolite structure. Selective nitration of toluene has been shown to occur in high-surface-area microcrystalline beta, not macrocrystalline beta.[27] This indicates that diffusion into the micropore system is required for selective *para* nitration to occur. A shape-selection explanation was also explored to explain the *para* selectivity where the pore network allowed for excellent diffusion of reagents into the network and a possible restriction of *ortho* species diffusion out of the system. The *ortho* species have a geometry that may make their diffusion more restrictive. However, Smith et al. and Prins and coworkers reported almost quantitative recovery of the nitrotoluene

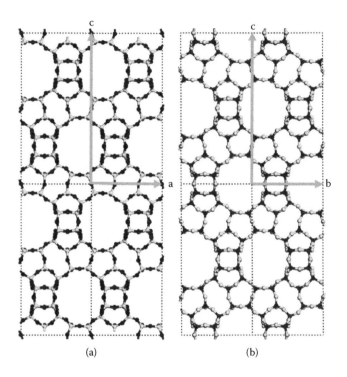

FIGURE 1.2 Polymorph A ac plane view (left). Polymorph A bc plane view (right). The dark balls are oxygen atoms, and the light balls are silicon atoms.

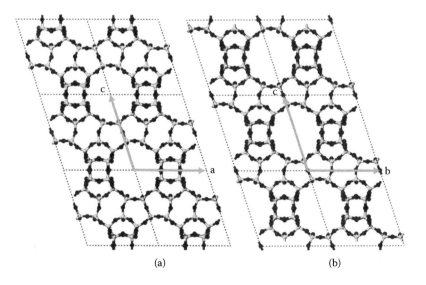

FIGURE 1.3 Polymorph B ac plane view (left). Polymorph B bc plane view (right). The dark balls are oxygen atoms, and the light balls are silicon atoms.

product, regardless of the isomer shape. Thus, the *para* selectivity of beta zeolite seems to be an intrinsic property of this zeolite that is not completely explained by diffusion properties. Other zeolites with large 12-T channel "plumbing" (e.g., mordenite) do not show the same *para* selectivity that beta exhibits. It is the aim of this study to provide in-depth mechanistic details of *para* nitration of toluene by *in situ* acetyl nitrate in H-beta zeolite via atomistic and electronic structure insight.

1.2 THEORETICAL METHOD

1.2.1 DENSITY FUNCTIONAL THEORY CALCULATIONS

All-electron periodic density functional theory (DFT) calculations were performed using the commercial version (Accelrys' Materials Studio® 4.0[28]) of DMol[3].[29,30] The generalized gradient approximation exchange-correlation functional of Perdew, Burke, and Ernzerhof was used for all DFT calculations.[31] All calculations employed the double numerical local atomic basis set including a polarization *d*-function on all nonhydrogen atoms and a polarization *p*-function on all hydrogen atoms. The atomic radius cutoff was set at 4.2 Å. Because the zeolite models considered here are large (>96 atoms), calculations at the Γ-point were sufficient. The energy convergence tolerance was set to 1×10^{-5} Ha. The geometry convergence tolerance was set to 2×10^{-5} Ha, and the force and displacement convergence thresholds were set to 4×10^{-3} Ha/Å and 5×10^{-3} Å, respectively. Thermal smearing was used to aid convergence and was set to 0.005 Ha. The delocalized internal option was invoked for geometry optimization.[32] For transition state searches along proposed reaction pathways, the generalized synchronous transit approach was used to locate saddle points along a path anchored by the reactant and product minima.[33]

1.2.2 STRUCTURE OF BETA ZEOLITE

Pure silica and aluminosilicate beta zeolite consists of an intergrowth of two polymorphs, A and B, having approximately a 40:60 ratio of the two polymorphs.[34] The ratio of A and B polymorphs can vary with crystallite size.[35] Figure 1.2 shows the structure of polymorph A, and Figure 1.3 shows the structure of polymorph B. The polymorphs of beta differ primarily in the stacking of tertiary building units in the *z*-direction. Polymorph A is characterized as a racemic mixture of two enantiomeric structures, having either left or right helical stacking of building units at the (001) plane about a fourfold screw axis. The enantiomeric space group pair for polymorph A is *P*4₁22 and *P*4₃22. On the other hand, polymorph B is achiral in character and has the space group *C*2/*c*. Newsam et al.[34] also described a hypothetical polymorph C with higher symmetry (space group *P*42/*mmc*), which is not found in the pure silica and aluminosilicate varieties of beta zeolite. However, germanosilicate varieties of beta zeolite bearing the polymorph C structure have been synthesized recently.[36] Interestingly, the disordered stacking sequences of the polymorphs do not greatly affect the accessible pore volume. However, the stacking disorder greatly determines the tortuosity of the 12-T pore system along the *z*-direction.[34] Incidentally, stacking faults can occur at the (001) stacking plane, creating Lewis acid-type undercoordinated aluminum sites and silanol functionality.[37]

In this work, we chose to use the beta primitive polymorph B cell lattice in all our calculations. The primitive cell of pure silica beta polymorph B has 96 atoms, which is a little less than half the size of the cell lattice of pure silica polymorph A (206 atoms). Thus, the beta polymorph B model is the least expensive beta model from a computational standpoint. Moreover, polymorph B tends to be present in slightly higher amounts compared with polymorph A (polymorph B being about 62.5%).[34] The lattice parameters of the primitive monoclinic beta polymorph B cell are $a = b = 12.66$ Å, $c = 14.33$ Å, $\alpha = \beta = 107.24°$, and $\Gamma = 90.06°$ (constructed from the conventional cell described in Newsam et al.[34]). These cell parameters were held fixed during structure optimization, but the atom positions were allowed to relax.

1.2.3 LOCATION OF BRØNSTED ACID SITES

The approach for studying the Brønsted acid sites in H-beta is similar to that of Hill et al.[38] and Govind et al.[39] For polymorph B of beta zeolite, one silicon atom in a pure silica beta B model was replaced with an aluminum atom, and the resulting positive charge deficit was compensated with a proton at one of the oxygen atoms coordinating the aluminum to ensure neutrality. This protocol was repeated for all crystallographically unique Brønsted acid sites in the original pure silica beta B model. According to the original paper by Newsam et al.[34] detailing the structure of beta zeolite, all polymorphs of pure silica beta have nine crystallographically unique silicon atoms. These nine silicon atoms have been labeled in the beta polymorph B shown in Figure 1.4. Because each silicon is coordinated by four oxygen atoms, there are $9 \times 4 = 36$ possible Brønsted acid sites. However, the silicon atoms labeled with numbers 8 and 9 in the work of Newsam et al. have only two unique oxygen atoms.[34] Thus, there are only 32 unique Brønsted acid sites. All 32 of these Brønsted acid site beta structures were optimized with DFT as detailed in Section 1.2.1. Relative energies for the optimized 32 structures were established based on the total energy of the optimized structure bearing the lowest total energy.

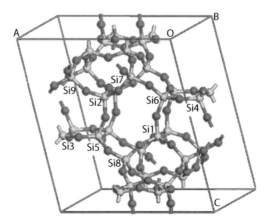

FIGURE 1.4 The nine crystallographically unique silicon atoms of beta polymorph B. Note that atom assignments in beta polymorph A are similar.

1.2.4 NITRATION OF TOLUENE BY ACETYL NITRATE AND THE BETA BRØNSTED ACID SITE

Once a likely Brønsted acid site was determined from the above protocol, a number of acetyl nitrate orientations around the acid site were generated and optimized. The lowest energy orientation was chosen to study the reaction of acetyl nitrate with toluene. The toluene was placed close to the acetyl nitrate NO_2 moiety for two scenarios: electrophilic *para* attack and electrophilic *ortho* attack. The H-beta zeolite, acetyl nitrate, and toluene models were then optimized to give the "reactant" minima for subsequent transition state searches. The mechanism for the reactant, intermediate, and product minima and transition states for *para* and *ortho* toluene nitration were partially inspired by the mechanistic work of Olah and coworkers[1,40] and the work of Silva and Nascimento[41] for the nitration of benzene in the gas phase and via formyl nitrate and a 5-T (pentameric) cluster carved from beta zeolite, respectively. From these minima, the appropriate transition states were found using the transition state algorithm described earlier to build potential energy profiles for the nitration of toluene by acetyl nitrate in the presence of the beta acid site catalyst.

1.3 RESULTS AND DISCUSSION

1.3.1 ALUMINUM AND BRØNSTED ACID SITES

Table 1.1 lists the relative energies based on our calculated DFT total energies for the 32 crystallographically unique Brønsted acid sites. The distribution of relative energies is fairly narrow (6.0 ± 3.5 kcal/mol), but a little wider than the 1.4 kcal/mol range previously reported by Pápai et al.[42] from DFT calculations on pentameric cluster models. Despite the location of the Brønsted acid site in a strained 4-T ring, the Al1–O3 acid site appears to be the most stable among the acid sites (see Figure 1.5 for an illustration of this site). Moreover, it is interesting that all four coordinated oxygen atoms around the Al1 site have fairly low energies relative to Al1–O3. The Al1–O2 site is only 2.0 kcal/mol higher than the Al1–O3 site. This is an important feature of this site, which will be explored shortly in the reaction mechanism proposed for nitration of toluene. Note also that much of the stability of sites Al1–O3 and Al–O2 is due to the noticeable intracage hydrogen bonding. These Brønsted acid sites are similar to the lowest energy sites Fujita et al.[43] have found in the oval-shaped 6-T rings of polymorph A using electronic structure methods.

Though nine unique silicon sites can be discerned from Fyfe et al.'s ^{29}Si MAS NMR spectra for highly siliceous beta, the conspicuous trimodal character of these spectra suggests that these nine unique silicon sites can be classified into three general categories.[44] An in-depth theoretical analysis of the early findings of Fyfe et al. was performed by Valerio et al.[45] using sum-over-states density functional theory to simulate NMR shifts. Their assumption of three primary categories for the nine unique Al site types based on inclusion/exclusion of the Al sites in 4-T rings (i.e., whether the Al is included in two, one, or zero 4-T rings) proved to agree well with the experiment. From their work, it was found that sites 3–6 comprise the group of T sites contained in two 4-T ring systems. Sites 7–9 comprise the group of T sites not

TABLE 1.1

Brønsted Acid Site Energies (in kcal/mol) for All Possible Sites in Beta B Polymorph

Aluminum Site[a]	Brønsted Site[a]	Relative Energy[b]
Al1	O1	5.8
	O2	2.0
	O3	0.0
	O4	4.1
Al2	O1	10.0
	O5	3.5
	O6	4.1
	O7	5.7
Al3	O8	6.2
	O9	5.0
	O10	7.4
	O11	5.6
Al4	O8	4.1
	O11	2.2
	O12	5.4
	O13	6.7
Al5	O5	3.9
	O9	6.6
	O14	5.1
	O15	10.6
Al6	O2	3.1
	O12	6.0
	O14	3.6
	O16	9.5
Al7	O4	12.4
	O6	3.8
	O10	4.0
	O16	7.9
Al8	O3	2.9
	O15	5.6
Al9	O7	15.0
	O13	7.3

[a] Numbering based on crystallographic ordering specified by Newsam et al.[34]

[b] Relative energy with respect to the lowest energy site found in this study (in this case, Al1–O3).

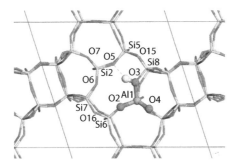

FIGURE 1.5 The most stable Brønsted acid site (Al1–O3) for the beta polymorph B system. Note that Al1–O2 and Al1–O4 are also relatively low in energy (2.0 kcal/mol and 4.1 kcal/mol, respectively).

contained in any 4-T ring system. Finally, sites 1 and 2 comprise the group of T sites contained in only one 4-T ring moiety. van Bokhoven et al.[17] have done further ^{27}Al MAS NMR studies of beta zeolite undergoing steam dealumination and found that sites 1 and 2 are the most stable aluminum sites in the structure, resisting dealumination up to 600°C. Moreover, they found that all sites, but sites 1 and 2, can readily undergo a tetrahedral to octahedral coordination change under steam conditions. This change in coordination is reversible under mild steaming conditions. This octahedral aluminum coordination has also been observed in the addition of concentrated nitric acid to beta zeolite, which is the first step in the synthesis of nitrotoluene in beta.[14,18,27]

The experimental ^{27}Al MAS NMR work of Abraham et al. on stepwise calcined H-beta[19] has shown Si/Al ratio dependence for Al site occupation, suggesting a potential nonrandom process for Al site selection (likely due to the crystallization kinetics process). Their experiments also suggest that two adjacent Al sites (obeying Löwenstein's rule of no Al–O–Al linkages[46]) are required for the hydrolysis of Al–O–Si linkages and, therefore, octahedral coordination of Al sites. Our results for aluminum site occupation in our model H-beta system agree with the observation of Abraham et al. that, for high Si/Al ratio (i.e., Si/Al = 110 and 215) or extremely low aluminum content, octahedral aluminum sites are absent. That is, the most stable aluminum sites will tend to be sites 1 and 2, which do not demonstrate an octahedral coordination shift of the aluminum on calcination, steaming, or, in the case of aromatic nitration, nitric acid treatment. However, the lack of octahedral species may also be due to a lack of close aluminum site pairs. The presence of aluminum acid site pairs in beta zeolite has also been proposed by Bortnovsky et al.[47] based on Fourier transform infrared spectroscopy with acetonitrile molecular probe experiments on Co^{2+}-exchanged beta. Indeed, the presence of two close-by Al sites is an appealing zeolite feature for divalent transition metal exchange ions (which require a twofold counterion compensation) for NO_x selective catalytic reduction catalysts, as has been explored recently in the theory works of Fujita et al.[43] and Fischer et al.[48]

Though aluminum site pairs like those seen in the work of Abraham et al. are likely to be prevalent with the Si/Al = ~13 beta used in the nitration experiments of Prins and coworkers[14,16,18,27] and Smith et al.,[4] our proposed mechanism for the nitration of toluene with the acetyl nitrate nitration agent in H-beta depends on a single acid site. Though our present mechanism adequately explains the basic mechanism for *para*-selective nitration, we are exploring the formation of the acetyl nitrate nitration agent and subsequent toluene nitration in an aluminum acid site pair model for an upcoming study.[49]

1.3.2 Adsorption of Acetyl Nitrate at Brønsted Acid Sites

We first searched for a nitrate–aluminum bound acetyl nitrate structure as proposed by Prins and coworkers. Prins and coworkers have used chemical anisotropy broadening arguments in their ^{15}N NMR results to claim that the NO_2 moiety is bound to the aluminum site.[14,18,27] We, however, were unsuccessful in finding such a species with our Al1–O3 acid site model. As mentioned previously, Al sites 1 and 2 are not as flexible as Al sites 3–9.[19] Thus, stable nitrate coordination with Al1 would not be expected. More of an interaction between the nitrate moiety of the acetyl nitrate and Al sites 3–9 may be possible because these sites participate in octahedral coordination because of their flexibility, which we are currently pursuing.[49] Prins and coworkers have noted that ZSM-5 also exhibits similar broadening, but octahedral aluminum coordination occurs negligibly in ZSM-5.[27] Thus, there appears to be more than one way to explain the broadening they observed. We show that chemical

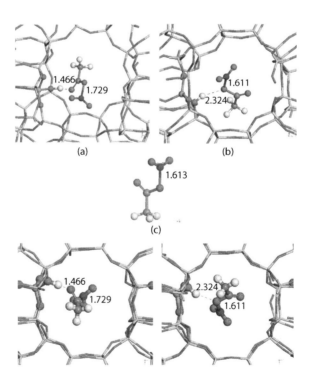

FIGURE 1.6   (a) Carbonyl oxygen hydrogen bonding of acetyl nitrate, (b) ester linkage oxygen hydrogen bonding of acetyl nitrate, and (c) gas phase acetyl nitrate. Distances and bond lengths are in angstroms.

anisotropy (i.e., distortion of the nitrate moiety) can occur through means other than nitrate binding to Al. For the acetyl carbonyl oxygen binding at the proton site shown in Figure 1.6, the optimized structure shows a sizeable elongation of the N–O (ester linkage) bond (1.729 Å). The hydrogen bonding of the zeolite cage proton and the carbonyl oxygen is short (1.466 Å). Both the long N–O (ester linkage) bond and the short H (cage)–O (carbonyl) hydrogen bond suggest that this mode of acetyl nitrate binding is going to be the most amenable to toluene nitration. That is, the N–O (ester linkage) will readily rupture to form a "nitronium" moiety to attack toluene, and the transfer of the cage proton to the acetyl carbonyl oxygen to form acetic acid will be facile.

We also considered the scenario where the ester linkage oxygen of acetyl nitrate interacts with the proton site. This manner of acetyl nitrate interaction was considered in the mechanism proposed by Silva and Nascimento[41] for the nitration of benzene by formyl nitrate in their DFT calculation on a pentameric cluster model. In this scenario, the distortion of the N–O (ester linkage) bond is not as pronounced (1.611 Å), and the hydrogen bond between the ester linkage oxygen and the zeolite proton appears weaker (2.324 Å). Thus, the scenario involving the acetyl carbonyl interaction with the zeolite cage proton may be more in line with the observation of anisotropy. Moreover, this interaction scenario is the most stable from the total energy calculations performed here (by 10.1 kcal/mol).

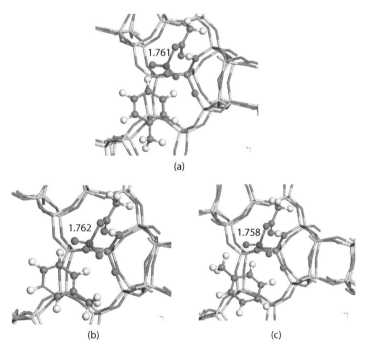

FIGURE 1.7 **(See color insert.)** Reactant structures (acetyl nitrate and toluene) in side cross section of 12-T pore around Brønsted acid site for (a) *para*, (b) *ortho* 1, and (c) *ortho* 2 attack orientations of toluene. Distance and bond lengths are in angstroms.

1.3.3   *Para* versus *Ortho* Nitration of Toluene in Beta

Para and *ortho* nitration was considered around the Al1 site. As mentioned earlier, the three oxygens surrounding Al1 in the 12-T pore wall are the most stable Brønsted acid sites, with the Al1–O3 site being the lowest in energy. The Al1–O3 is assumed to be protonated at the beginning of the series of reactions leading to toluene nitration with the carbonyl oxygen of the acetyl nitrate hydrogen bonding with the proton of the Al1–O3 site. However, there are likely to be other potential Brønsted acid sites. The calculations considered one *para* site and the two possible *ortho* sites on toluene. Figure 1.7 shows the starting point for the reactants, toluene and acetyl nitrate, in our calculations.

Nitration of the toluene appears to be a three-step process, as demonstrated with DFT calculations of benzene nitration in the gas phase by Olah and coworkers.[40] The acetyl nitrate readily transfers a "nitronium"-like moiety to the π-system of the toluene close to the *para* or *ortho* sites. This is a π-complex with no real bonding interaction. An sp^2 to sp^3 hybridization of the *para* or *ortho* carbon must then occur to create a σ-complex (aka Wheland intermediate or arenium cation). Finally, this σ-complex must transfer the proton from the *para* or *ortho* site to a Brønsted acid site in the zeolite wall. Upon optimizing the H-beta structure with the acetyl nitrate site and toluene, it became clear that the most likely acid transfer site after creation of the σ-complex is the Al1–O2 site in our model, which, as we already

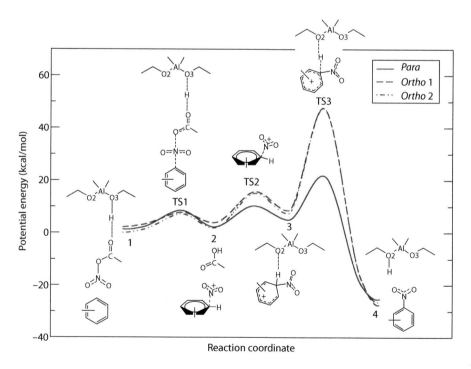

FIGURE 1.8   Potential energy plots for the *para* and two *ortho* nitrations according to the calculations done here.

TABLE 1.2
Energetics (in kcal/mol) for the *Para* and *Ortho* Electrophilic
Nitration by Acetyl Nitrate in the Polymorph B H-Beta
Structure at the Acid Site

Attack Site	Para	Ortho-1	Ortho-2
	Reaction Energetics		
1→2	1.0	1.6	−0.9
2→3	2.7	4.5	5.6
3→4	−30.3	−36.0	−33.5
	Transition State Barriers		
1→TS1	7.2	5.4	4.4
2→TS2	7.8	11.5	13.0
3→TS3	16.6	38.9	40.0

Note: See Figure 1.8 for corresponding product, intermediate, and transition states.

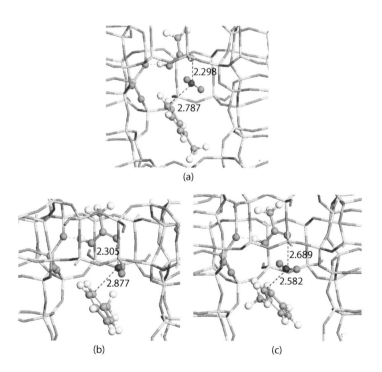

FIGURE 1.9 The transition state for the "nitronium" transfer from the acetyl nitrate to the toluene π-system for (a) *para*, (b) *ortho* 1, and (c) *ortho* 2 attack. Distances and bond lengths are in angstroms.

mentioned previously, has comparable stability to Al1–O3. The presence of two nearby Brønsted acid sites that participate in the necessary proton transfer steps is an important feature of the selectivity mechanism of our model. A similar cluster model of two oxygens with one aluminum acid site was proposed by Silva and Nascimento[41] in their mechanism for benzene nitration by formyl nitrate. Two nearby Brønsted acid sites may be achieved with a nearby aluminum pair model (Al–O–Si–O–Al in accordance with Löwenstein's rule). We are currently studying this model.[49]

Figure 1.8 shows the reaction path for the nitration of toluene in the beta zeolite pore system and for the three-site nitration scenarios. Table 1.2 shows the energetics for the three reactions and the barriers to reaction. The energetics have not been corrected for temperature, zero point vibrational energies, or solvent effects. However, such corrections are not expected to change the mechanistic message of our work. Figure 1.9 shows the structures of the transition state for the *para* and two *ortho* nitration scenarios, and Figure 1.10 shows the structures of the optimized π-complex intermediates for the *para* and two *ortho* nitration scenarios. The C–N distances in the three π-complexes of our calculations are between the 1.997 Å and 2.434 Å values reported by Esteves et al.[40] for the addition of NO_2^+ to benzene with two different NO_2^+ rotamer configurations in the gas phase calculated at the DFT-B3LYP/6-31++G** level of theory. Figure 1.8 shows that the addition of a "nitronium" moiety to the toluene system from the acetyl nitrate is almost the same for all three nitration site scenarios and requires little energy.

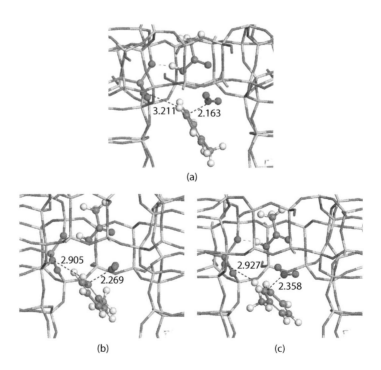

FIGURE 1.10   (See color insert.) The π-complex for the (a) *para*, (b) *ortho* 1, and (c) *ortho* 2 attack. Distances and bond lengths are in angstroms.

However, the second step where the π-complex converts to a σ-complex appears to be energetically prohibitive for the *ortho* nitration scenarios compared with the *para* nitration scenario. The barrier to π–σ conversion is 4 kcal/mol–5 kcal/mol for *para* nitration compared with the two *ortho* nitration scenarios. The *para* nitration reaction for the π–σ shift is 2 kcal/mol–3 kcal/mol more exothermic compared with the two *ortho* nitration cases. Figure 1.11 shows the transition state structures for the *para* and two *ortho* nitration scenarios. The reason for this difference between the *para* and *ortho* cases is likely due to an increased steric hindrance between the toluene methyl moiety and the zeolite pore wall in the *ortho* cases. In Figure 1.12, the *para* nitration intermediate (Figure 1.12a) is closer to the closest open acid site (Al1–O2) with the sp³ hydrogen of the σ-complex at the *para* site 0.350 Å closer than that of the *ortho* 1 nitration case (Figure 1.12b) and 0.267 Å closer than that of the *ortho* 2 nitration case (Figure 1.12c). Moreover, the C–N bond length of the σ-complex is shorter in the *para* case by 0.157 Å and 0.016 Å compared with the *ortho* 1 and *ortho* 2 nitration cases, respectively. These observations from Figure 1.12a through c suggest that the σ-complex requires stabilization by the closest open acid site (Al1–O2) of the zeolite pore wall, and the methyl moiety of the toluene hinders this stabilization in the two *ortho* site nitration scenarios. The C–N distances in the three σ-complexes of our calculations are in reasonable agreement with the 1.512, 1.587, and 1.747 Å values reported by Esteves et al.[40] for the addition of NO_2^+ to benzene

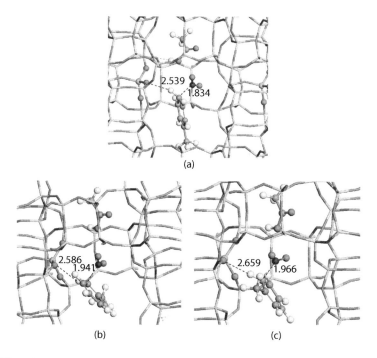

FIGURE 1.11 The transition state from the π-complex to the σ-complex for (a) *para*, (b) *ortho* 1, and (c) *ortho* 2 attack. Distances and bond lengths are in angstroms.

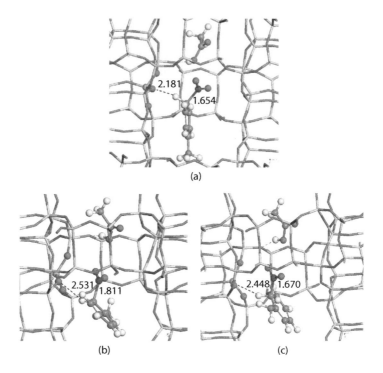

FIGURE 1.12   The σ-complex for (a) *para*, (b) *ortho* 1, and (c) *ortho* 2 attack. Distances and bond lengths are in angstroms.

with three different NO_2^+ rotamer configurations in the gas phase calculated at the DFT-B3LYP/6-31++G** level of theory.

The steric hindrance of the methyl moiety of the toluene for the *ortho* site nitration is even more profound in the final reaction with the transfer of the sp³ hydrogen of the σ-complex to the open acid site of the zeolite cage (Al1–O2). This final step has a barrier to reaction that is 22 kcal/mol–23 kcal/mol higher in the *ortho* cases compared with that of the *para* case. However, the reaction is 4 kcal/mol–6 kcal/mol more exothermic in the *ortho* cases compared with the *para* case. Figure 1.13 shows the distances that the transferring "proton" is from open acid site (Al1–O2) and the nitration site of the toluene in each of the transition state structures of the three nitration scenarios (the final product structures are shown in Figure 1.14). As expected, the distances for the "proton" in the *para* case are noticeably shorter compared with those of the *ortho* scenarios [by 0.145 Å and 0.426 Å for the H–O2(Al1) distance in the *ortho* 1 and *ortho* 2 cases, respectively, and 0.140 Å and 0.296 Å for the H–C in the *ortho* 1 and *ortho* 2 cases, respectively]. This suggests that the steric hindrance of the methyl moiety of the toluene in the *ortho* nitration cases is making the "proton" transfer of the final step difficult compared with the *para* nitration case. From our analysis, it appears that the *ortho* nitration of toluene becomes increasingly difficult with the second step (conversion from the π to σ complex) and third step ("proton" transfer from the nitration site to the open acid site in the zeolite pore

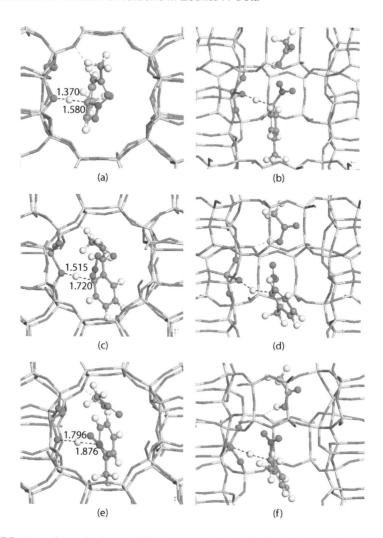

(a)

(b)

(c)

(d)

(e)

(f)

FIGURE 1.13 **(See color insert.)** TS3 for *para* nitration (a) viewing down the 12-T pore and (b) viewing a side cross section of the 12-T pore, *ortho* 1 nitration (c) viewing down the 12-T pore, and (d) viewing a side cross section of the 12-T pore, and *ortho* 2 nitration (e) viewing down the 12-T pore and (f) viewing a side cross section of the 12-T pore. Distances shown are in angstroms.

wall) primarily because of the toluene methyl–zeolite pore wall steric hindrance. Our mechanism appears to be at odds with the mechanism that Prins and coworkers proposed (illustrated in Figure 1.1) to explain H-beta's *para*-selective nitration of toluene.[14] They propose that the acetyl methyl group provides the steric hindrance required for the *para* selectivity and not the zeolite pore walls. Conceivably, chemical kinetics experiments exploring the isotopic effect of deuterated H-beta could be used to give credence to the mechanism presented here.

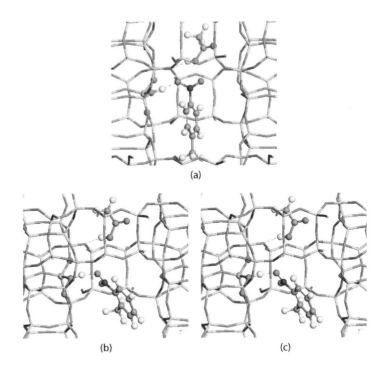

FIGURE 1.14   The nitration product resulting from (a) *para*, (b) *ortho* 1, and (c) *ortho* 2 attack.

1.4   CONCLUSION

Through DFT calculations, we have explored the likely Brønsted acid sites of H-beta polymorph B and applied this work to exploring the origin of H-beta's propensity toward high *para* selection in the nitration of toluene by acetyl nitrate. According to our calculations, assuming a single Brønsted acid site, the likely mode of binding of the acetyl nitrate is via the carbonyl oxygen of the acetyl moiety rather than via octahedral coordination of Al with two nitrate oxygens or the ester linkage oxygen between the acetyl and nitrate moieties. A stable octahedral coordination of the Al was not found with the theory methods used here, and the ester linkage oxygen interaction with the acid site proton was weaker than that of the carbonyl interaction. Moreover, the carbonyl interaction with the cage proton resulted in a longer O–N bond at the ester linkage compared with that of the ester linkage oxygen interaction. This suggests a more labile NO_2^+ moiety for electrophilic attack of the toluene π-system.

With this mode of acetyl nitrate interaction with the acid site proton, further calculations were done to determine stable minima and transition states for the nitration of toluene at the *para* site and both of the *ortho* sites. Although the energetics suggest that addition of the "nitronium" moiety to the toluene to form a "π-complex" appears to be equally likely at all three electrophilic attack sites, the stabilization of the "σ-complex" and the subsequent proton transfer to a zeolite acid site to ultimately form nitrotoluene are more favorable in the *para* attack case than it is for either of the two *ortho* attack cases. In the *ortho* attack cases, steric hindrance caused by the interaction of the methyl group of the toluene with the wall of the zeolite 12-T pore system appears to prevent the stabilization of the σ-complex slightly and the final transfer of the complex's proton to one of the acid site oxygens greatly (the barrier to reaction in either of the *ortho* cases is nearly 2.5 times greater than that of the *para* case). With this, we have shown that a more likely mechanism for the *para* selectivity in beta zeolite involves the steric hindrance of the toluene methyl moiety with the zeolite cage wall, not the steric hindrance from the acetyl methyl moiety as suggested by Prins and coworkers.

We also acknowledge that the favorable diffusion within microcrystalline beta and the highly flexible framework of beta as well as an appropriate Si/Al ratio (e.g., ~13) and distribution of Al sites must also be partially credited with this zeolite's selective nature in the nitration of monosubstituted benzene compounds. Evidence for this has been provided by the work of Prins and coworkers who have shown a decrease in selectivity on considerable dealumination (i.e., decrease in acid sites to give a high Si/Al ratio) of beta.[16] The favorable diffusion provided by the large 12-T pore channels allows for transport of the reagents to the internal acid sites and for transport of the products out of the pore system. The flexible framework possibly allows for prolonged sequestering of the nitrate ions at the Lewis acid sites (i.e., effectively trapping nitrate ions at octahedral aluminum acid sites). This may allow for creation of the acetyl nitrate close to the internal acid sites for subsequent reaction with toluene. The flexible acid sites may also create traps for acetic acid species (cage-bound acetate ions at octahedral Lewis acid aluminum sites and cage-bound protons at nearby Brønsted acid sites) that prevent protons from participating in nonselective nitration through homogeneous free acid species in the large pore channels. Prins and coworkers very briefly mentioned that trapping of acetic acid by-product may lead to selectivity but did not give any mechanism on how this may lead to selectivity.[14] We believe that the dual Lewis and Brønsted acid character of many aluminum sites in beta may lead to anchoring both the acetate anion and the proton from acetic acid onto the cage structure. With very few or no "free" acetic acid protons in the spacious 12-T channels, there is very little unselective solution-phase proton transfer from unbound acetic acid taking place. Thus, only heterogeneous proton transfer with the zeolite cage can take place, which is where our mechanism for sterically restrictive selectivity comes into play. We will be exploring this in our future work on this elusive zeolite system.

ACKNOWLEDGMENTS

The authors thank Accelrys, Inc. for the computational resources to perform this work. J. T. Hynes is also acknowledged for his comments regarding potential hydrogen isotopic experiments that could be explored to determine the validity of our mechanism.

REFERENCES

1. G. A. Olah, R. Malhotra and S. C. Narang. *Nitration: Methods and Mechanisms*. New York: VCH, 1989.
2. K. Schofield. *Aromatic Nitration*. Cambridge: Cambridge University Press, 1980.
3. R. Taylor. *Electrophilic Aromatic Substitution*. Chichester: John Wiley & Sons, 1990.
4. K. S. Smith, A. Musson and G. A. DeBoos. A novel method for the nitration of simple aromatic compounds. *J. Org. Chem.* **63**, 1998, 8448–8454.
5. L. G. Wade. *Organic Chemistry*, 3rd edition. Englewood Cliffs, NJ: Prentice-Hall, 1995.
6. F. J. Waller, A. G. M. Barrett, D. C. Braddock and D. Ramprasad. Lanthanide(III) triflates as recyclable catalysts for atom economic aromatic nitration. *Chem. Commun.* 1997, 613–614.
7. G. A. Olah, R. Malhotra and S. C. Narang. Aromatic substitution. 43. Perfluorinated resinsulfonic acid catalyzed nitration of aromatics. *J. Org. Chem.* **43**, 1978, 4628–4630.
8. H. A. Muathen. Selective nitration of aromatic compounds with bismuth subnitrate and thionyl chloride. *Molecules* **8**, 2003, 593–598.
9. A. Cornélis, L. Delaude, A. Gerstmans and P. Laszlo. A procedure for quantitative regio-selective nitration of aromatic hydrocarbons in the laboratory. *Tetrahedron Lett.* **29**, 1988, 5909–5912.
10. K. Smith, K. Fry, M. Butters and B. Nay. *Para*-selective mononitration of alkylbenzene under mild conditions by use of benzoyl nitrate in the presence of a zeolite catalyst. *Tetrahedron Lett.* **30**, 1989, 5333–5336.
11. S. M. Nagy, K. A. Yarovoy, M. M. Shakirov, V. G. Shubin, L. A. Vostrikova and K. G. Ione. Nitration of aromatic compounds with benzoyl nitrate on zeolites. *J. Mol. Catal.* **64**, 1991, L31.
12. T. J. Kwok, K. Jayasuriya, R. Damavaparu and B. W. Brodman. Application of H-ZSM-5 zeolite for regioselective mononitration of toluene. *J. Org. Chem.* **59**, 1994, 4939–4942.
13. T. Esakkidurai, M. Kumarraja and K. Pitchumani. Regioselective nitration of aromatic substrates in zeolite cage. *Proc. Indian Acad. Sci.* **115**, 2003, 113–121.
14. M. Haouas, S. Bernasconi, A. Kogelbauer and R. Prins. An NMR study of the nitration of toluene over zeolites by HNO_3–AC_2O. *Phys. Chem. Chem. Phys.* **3**, 2001, 5067–5075.
15. S. Bernasconi. Liquid phase nitration of toluene and 2-nitrotoluene using acetyl nitrate: Zeolite BEA as *para*-selective catalyst. PhD thesis, Swiss Federal Institute of Technology Zurich, 2003.
16. S. Bernasconi, G. D. Pirngruber and R. Prins. Influence of the properties of zeolite BEA on its performance in the nitration of toluene and nitrotoluene. *J. Catal.* **224**, 2004, 297–303.
17. J. A. van Bokhoven, D. C. Koningsberger, P. Kunkeler, H. van Bekkum and A. P. M. Kentgens. Stepwise dealumination of zeolite beta at specific T-sites observed with ^{27}Al MAS and ^{27}Al MQ MAS NMR. *J. Am. Chem. Soc.* **122**, 2000, 12842–12847.
18. M. Haouas, A. Kogelbauer and R. Prins. The effect of flexible lattice aluminium in zeolite beta during the nitration of toluene with nitric acid and acetic anhydride. *Catal. Lett.* **70**, 2000, 61–65.
19. A. Abraham, S.-H. Lee, C.-H. Shin, S. B. Hong, R. Prins and J. A. van Bokhoven. Influence of framework silicon to aluminium ratio on aluminium coordination and distribution in zeolite beta investigated by ^{27}Al MAS and ^{27}Al MQ MAS NMR. *Phys. Chem. Chem. Phys.* **6**, 2004, 3031–3036.
20. J. A. van Bokhoven, A. M. J. van der Eerden and D. C. Koningsberger. Three-coordinate aluminum in zeolites observed with *in situ* x-ray absorption near-edge spectroscopy at the Al K-edge: Flexibility of aluminum coordinations in zeolites. *J. Am. Chem. Soc.* **125**, 2003, 7435–7442.

21. J. A. van Bokhoven, D. C. Koningsberger, P. Kunkeler and H. van Bekkum. Influence of steam activation on pore structure and acidity of zeolite beta: An Al K edge XANES study of aluminum coordination. *J. Catal.* **211**, 2002, 540–547.

22. A. Omegna, J. A. van Bokhoven and R. Prins. Flexible aluminum coordination in alumino-silicates. Structure of zeolite H-USY and amorphous silica–alumina amorphous silica–alumina. *J. Phys. Chem. B* **107**, 2003, 8854–8860.

23. A. Omegna, M. Haouas, A. Kogelbauer and R. Prins. Realumination of dealuminated HZSM-5 zeolites by acid treatment: A re-examination. *Microporous Mesoporous Mater.* **46**, 2001, 177–184.

24. A. Omegna, M. Vasic, J. A. van Bokhoven, G. Pirngruber and R. Prin. Dealumination and realumination of microcrystalline zeolite beta: An XRD, FTIR and quantitative multinuclear (MQ) MAS NMR study. *Phys. Chem. Chem. Phys.* **6**, 2004, 447–452.

25. A. Omegna, R. Prins and J. A. van Bokhoven. Effect of temperature on aluminum coordination in zeolites H-Y and H-USY and amorphous silica–alumina: An *in situ* Al K edge XANES study. *J. Phys. Chem. B* **109**, 2005, 9280–9283.

26. C. Jia, P. Massiani and D. Barthomeuf. Characterization by infrared and nuclear magnetic resonance spectroscopies of calcined beta zeolite. *J. Chem. Soc., Faraday Trans.* **89**, 1993, 3659–3665.

27. S. Bernasconi, G. D. Pirngruber, A. Kogelbauer and R. Prins. Factors determining the suitability of zeolite BEA as *para*-selective nitration catalyst. *J. Catal.* **219**, 2003, 231–241.

28. Materials Studio, Version 4.0, Accelrys Software, Inc., San Diego, 2005.*

29. B. J. Delley. An all-electron method for solving the local density functional for polyatomic molecules. *J. Chem. Phys.* **92**, 1990, 508–517.

30. B. J. Delley. From molecules to solids with the DMol^3 approach. *J. Chem. Phys.* **113**, 2000, 7756–7764.

31. J. P. Perdew, K. Burke and M. Ernzerhof. Generalized gradient approximation made simple. *Phys. Rev. Lett.* **77**, 1996, 3865–3868.

32. J. Andzelm, R. D. King-Smith and G. Fitzgerald. Geometry optimization of solids using delocalized internal coordinates. *Chem. Phys. Lett.* **335**, 2001, 321–326.

33. N. Govind, M. Petersen, G. Fitzgerald, D. King-Smith and J. Andzelm. A generalized synchronous transit method for transition state location. *Comput. Mater. Sci.* **28**, 2003, 250–258.

34. J. M. Newsam, M. M. J. Treacy, W. T. Koetsier and C. B. de Gruyter. Structural characterization of zeolite beta. *Proc. R. Soc. London, Ser. A* **420**, 1988, 375–409.

35. B. Mihailova, V. Valtchev, S. Mintova, A.-C. Faust, N. Petkova and T. Bein. Interlayer stacking disorder in zeolite beta family: A Raman spectroscopic study. *Phys. Chem. Chem. Phys.* **7**, 2005, 2756–2763.

36. A. Corma, M. T. Navarro, F. Rey, J. Rius and S. Valencia. Pure polymorph C of zeolite beta synthesized by using framework isomorphous substitution as a structure-directing mechanism. *Angew. Chem., Int. Ed.* **40**, 2001, 2277–2280.

37. J. C. Jansen, E. J. Creyghton, S. L. Njo, H. van Koningsveld and H. van Bekkum. On the remarkable behavior of zeolite beta in acid catalysis. *Catal. Today* **38**, 1997, 205–212.

38. J. Hill, C. M. Freeman and B. Delley. Bridging hydroxyl groups in faujasite: Periodic vs cluster density functional calculations. *J. Phys. Chem. A* **103**, 1999, 3772–3777.

39. N. Govind, J. Andzelm, K. Reindel and G. Fitzgerald. Zeolite-catalyzed hydrocarbon formation from methanol: Density functional simulations. *Int. J. Mol. Sci.* **3**, 2002, 423–434.

---

* Three-dimensional graphical images appearing in this work were generated using the Materials Studio visualizer.

40. P. M. Esteves, J. W. de M. Carneiro, S. P. Cardoso, A. G. H. Barbosa, K. K. Laali, G. Rasul, G. K. S. Prakash and G. A. Olah. Unified mechanistic concept of electrophilic aromatic nitration: Convergence of computational results and experimental data. *J. Am. Chem. Soc.* **125**, 2003, 4836–4849.

41. A. M. Silva and M. A. C. Nascimento. A DFT study of nitration of benzene by acyl nitrate catalyzed by zeolites. *Chem. Phys. Lett.* **393**, 2004, 173–178.

42. I. Pápai, A. Goursot, F. Fajula and J. Weber. Density functional calculations on model clusters of zeolite-beta. *J. Phys. Chem.* **98**, 1994, 4654–4659.

43. H. Fujita, T. Kanougi and T. Atoguchi. Distribution of Brønsted acid sites on beta zeolite H-BEA: A periodic density functional theory calculation. *Appl. Catal., A* **313**, 2006, 160–166.

44. C. A. Fyfe, H. Strobl, G. T. Kokotailo, C. T. Pasztor, G. E. Barlow and S. Bradley. Correlations between lattice structures of zeolites and their ^{29}Si MAS N.M.R. spectra: Zeolites KZ-2, ZSM-12 and beta. *Zeolites* **8**, 1988, 132–136.

45. G. Valerio, A. Goursot, R. Vetrivel, O. Malkina, V. Malkin and D. R. Salahub. Calculation of ^{29}Si and ^{27}Al MAS NMR chemical shifts in zeolite-beta using density functional theory: Correlation with lattice structures. *J. Am. Chem. Soc.* **120**, 1998, 11426–11431.

46. W. Löwenstein. The distribution of aluminum in the tetrahedra of silicates and aluminates. *Am. Mineral.* **39**, 1954, 92–96.

47. O. Bortnovsky, Z. Sobalík and B. Wichterlová. Exchange of Co(II) ions in H-BEA zeolites: Identification of aluminum pairs in the zeolite framework. *Microporous Mesoporous Mater.* **46**, 2001, 265–275.

48. G. Fischer, A. Goursot, B. Coq, G. Delahay and S. Pal. Theoretical study of N_2O reduction by Co in Fe–BEA zeolite. *Chem. Phys. Chem.* **7**, 2006, 1795–1801.

49. A. Andersen, N. Govind and E. J. Bylaska, in progress.*

---

* *Note:* Around the time this work was accepted for publication, our preliminary DFT results show that aluminum sites at the flexible 3–9 positions will not bind with the nitrate moiety of acetyl nitrate to form octahedrally coordinated aluminum. However, nitrate and acetate moieties will bind to these sites to form octahedral aluminum. These promising results further support our mechanism, which does not rely on an Al-bound acetyl nitrate for *para*-selective nitration.

2 Computational Designing of Gradient-Type Catalytic Membrane
Application to the Conversion of Methanol to Ethylene

Abhijit Chatterjee and Maya Chatterjee

CONTENTS

2.1 INTRODUCTION

Zeolites are nanometer-sized crystalline porous materials. Their small pore size and the possibility of having different chemical compositions can give rise to a very selective interaction with adsorbed molecules depending on size, shape, and chemical compositions. A zeolite's surface can be changed from hydrophobic to hydrophilic and nonacidic to acidic simply by substitution of Si by Al, which significantly affects structural property-related phenomena such as diffusion. During the last 10 years, interest in developing thin zeolite films or zeolite membranes has grown enormously [1,2]. Zeolite membranes are generally used as a catalytic membrane reactor, as it can combine separation [3–7] and catalytic activity [8] efficiently due to the significantly different diffusivities in the uniform, molecular-sized pores and the presence of its catalytic sites. The selective sorption properties with their catalytic activity and thermal stability make zeolite as an ideal candidate for the inorganic catalytic membrane. Most of the zeolite membranes reported in the literature are of the mobil five (MFI) type because of their pore architecture, which consists of straight channels interconnected by zigzag channels. This amazing pore structure leads to highly anisotropic diffusivity, and MFI membranes were widely used in gas separation [9] and catalytic reactors [10,11]. Silicalite is an MFI-type zeolite comprising only Si and O, is hydrophobic and inert, and is used in hydroisomerization [12], pervaporation [13], etc.

Despite the enormous success of the zeolite membrane, practical application in a larger scale is still limited. Several factors such as cost of membrane development, reproducibility, long-term stability, and the method for the preparation of the defect-free membrane restrict its implementation in the industry. Molecular simulations become a powerful tool to predict the catalytic behavior of zeolite [14]. Compared to the experimental process, it is rapid and convenient, is cost effective, can handle more complex systems within a reasonable period of time, and results to better understanding of the system. Many computer simulation methodologies have been employed to understand the physicochemical properties of zeolite such as adsorption characteristics [15], diffusion and permeation [16], catalytic reaction [17], and also the nature of the acidic site [18,19]. The main concern of this work is to design a membrane using computer simulation methodology.

A gradient-type membrane is defined as a membrane when there is a regular or graded ascent or descent in terms of any measurable parameter such as adsorption and diffusion. The most important aspect of a gradient membrane is its compatibility with its constituents, which can be measured by the diffusivity as well as from the mismatch of the components. Experimentally, the synthesis of a zeolitic gradient-type membrane is very critical and challenging due to surface matching. Moreover, complete interpretation of adsorption and diffusion processes of guest molecules in zeolites only by experiments is quite difficult or impossible. Thus, computer simulation is highly desirable to get the fundamental insight into the nature of the adsorption and diffusion behavior of the zeolite and the adsorbate system.

Here, we will describe the designing of functionally gradient material for a catalytic membrane reactor depending on adsorption and diffusion with varying Si/Al ratio to find the mismatch. To test the efficiency of the designed material, transition state and diffusivity calculations were performed to track the feasibility of the reaction and to monitor the diffusion path of the reactant and product to validate the design of the gradient-type membrane, respectively.

2.2 COMPUTATIONAL METHODOLOGY

Ab initio total energy pseudopotential geometry optimization calculations of zeolite unit cell were performed using Cambridge Serial Total Energy Package (CASTEP) and associated programs of Accelrys, with imposed periodic boundary conditions [20,21]. For the calculation, the unit cell of varying Si/Al ratio followed by a combination of high Al and low Al content was taken. Nonlocal ultrasoft pseudopotentials [22] and Becke–Perdew parameterization [23,24] of the exchange correlation functional including gradient correction generalized gradient approximation were used. The k-point set used in a calculation defines the accuracy of the Brillouin zone sampling. The Monkhorst–Pack [25] k-points used in CASTEP are characterized by divisions along three reciprocal space axes and by an optional origin shift. Thus, the k-point setup procedure simply involves selecting three integer divisions. The quality of the k-point set can be quantified in a number of ways; CASTEP uses the distance between the points in reciprocal space as a numerical measure. Sampling of reciprocal space has been performed using $2 \times 2 \times 2$ k-point in the Brillouin zone with kinetic energy cutoff of 350 eV using the Monkhorst–Pack scheme [25]. The entries 2, 2, and 2 are R_i, R_j, R_k, which are the fractions of the reciprocal space lattice vector that give the value of the offset in the three directions. ZSM-5 (silicalite) exists in three distinct forms: monoclinic, *Pnma*, and $P2_12_12_1$ orthorhombic. We have used a unit cell with *Pnma* symmetry containing 288 atoms and the starting composition was $O_{192}Si_{96}$, where Si atoms were substituted by Al to change the Si/Al ratio. The charge after Al replacement is saturated by attaching the hydrogen atom at the bridging oxygen connecting the Si and Al. All the calculations related to the transition state were performed with density functional theory (DFT) [26] using $DMol^3$ [27–29] code of Accelrys, DNP basis set [30], and BLYP exchange correlation functional [31,32]. The transition state calculations were performed using the synchronous transit methods as included in the $DMol^3$ module of Accelrys. Complete linear synchronous transit (LST)/quadratic synchronous transit (QST) begins by performing an LST/optimization calculation. The transition state approximation obtained in that way is used to perform a QST maximization. From that point, another conjugate gradient minimization is performed. The cycle is repeated until a stationary point is located or the number of allowed QST steps is exhausted. $DMol^3$ uses the nudged elastic band (NEB) method for minimum energy path calculations. The NEB method introduces a fictitious spring force that connects neighboring points on the path to ensure continuity of the path and projection of the force so that the system converges to the minimum energy path, which is also called an intrinsic reaction coordinate if the coordinate system is mass weighted. We also calculated the vibration mode to identify the negative frequency to confirm the transition state.

The "Free Volume" routine within the "Visualizer" capability of Materials Studio® was used to generate and calculate internal surfaces. The surfaces are created by systematically (algorithmically) passing a spherical probe-atom through the periodic structures (and using periodic boundary conditions). The centroid of the probe-atom is used to generate the surface rather than the actual "contact points" between the probe-atom and the framework atoms. There are three distinct computational modes within the Free Volume routine: "Total," "Occupiable," and "Accessible." The Total mode does not require the use of a probe-atom and calculates the internal surface area by summing the surfaces of all the atoms (represented by hard spheres) that reside within a particular unit cell. Thus, the internal surface area for each and every structure, on a molar basis, is exactly the same. We want to generate solvent-accessible surfaces (i.e., surfaces that are functions of the framework's relative geometry, porosity, and spatial arrangement of atoms) and the Total mode was not used in this study. Materials Studio characterizes spatial domains in model spaces as (a) "occupiable" or "not occupiable," and (b) "accessible" or "inaccessible."

Absolute adsorption isotherms were computed using a grand canonical Monte Carlo (GCMC) calculation algorithm, which allows displacements (translations and rotations), creations, and destructions. The GCMC method was convenient to simulate adsorption [33,34] and carried out by the Sorption module in Materials Studio software of Accelrys, which can execute a series of fixed pressure simulations at a set temperature in a single step. At the start of the Sorption run, a fixed pressure simulation is performed for a specified number of steps between the start and end fugacity set. In the default setup, the start fugacity of each component is increased linearly to the end fugacity with an equidistant step. However, it is also possible to regulate those steps to space logarithmically; in that case, the logarithm of the start fugacity of each component is increased linearly to the logarithm of the end fugacity in equidistant steps (the fugacity thus increases exponentially). Sorption result reports the sum of component fugacities as the total fugacity. This is equal to the pressure if the reservoir is an ideal gas. The equilibration and production steps were set to 1 million and 1 billion Monte Carlo moves, respectively. We have saved the frames for every 1000 steps. Materials Studio database provides various types of force fields. A cvff_aug force field was applied in this study. The zeolite structure was assumed to be rigid during the sorption process. This assumption is not as drastic because the flexibility of the lattice more significantly influences the diffusion properties [35].

2.3   RESULTS AND DISCUSSIONS

To design a gradient-type membrane, the model structure of ZSM-5 with different Si/Al ratio such as 1, 10, 30, 60, 90, and 300 and only silica were generated. An all-silica model is shown in Figure 2.1. We have optimized the individual bulk structure of one unit cell with CASTEP and the Si/Al ratio of 300 was chosen to mimic a situation of very low Al content. It has to be mentioned that for a gradient-type membrane, the compatibility of its component, which means the combination of two MFI layers with varied Si/Al ratio, is an integral part to achieving a stable membrane.

2.3.1   AB INITIO FIRST PRINCIPLE CALCULATION

Ab initio first principle calculation was performed to find out the mismatch of the component from the best combination of two bulk structures with different Si/Al ratios, which will be energetically stable. We have first optimized the individual zeolite structure with a specific Si/Al ratio and then designed a layer with the two possible matrices. Possible combinations of low and high Si/Al ratio were tried and then the high Si/Al ratio was fixed to 300. After that, the other entire low Si/Al ratios (1, 10, 30, 60, and 90) are combined with the high Si/Al ratio (300) and each of these combined structures with high Si/Al and low Si/Al ratio was optimized for the mismatch and stabilization of the structure. The stabilization energy was calculated using the following formula:

$$E_{stabilization} = E_{complex} - [E_{component1} + E_{component2}],$$

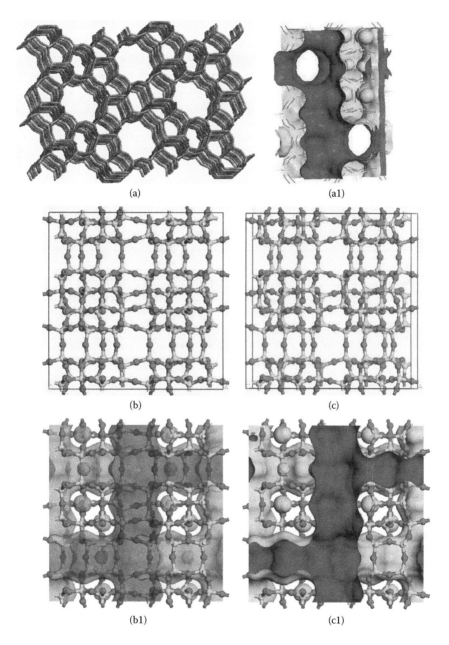

(a)   (a1)

(b)   (c)

(b1)   (c1)

FIGURE 2.1   Structural model of MFI (a) all-silica, (b) Si/Al = 30, and (c) Si/Al = 300. Void volume and occupied volume of MFI (a1) all-silica, (b1) Si/Al = 30, and (c1) Si/Al = 300, with void shown in darker gray and occupied volume shown in white.

TABLE 2.1

Stabilization Energy and Percentage Mismatch for Different Combinations of Zeolite Matrix with Varying Si/Al Ratio

Membrane Matrix with Different Si/Al Ratio Combination[a]	Stabilization Energy (kcal/mol)	Percentage Mismatch
0/300	−6.19	19.23
1/300	−15.2	12.14
10/300	−22.56	6.56
30/300	−53.34	1.86
60/300	−23.23	12.47
90/300	−5.19	21.67

[a] We have tested here the lowest and highest Si/Al ratios.

where $E_{complex}$ is the total energy of the complex resulting from the combination of two different sets of Si/Al ratio. $E_{component1}$ and $E_{component2}$ are the energies of the optimized individual zeolite framework used to make the complex. According to the stabilization energy, it was possible to predict the energetically most stable combination depending on the Si/Al ratio. Table 2.1 shows the stabilization energy and percentage mismatch of the different combinations of Si/Al ratio. The highest and the lowest stabilization energy of −6.19 kcal/mol and −53.34 kcal/mol were obtained for 90/300 and 30/300 combination, respectively. Hence, in terms of stabilization energy, Si/Al 30 and Si/Al 300 were chosen as the best combination for a gradient-type membrane. A reasonable relative energy difference of ~30 kcal/mol was obtained between the most stable composition (Si/Al 300 and Si/Al 30) and others (Table 2.1). To calculate the mismatch, a little quench was observed in the structures with higher Al content. However, Si/Al 300 and Si/Al 30 exhibit a mismatch of less than 2%. Thus, considering the stability and mismatch, an optimum combination of Si/Al 300 (hydrophobic) and Si/Al 30 (hydrophilic) was obtained for developing the gradient type of membrane. Figure 2.1b and 2.1c show the distribution of the Al centers within the membrane matrix of the Si/Al 300 and Si/Al 30 structure, respectively. Particular cell parameters and the cell volumes were compared to find the best match among the candidates.

2.3.2 ADSORPTION PROCESS

This contains two steps: first, we look into the zeolite membrane to calculate the void volume accessible to the guest molecules or, in other words, to calculate the internal surface area of the individual zeolite components to be used for the gradient-type zeolitic membrane. To do this, we have taken the all-silica zeolite of MFI type and exchanged the Si with Al to a desired Si/Al ratio. Then the void volume and occupied volume were calculated along with the surface area for all the structures as shown in Figure 2.1a all-silica, Figure 2.1b Si/Al 300, and Figure 2.1c Si/Al 30. The results are shown in the adjacent figures. The void and occupied volume are tabulated in Table 2.2. The results show that Si/Al ratio 30 has the largest free volume and all Si has

TABLE 2.2
Void Volume, Occupied Volume, and Surface Area for the Zeolite Matrix of MFI Type with Varying Si/Al Ratio

Composition of MFI Zeolite	Void Volume $(\text{Å})^3$	Occupied Volume $(\text{Å})^3$	Surface Area $(\text{Å})^2$
All Si	1809.10	3522.92	1361.96
Si/Al = 300	1810.33	3521.69	1363.04
Si/Al = 30	1813.47	3518.55	1363.19

the lowest void. The trend is similar for surface area values, but the occupied volume trend is opposite as expected, justifying the role of aluminum.

Adsorption and diffusion of the guest molecule in the zeolite host strongly influence the product distribution of catalytic reaction. For the above-mentioned gradient-type membrane, the conversion of methanol to ethylene was chosen as the target reaction. Hence, the adsorption behavior of methanol (reactant) and ethylene (product) on Si/Al 30 and Si/Al 300 was studied. Figure 2.2a and 2.2b (methanol) and Figures 2.2c and 2.2d (ethylene) show adsorption isotherms for each adsorbed species at 298 K using standard GCMC methods. From the results, it was encouraging to note that the loading of methanol (reactant) and ethylene (product) on the membrane varies greatly with Si/Al ratio. Methanol shows maximum loading at higher Al content part, whereas ethylene barely has any effect. It is well known that the low Si/Al ratio (high Al content) of zeolite prefers polar molecule to adsorb, whereas at high Si/Al ratio (low Al content), as the surface property of zeolite changes from hydrophilic to hydrophobic, the adsorption affinity toward the polar molecule decreased [36]. For instance, the hydrophilic part of the membrane, which is associated with high Al content, was more polar, adsorbed methanol easily, and hindered the diffusion, whereas ethylene can pass easily. Hence, the difference in loading of the reactant and the product depending on Si/Al ratio could be a perfect gradation to design a gradient-type zeolite membrane. With regard to the conversion of methanol to ethylene, methanol needs to be adsorbed in the zeolite matrices followed by the formation of ethylene, which could be separated easily because of its smooth diffusion through the membrane. Thus, considering the adsorption results of methanol and ethylene, the best possible combination will be hydrophilic (high Al content) entrance through which methanol could adsorb and ethylene formed can pass easily through the hydrophobic (low Al content) environment, which proposes its easy separation after the reaction. The energy distribution calculation shows that the adsorption energy of ethylene remains lower compared with that of methanol throughout the process, and it is even lowest in the higher Si/Al ratio where methanol shows stronger adsorption behavior.

2.3.3 TRANSITION STATE CALCULATIONS

The conversion of methanol to ethylene consists of the dehydration of methanol on acidic zeolite with the formation of an equilibrium mixture containing methanol and

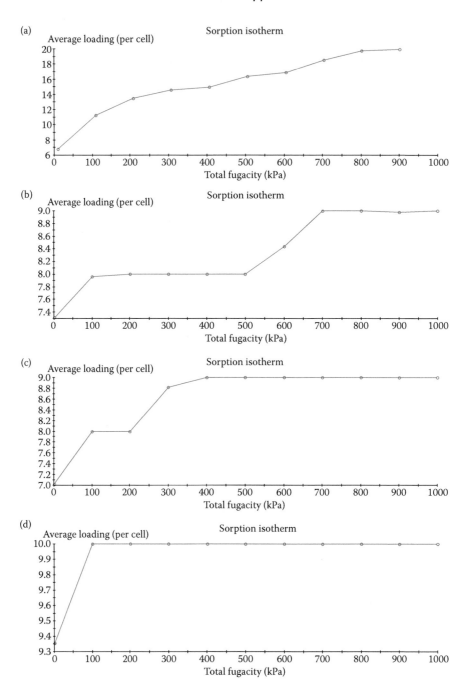

FIGURE 2.2   Adsorption isotherm of the reactant methanol (a) Si/Al = 30, (b) Si/Al = 300, and the product ethylene (c) Si/Al = 30, (d) Si/Al = 300.

TABLE 2.3
Barrier Height of Dimethyl Ether and Ethylene as Obtained by DFT Calculations

System	Barrier Height (kJ/mol)	
	No Membrane	**Gradient-Type Membrane**
Dimethyl ether formation	45.3	30.2
Ethylene formation	35.2	25.1

dimethyl ether (DME) followed by the formation of olefins through the carbon–carbon bond formation. In a transition state calculation, first we will identify the intermediate and locate the normal mode to rationalize the intrinsic reaction coordinate. In this study, a transition state search calculation using DFT formalism was performed to check the applicability of the designed gradient-type membrane reactor to the conversion of methanol to ethylene. Table 2.3 shows the barrier height of DME and the product formation, in the presence and in the absence of the gradient-type catalytic membrane. The result indicates that irrespective of the presence of the membrane, the barrier height of DME is higher compared with that of ethylene and this is attributed to the formation of ethylene as a primary product from DME. Moreover, in the presence of the membrane, the barrier height of DME and the product is decreased, suggesting the increase in conversion due to the easy removal of the product, forcing the equilibrium of the reaction "to the right" (according to Le Chatelier's Principle). We have mentioned earlier that diffusion of the guest molecule in the zeolite host determines the utility of the zeolite in the product distribution. More exhaustive study with the reaction mechanism is under progress; the goal for the transition state calculation here is to identify the barrier height and the dependence of the membrane composition over the reaction. The composition is detrimental for the gradient-type membrane. Once the product formation is confirmed, then it is necessary to confirm the easy separation of the product and also to check the reaction pathway at the boundary of the hydrophobic and hydrophilic domain. This will justify if the material can be usable for the desired separation type reaction.

2.3.4 DIFFUSION PROCESS

To validate the adsorption of methanol on the hydrophilic site (low Si/Al ratio), the diffusion behavior of the methanol is tested. The direct contribution of the gradient was tested by moving the methanol molecule along the channel direction. Generally, the diffusivity behavior of a molecule in the zeolite channel is strongly influenced by the specific location of the molecule in the pore system and some energy is needed to move from one location to another. We moved the molecules at a rate of 1.5 Å and then calculated the energy at that point, which will allow tracing a diffusion path through the specified channel. This is an indirect way to measure the diffusion path. We have optimized the energy for each of the structures using the DMol^3 of Accelrys, as described in Section 2.2. The increase and decrease in energy from the

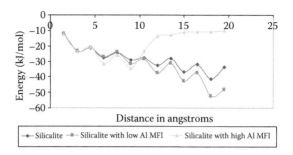

FIGURE 2.3 Diffusion path of methanol through silicalite and the gradient-type membrane.

previous step would correspond to the instability and the smooth diffusion of the molecule in that direction, respectively. Figure 2.3 shows the diffusion behavior of methanol through the low and high Al content ZSM-5. The results clearly indicate that the zeolite matrix with low Al content partly allows the diffusion of methanol quite smoothly similar to silicalite, whereas a steep rise in energy was observed

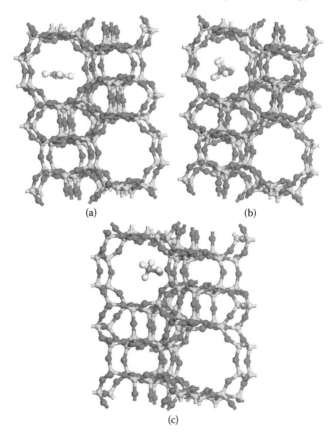

FIGURE 2.4 The optimized geometry of the (a) reactant methanol and two products, (b) ethylene, and (c) DME within the MFI all-silica matrix.

when methanol diffuses through the high Al content part of the membrane matrix and suggested that the diffusion of methanol was prohibited. At the high Al content part, methanol gets adsorbed close to the Al site, thus favoring the DME formation because of the increase in acidity of the membrane [37]. On the other hand, independent from the Si/Al ratio, ethylene molecules continue to pass through the membrane and separate easily after the formation. The individual structures at their minimum energy conformation are shown in Figure 2.4a, 2.4b, and 2.4c (methanol, ethylene, and DME, respectively). The minimum energy conformations are calculated using DMol^3 as well to show the affinity of the molecule with the MFI structure. The diffusivity through a gradient membrane is a combination of three phenomena: (1) physisorption followed by (2) chemisorptions, and (3) catalytic reaction. Once the product is generated, it will follow the same order and this sequential phenomenon is detrimental in product separation. From the above observation, it can be predicted that for the reactant (methanol), Si/Al ratio significantly influences the binding, but for the product, Si/Al ratio does not have any considerable effect on the diffusion process. In other words, the chemisorption is more detrimental in the catalytic reaction to undergo and once the product is formed, it does not need any favor from the lattice because it needs to go out of the membrane pore. Therefore, a zeolite membrane with hydrophilic entrance modeled by increasing Al content and a hydrophobic exit with lower Al content significantly controls the conversion of methanol to ethylene along with the product separation.

2.4   CONCLUSIONS

In summary, this work successfully describes the designing of gradient-type zeolite membrane based on Si/Al ratio by computer simulation. This study exhibits high relevance to the understanding of structure–function relationships that govern the applications of these materials as highly efficient catalytic membrane reactors from the reaction to the product separation. We have also shown that it is possible to propose the best composition for a zeolite membrane and looked into the feasibility of the gas adsorption within the zeolite domain using first principle structural optimization. Moreover, the transition state calculation suggested that the conversion of methanol to ethylene through the membrane reactor, followed by the product separation, was effectively done depending on the Si/Al ratio (hydrophobicity and hydrophilicity). We then intend to apply this methodology to predict the various chemical compositions to design catalytic membrane reactors depending on the nature of the reaction and also by varying the zeolite type.

REFERENCES

1. J. Caro, M. Noack, P. Kolsch and R. Schafer. Zeolite membranes—State of their development and perspective. *Microporous Mesoporous Mater.* **38**, 2000, 3–24.
2. A. Tavolaro and E. Drioli. Zeolite membranes. *Adv. Mater.* **11**, 1999, 975–996.
3. H. H. Funke, M. G. Kovalchick, J. L. Falconer and R. D. Noble. Separation of hydrocarbon isomer vapors with silicalite zeolite membranes. *Ind. Eng. Chem. Res.* **35**, 1996, 1575–1582.

4. H. Kita, T. Inoue, H. Asamura, K. Tanaka and K. Okamoto. NaY zeolite membrane for the pervaporation separation of methanol–methyl *tert*-butyl ether mixtures. *Chem. Commun.* 1997, 45–46.

5. K. Kusakabe, T. Kuroda and S. Morooka. Separation of carbon dioxide from nitrogen using ion-exchanged faujasite-type zeolite membranes formed on porous support tubes. *J. Membr. Sci.* **148**, 1998, 13–23.

6. G. Xomeritakis and M. Tsapatsis. Permeation of aromatic isomer vapors through oriented MFI-type membranes made by secondary growth. *Chem. Mater.* **11**, 1999, 875–878.

7. J. Coronas, J. L. Falconer and R. D. Noble. Separations of C-4 and C-6 isomers in ZSM-5 tubular membranes. *Ind. Eng. Chem. Res.* **37**, 1998, 166–176.

8. M. P. Bernal, J. Coronas, M. Menendez and J. Santamaria. Coupling of reaction and separation at the microscopic level: Esterification processes in a H-ZSM-5 membrane reactor. *Chem. Eng. Sci.* **57**, 2002, 1557–1562.

9. Z. L. Cheng, Z. S. Chao and H. L. Wan. Progress in the research of zeolite membrane on gas separation. *Prog. Chem.* **16**, 2004, 61–67.

10. S. Haag, M. Hanebuta, G. T. P. Mabande, A. Avhale, W. Schweiger and R. Dittmeyer. On the use of a catalytic H-ZSM-5 membrane for xylene isomerization. *Microporous Mesoporous Mater.* **96**, 2006, 168–176.

11. E. Piera, C. Tellez, J. Coronas and M. Menendez. Use of zeolite membrane reactors for selectivity enhancement: Application to the liquid-phase oligomerization of *i*-butene. *Catal. Today* **67**, 2001, 127–138.

12. E. E. Mcleary, E. W. J. Buijsse, L. Gora, J. C. Jansen and T. Maschmeyer. Membrane reactor technology for C-5/C-6 hydroisomerization. *Philos. Trans. R. Soc., A* **363**, 2005, 989–1000.

13. H. Ahn and Y. Lee. Pervaporation of dichlorinated organic compounds through silicalite-1 zeolite membrane. *J. Membr. Sci.* **279**, 2006, 459–465.

14. A. Chatterjee. Lecture Notes in Computer Science (LNCS), 77, 2006, 3993 [Volume editor V. N. Alexandrov et al. for ICCS 2006, Part III].

15. H. Takaba, R. Koshita, K. Mizukami, Y. Oumi, N. Ito, M. Kubo, A. Fahmi and A. Miyamoto. Molecular dynamics simulation of *iso*- and *n*-butane permeations through a ZSM-5 type silicalite membrane. *J. Membr. Sci.* **134**, 1997, 127–139.

16. J. Z. Yang, Q. L. Liu and H. T. Wang. Analyzing adsorption and diffusion behaviors of ethanol/water through silicalite membranes by molecular simulation. *J. Membr. Sci.* **291**, 2007, 1–9.

17. S. Jakobtorweihen, N. Hansen and F. J. Keel. Combining reactive and configurational-bias Monte Carlo: Confinement influence on the propene metathesis reaction system in various zeolites. *J. Chem. Phys.* **125**, 2006, 224709.

18. J. Sauer, P. Uglieno, E. Garrone and V. R. Saunders. Theoretical-study of van der Waals complexes at surface sites in comparison with the experiment. *Chem. Rev.* **94**, 1994, 2095–2160.

19. S. P. Yuan, J. G. Wang, Y. W. Li and H. Jiao. Bronsted acidity of isomorphously substituted ZSM-5 by B, Al, Ga, and Fe. Density functional investigations. *J. Phys. Chem. A* **106**, 2002, 8167–8172.

20. M. P. Teter, M. C. Payne and D. C. Allen. Solution of Schrodinger equation for large system. *Phys. Rev. B* **40**, 1989, 12255–12263.

21. M. C. Payne, M. P. Teter, D. C. Allen, T. A. Arias and J. D. Johannopoulos. Iterative minimization techniques for *ab initio* total energy calculations molecular dynamics and conjugate gradients. *Rev. Mod. Phys.* **64**, 1992, 1045.

22. D. Vanderbilt. Soft self consistent pseudopotentials in a generalized Eigen value formalism. *Phys. Rev. B* **41**, 1990, 7892–7895.

23. J. P. Perdew. Density functional approximation for the correlation energy of the inhomogeneous electron gas. *Phys. Rev. B* **33**, 1986, 8822–8824.
24. A. D. Becke. Density functional exchange energy approximation with correct asymptotic behavior. *Phys. Rev. A* **38**, 1988, 3098–3100.
25. H. J. Monkhorst and J. D. Pack. Special points for Brillouin-zone integrations. *Phys. Rev. B* **13**, 1976, 5188–5192.
26. W. Kohn and L. J. Sham. Self-consistent equations including exchange and correlation effects. *Phys. Rev. A* **140**, 1965, A1133–A1138.
27. B. Delley. An all electron numerical method for solving the local density functional for polyatomic molecules. *J. Chem. Phys.* **92**, 1990, 508–517.
28. B. Delley. Analytical energy derivatives in the numerical local density functional approach. *J. Chem. Phys.* **94**, 1991, 7245–7250.
29. B. Delley. From molecules to solids with the DMol(3) approach. *J. Chem. Phys.* **113**, 2000, 7756–7764.
30. B. Delley. Fast calculation of electrostatics in crystals and large molecules. *J. Phys. Chem.* **100**, 1996, 6107–6110.
31. A. D. Becke. A multicenter numerical integration scheme for polyatomic molecules. *J. Chem. Phys.* **88**, 1988, 2547–2553.
32. C. T. Lee, W. T. Yang and R. G. Parr. Development of the Colle–Salvetti correlation energy formula into a functional of the electron density. *Phys. Rev. B* **37**, 1988, 785–789.
33. M. Fleys and R. W. Thompson. Monte Carlo simulations of water adsorption isotherms in silicalite and dealuminated zeolite Y. *J. Chem. Theory Comput.* **1**, 2005, 453–458.
34. T. J. Hou, L. L. Zhu and X. J. Xu. Adsorption and diffusion of benzene in ITQ-1 type zeolite: Grand canonical Monte Carlo and molecular dynamics simulation study. *J. Phys. Chem. B* **104**, 2000, 9356–9364.
35. I. Stara, D. Zeze, V. Matolin, J. Pavluch and B. Gruzza. AES and EELS study of alumina model catalyst supports. *Appl. Surf. Sci.* **115**, 1997, 46–52.
36. B. H. Engler, D. Lindner, E. S. Lox, A. Schafer-Sinlinger and K. Ostgathe. Development of improved Pd-only and Pd/Rh three-way catalysts. *Stud. Surf. Sci. Catal.* **96**, 1995, 441–460.
37. P. L. Benito, A. G. Gayubo, A. T. Aguyao, M. Olazar and J. Bilbao. Effect of Si/Al ratio and of acidity of H-ZSM5 zeolites on the primary products of methanol to gasoline conversion. *J. Chem. Technol. Biotechnol.* **66**, 1996, 183–191.

3 Wetting of Paracetamol Surfaces Studied by DMol³-COSMO Calculations

Theodora Todorova and Bernard Delley

CONTENTS

3.1   INTRODUCTION

The energetics and wetting properties of solid-state materials are of great importance in the performance of pharmaceutical and chemical materials. A detailed knowledge on the surface chemical behavior will assist in predicting surface properties such as solubility, adhesion, surfactant adsorption, and many others.

Paracetamol is a compound with a significant usage in the pharmaceutical industry. It is already known that the paracetamol crystal exhibits a polymorphism. Three polymorphs of paracetamol have been reported[1]: a thermodynamically stable monoclinic form[2] I, a metastable orthorhombic form[3] II, and a very unstable form III. The monoclinic form I is the form used commercially due to its thermodynamic stability at room temperature. In the current work, we investigate both forms I and II to underline the difference in their chemical behavior and to point out that it is not possible to gain understanding about one form based upon the knowledge on another form.

Duncan-Hewitt and Nisman[4] examined the wettability of paracetamol using several experimental techniques such as Washburn capillary rise method, sedimentation volume methods, and the sessile drop determinations on compacts, films, and single crystals. It was concluded that these powder characterization techniques have difficulties concerning the sample preparation, data interpretation, and methodology imprecision.

Semiempirical methods have been applied to evaluate surface free energy components of solid materials and polymer surfaces.[5–10] Owens and Wendt[6] developed a method for measuring the surface energy of solids and for resolving the surface energy into contributions from dispersion and dipole–hydrogen bonding forces. By measuring a contact angle of two different liquids against a solid, the components of surface free energy due to hydrogen bonding (γ^h) and dispersive forces (γ^d) can be estimated (the sum of these components should yield an approximation of the total solid surface energy). Based on this model, the surface energy could be predicted for other solvents with known γ^h and γ^d. Theoretical considerations for determination of surface energies and semiempirical methods are reviewed in Ref. 11.

Recent years have shown significant progress in the use of computation-based approaches for predicting surface energies and wettability. Continuum solvation models, in particular, the conductor-like screening model (COSMO), are well-established methods to incorporate solvation effects into quantum chemical calculations. COSMO takes into account the detailed atomic structure of the interface interacting with the dielectric liquid.

In this work, the DMol[3]-COSMO method[12,13] is applied with periodic boundary conditions. An extension to a wide class of liquids could be made by semiempirical "real solvent" parameterization as it is done routinely for solvent modeling.[14] This surface COSMO approach yields first principle prediction for the solid–liquid interfacial energy of a specific surface with an idealized high dielectric liquid.

3.2   METHODOLOGY

The calculations reported in this work are performed by the all-electron density functional theory DMol[3] code.[15,16] Double numerical polarized (DNP) basis set that includes all occupied atomic orbitals plus a second set of valence orbitals plus polarized d-valence orbitals is used. Atom element-dependent cutoff radii with a medium size of 8.0 Å are applied. Perdew–Burke–Ernzerhof (PBE)[17] exchange-correlation potential is used. DMol[3] PBE calculations have been demonstrated to give a very successful account of reaction enthalpies of molecules in the gas phase.[18] A sufficient level of convergence for the COSMO solvent accessible surface (SAS) is reached using a 110-point scheme[19] for all atoms except hydrogen, where the 50-point scheme is used.

The lattice parameters that we used for the bulk geometry optimization of the crystal structures are those reported by Haisa et al.[2,3] The optimized bulk crystal structures are used for cleaving crystal surfaces, characterized by their Miller indices. Crystal surfaces are kept fixed in the calculations and no relaxation of the surface structures is considered. The unrelaxed surface slab is constructed via a bulk primitive cell containing sides with the desired Miller index for the surface. Molecules are assigned entirely to the cell where the molecular center of mass resides. The slab model here is one or two layers of such primitive cells. The choice of origin normal to the surface plane leaves a free parameter for constructing possibly different surfaces with the same index. The presence of an inversion symmetry element in both orthorhombic and monoclinic paracetamol leaves automatically symmetry equivalent faces on both sides when the slabs are composed of an integer number of primitive cell layers.

Our present minimalistic slab models contain 80 atoms per cell, respectively, 160 atoms per cell in the case of the 201 surface and the orthorhombic form. In some cases, the back and the front surfaces of the slab derive from the same molecule. Because the molecules are stiff as compared to intermolecular bonding interactions, geometry optimization should mainly adjust the relative geometry of molecular constituents. The authors feel that for thin slab models as the present ones, a geometry optimization does not improve the realism of the solid–liquid interface. The molecules are placed according to the bulk crystal, but there is no bulk substrate constraining the position of surface molecules in the present thin slab models. Adsorbed molecules and hydrogen bonds to the surface can be studied with the present approach by adding such extra molecules explicitly, but this has not been tried yet. The present models turn out to contain a smaller number of atoms than a cluster construct for the surface, where additional molecules must be added to surround the surface cell of interest. Not surprisingly, the present calculations are not only more pleasing but also faster than the cluster model for the corresponding surface.

Wetting of the surface is simulated by means of the COSMO method. The detailed methodology using the COSMO approach is given in a previous publication,[12] but we give a brief overview of the SAS and the calculation of the solvation energy.

The DMol³ is a method[15,16] for density functional theory calculations of molecular clusters in the gas phase. It is generalized[13] to model a solvent environment of a molecule via the COSMO.[20] This method involves the construction of a SAS and the solution of electrostatics. The SAS is defined via element-dependent radii, with values typically about 17% larger than the van der Waals radii. The SAS is a model of a surface where the induced charges are located. The detailed construction is discussed in a previous publication.[12] A new SAS grid construction[12] is tested, where the grid points and weights are a continuous function of all atomic geometries. The calculated solvation energy is also continuous by consequence, which is useful for all calculations that involve geometry changes of the atomic framework. The basic COSMO approach assumes that the screening takes place exactly on that surface, hence the name *conductor-like screening model*. The distribution of the screening charges contains information about the hydrophilicity of the surface (models can be generalized to take into account the different nature of the solvents). This information can, in principle, be used for improved semiempirical analyses such as the COSMO-RS method.[14]

The total solvation energy at the COSMO level is $E_{solv} = E_t(q) - E_t(q = 0) + E_{diel}$, where E_t is the total energy in the presence/absence of COSMO screening, q is the vector of the screening charges on the surface of the cavity, and E_{diel} is the dielectric energy.

3.3   RESULTS

3.3.1   SURFACE ENERGY

The surface energy is the energy gained by wetting the surface. At the most basic level, static wetting is described by Young's law: $\gamma_{SL} - \gamma_{SV} + \gamma \cos(\theta) = 0$, where γ_{SV} is the surface energy of the solid against the gas phase, and γ_{SL} against the liquid. γ is the surface tension of the liquid. At the COSMO level, $\Delta\gamma = \gamma_{SL} - \gamma_{SV} = E_{solv}/A_c$, the "solvation energy" per planar surface area A_c. In a slab model, A_c usually refers to the combined area from both sides of the slab. On the experimental side, contact angle measurements are an important tool to study liquid–solid interfacial energies. With Young's law, one has $\Delta\gamma = -\gamma \cos(\theta)$. To convert the experimental contact angles from Ref. 21 to $\Delta\gamma$, the experimentally well-known surface tension of water $\gamma = 72.8$ mN/m is used.

3.3.2   PARACETAMOL FORM I

The crystal structure parameters reported by Haisa et al.[2] are $a = 12.93$, $b = 9.40$, $c = 7.10$, and $\beta = 115.9$ (P21/a group), whereas Cambridge Database[22] gives values of $a = 7.0941$, $b = 9.2322$, $c = 11.6196$, and $\beta = 97.821$ (P21/a group, a–c interchanged vs. the crystal structure resolved by Haisa et al.[2]). We report calculations by means of both crystal structures, but the discussion of the results is based on the crystal structures proposed by Haisa et al.[2] Facet (010) exposes the hydrophobic methyl groups together with the amine and carbonyl ones. At the (010) surface, the lack of polar groups determines the low attachment energy of this facet. Facets (201), (001), and (011) have similar functional group contributions and the differences are due to the different densities of the methyl, amine, and carbonyl groups and their orientation at the surface. These differences give rise to different interactions. In line with the COSMO calculations presented here, surface energies, calculated from the measured advancing contact angle,[23,24] indicated relatively high energy contributions on facets (201), (001), and (011). The surface energy on facet (110) was lower. The surface energy on the facet (010) was even lower and confirmed that the surface

---

TABLE 3.1

Wetting Energies (mN/m) for Paracetamol Form I (Molecules Optimized in Bulk Structure)

Surface	Experiment[23,24]	Haisa et al.[2,3]	Cambridge Database[22]	A_c
(201)	57.3	110.7[26]	100.2	232.70
(001)	70.0	113.7	93.6	107.25
(011)	63.2	82.0	81.3	134.80
(110)	46.0	67.3	98.2	107.25
(010)	27.6	29.5	29.3	81.65

Note: A_c is the surface area (in Å^2).

TABLE 3.2

Wetting Energies (mN/m) for Paracetamol Form II (Molecules Optimized in Bulk Structure)

Surface	Experiment[23,24]	Haisa et al.[2,3]	Cambridge Database[22]	A_c
(001)	31.3	30.1	22.6	207.15
(110)	69.8	126	116.8	150.15
(010)	69.3	111	124.5	84.95

Note: A_c is the surface area (in Å²).

is hydrophobic and possesses minimal potential for hydrogen bond interactions. Wetting energies and surface areas for paracetamol form I (molecules optimized in bulk structures) are given in Table 3.1.

3.3.3 PARACETAMOL FORM II

The crystal structure parameters reported by Haisa et al.[3] are $a = 11.805$, $b = 17.164$, and $c = 7.393$ (Pcab group), whereas Cambridge Database[22] proposes values of $a = 17.1657$, $b = 11.7773$, and $c = 7.212$ (Pbca group, a–b interchanged vs. the crystal structure resolved by Haisa et al.[3]).

For the form II, the trend in the surface energy is the opposite of that of form I, that is, the hydrophobic (001) facet with low surface energy and hydrophilic (110) and (010) facets with similar energy. Whereas the COSMO calculations and the experiments agree well on the surface energy for the (001) surface, there is almost a factor of 2 disagreement for the (010) and (110) facets. The maximum value of the calculated wetting energy, based on the advancing contact angle measurements, is 72.8 mJ/m² [$\gamma \cos(\theta)$]. For surfaces (110) and (010), this maximum wetting energy value is approached. On the contrary, the COSMO method is not limited to maximum wetting energy value of 72.8 mJ/m² and could get any positive number. Quantitatively, the two methods agree in distinguishing the hydrophilic (110) and (010) surfaces from the hydrophobic (001) surface of the orthorhombic polymorphic form of the paracetamol crystal. Wetting energies and surface areas for paracetamol form II (molecules optimized in bulk structures) are given in Table 3.2.

3.4 DISCUSSION

Our calculations show that the dependence of the wetting behavior of the polymorphic crystalline paracetamol forms I and II on their Miller indices is not obvious. For the forms I and II, the facets (001) and (201) are the most hydrophilic, respectively (Table 3.1), whereas the facet (001) is the most hydrophobic (Table 3.2). These results show that the hydrophilicity trends are opposite for forms I and II. Similar conclusions are drawn from the experimental studies of Heng et al.[23,24] From the measurements of advancing contact angles, the orders of hydrophilicity for the specific facets of forms I and II are

$$(001) > (011) > (201) > (110) > (010) \text{ (Form I)}$$

$$(010) \approx (110) > (001) \text{ (Form II)},$$

whereas COSMO orders them in the following sequence (based on the crystal structures proposed by Haisa et al.[2,3]):

$$(001) > (201) > (011) > (110) > (010) \text{ (Form I)}$$

$$(110) > (010) > (001) \text{ (Form II)}.$$

As mentioned before, the calculations are done with optimized bulk coordinates. The calculation of the surface structures, however, does not include relaxation and reconstruction. A possible relaxation/reconstruction of the surface lowers the wetting energies significantly for the higher energy surfaces.[25] To give room for intermolecular relaxations, it is advisable to consider thicker slab models than presented here. The SAS for the (201) surface of the monoclinic paracetamol (form I), shown to be highly hydrophilic, and for the most hydrophobic (001) surface of the orthorhombic paracetamol (form II) are shown in Figures 3.1 and 3.2, respectively. The SAS is strongly corrugated for the (201) surface of the monoclinic paracetamol (form I) and not so much for the (001) surface of the orthorhombic paracetamol (form II).

Figure 3.3 gives a comparison of the COSMO results and H_2O experimental wetting energies from the work of Heng et al.[23,24] using the structures proposed by Haisa

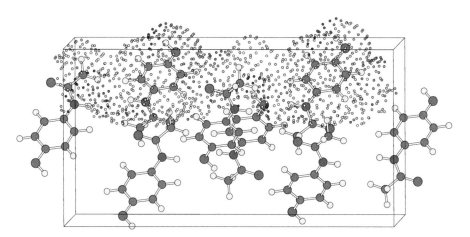

FIGURE 3.1 (See color insert.) COSMO charges for the wetted monoclinic (201) paracetamol surface.[26] Green balls are the carbon atoms, red—nitrogen atoms, blue—oxygen atoms, and dark blue—hydrogen atoms. Color code for COSMO charges: blue, −2 e/nm²; red, +2 e/nm²; and white, zero. Upper half of the COSMO surface is shown; face of cell indicates atomic surface plane.

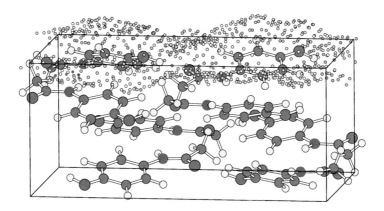

FIGURE 3.2   (See color insert.) COSMO charges for the wetted orthorhombic (001) paracetamol surface. Green balls are the carbon atoms, red—nitrogen atoms, blue—oxygen atoms, and dark blue—hydrogen atoms. Color code for COSMO charges: blue, −2 e/nm²; red, +2 e/nm²; and white, zero. Upper half of the COSMO surface is shown; face of cell indicates atomic surface plane.

et al.[2,3] and Cambridge Database structures[22] with reference numbers 135451/135452 and interpreting Cambridge results in Haisa crystal setting. A similar (relatively good) correlation is found between the theoretical results calculated with both structures (proposed by Haisa et al.[2,3] and Cambridge Database structures[22]) and the experimental results based on the measurements of the advancing contact angle.[23,24]

The orthorhombic (001) and the monoclinic (010) surfaces are the most stable ones in the dry and wet states. The COSMO approach reproduces the experimental

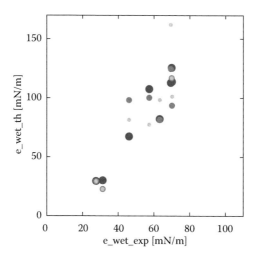

FIGURE 3.3   Comparison of the COSMO results and H_2O experimental.[23,24] Black—using crystal structures proposed by Haisa et al.,[2,3] dark gray—using Cambridge structures[22] 135451/135452, and light gray—interpreting Cambridge results in Haisa crystal setting.

values of the wetting energies for these two surfaces very well. The calculated COSMO wetting energies for the other, more polar, surfaces are higher and get over-estimated by the model. It is difficult to estimate how the wetting energy of the more polar and naturally less stable surfaces would be reduced to approach experimental values more closely. One may speculate that, because these surfaces are not very stable, significant surface relaxation and reconstruction, perhaps several molecular layers deep, might take place. Geometry optimization of sufficiently thick slab models ought to show this effect if it is real.

The COSMO involves a fictitious conductor with no limitation of the screening charge density. Even when scaled to the proper dielectric constant, this remains true. However, a real solvent has a fairly well-defined set of available surface charge densities, as evidenced in the so-called "sigma profile."[14] A postprocessing of COSMO results as the present ones with the real solvent model is expected to make these wetting energies more realistic.

3.5   SUMMARY AND CONCLUSION

This study presents a hydrophilicity ranking for several facets of paracetamol forms I and II based on the calculated wetting behavior by means of the COSMO approach. The local surface chemistry for the different facets is defined by the methyl, amine, and carbonyl groups present on the surface. Surfaces with the same Miller indices for forms I and II paracetamol show entirely opposite trends in their surface chemical behavior. The most hydrophilic surface for form I, (001), is the most hydrophobic one for form II, whereas one of the most hydrophilic surfaces for form II, (010), is the most hydrophobic for form I. Therefore, knowledge on one of the polymorphic forms of paracetamol cannot be directly applicable to another polymorphic form.

ACKNOWLEDGMENTS

The authors thank A. Klamt for the stimulating discussions. Swiss NSF grant 200021-109358 is gratefully acknowledged.

REFERENCES

1. P. DiMartino, P. Conflant, M. Drache, J.-P. Huvenne and A.-M. Guyot-Hermann. Preparation and physical characterization of forms II and III of paracetamol. *J. Therm. Anal. Calorim.* **48**, 1997, 447.
2. M. Haisa, S. Kashino, R. Kawai and H. Maeda. The monoclinic form of *p*-hydroxyacet-anilide. *Acta Crystallogr.* **B32**, 1976, 1283.
3. M. Haisa, S. Kashino and H. Maeda. The orthorhombic form of *p*-hydroxyacetanilide. *Acta Crystallogr.* **B30**, 1974, 2510.
4. W. Duncan-Hewitt and R. Nisman. Investigation of the surface free energies of phar-maceutical materials from contact angle, sedimentations, and adhesion measurements. In *Contact Angle, Wettability and Adhesion*, edited by K. L. Mittal. Utrecht, The Netherlands: VSP, 1993, 791–811.
5. F. M. Fowkes. Attractive forces at interfaces. *Ind. Eng. Chem.* **56**, 1964, 40.

6. D. K. Owens and R. C. Wendt. Estimation of the surface free energy of polymers. *J. Appl. Polym. Sci.* **13**, 1969, 1741.
7. S. Wu. *Polymer Interface and Adhesion*. New York: Marcel Dekker, 1982.
8. F. M. Fowkes and M. A. Mostafa. Acid–base interactions in polymer adsorption. *Ind. Eng. Chem. Prod. Res. Dev.* **17**, 1978, 3.
9. C. J. van Oss, R. J. Good and M. K. Chaudhury. Additive and nonadditive surface tension components and the interpretation of contact angles. *Langmuir* **4**, 1988, 884.
10. F. Chen and W. V. Chang. Applicability study of a new acid base interaction model in polypeptides and polyamides. *Langmuir* **7**, 1991, 2401.
11. F. M. Etzler. In *Contact Angle, Wettability and Adhesion*, edited by K. L. Mittal. Utrecht, The Netherlands: VSP, 1993, 1–46.
12. B. Delley. The conductor-like screening model for polymers and surfaces. *Mol. Simul.* **32**, 2006, 117.
13. J. Andzelm, Ch. Kölmel and A. Klamt. Incorporation of solvent effects into density functional calculations of molecular energies and geometries. *J. Chem. Phys.* **103**, 1995, 9312.
14. A. Klamt. *COSMO-RS from Quantum Chemistry to Fluid Phase Thermodynamics and Drug Design*. The Netherlands: Elsevier B.V., 2005.
15. B. Delley. An all-electron numerical method for solving the local density functional for polyatomic molecules. *J. Chem. Phys.* **92**, 1990, 508.
16. B. Delley. From molecules to solids with the DMol3 approach. *J. Chem. Phys.* **113**, 2000, 7756.
17. J. P. Perdew, K. Burke and M. Ernzerhof. Generalized gradient approximation made simple. *Phys. Rev. Lett.* **77**, 1996, 3865.
18. B. Delley. Ground-state enthalpies: Evaluation of electronic structure approaches with emphasis on the density functional method. *J. Phys. Chem. A* **110**, 2006, 13632.
19. B. Delley. High order integration schemes on the unit sphere. *J. Comp. Chem.* **17**, 1996, 1152.
20. A. Klamt and G. Schüürmann. COSMO: A new approach to dielectric screening in solvents with explicit expressions for the screening energy and its gradient. *J. Chem. Soc., Perkin Trans.* **2**, 1993, 799.
21. A. Sklodowska, M. Wozniak and R. Matlakowska. The method of contact angle measurements and estimation of work of adhesion in bioleaching of metals. *Biol. Proc. Online* **1**, 1999, 114.
22. D. A. Fletcher, R. F. McMeeking and D. Parkin. The United Kingdom chemical database service. *J. Chem. Inf. Comput. Sci.* **36**, 1996, 746.
23. J. Y. Y. Heng, A. Bismarck, A. F. Lee, K. Wilson and D. R. Williams. Anisotropic surface energetics and wettability of macroscopic form I paracetamol crystals. *Langmuir* **22**, 2006, 2760.
24. J. Y. Y. Heng and D. R. Williams. Wettability of paracetamol polymorphic forms I and II. *Langmuir* **22**, 2006, 6905.
25. B. Delley, unpublished results.
26. Erratum: Results for the (102) surface instead of (201) were shown in: T. Todorova and B. Delley. Wetting of paracetamol surfaces studied by DMol3-COSMO calculations. *Mol. Simul.* **34**, 2008, 1013–1017.

4 Molecular Dynamic Studies of the Compatibility of Some Cellulose Derivatives with Selected Ionic Liquids

Bela Derecskei and Agnes Derecskei-Kovacs

CONTENTS

4.1 INTRODUCTION

Mutual compatibility of polymers and solvents is one of the important issues of materials engineering not only in the polymer industry but also for pharmaceuticals, where drug delivery and release is closely related to the properties of the individual components. In simple terms, "like dissolves like," a principle that can be quantified through the Hildebrand solubility parameter, which is defined as the square root of the cohesive energy density. The basic concept has been further refined by Hansen [1], who introduced individual terms related to the van der Waals dispersion forces, dipole interaction, and hydrogen bonding. The total solubility parameters and the individual contributions are determined experimentally from intrinsic viscosity measurements and, for polymers, from swelling measurements.

Cellulose and its derivatives are widely used in the pharmaceutical, food, plant, oil, and other industries. It is a biorenewable material for which further expansion of the large industrial application is somewhat hindered by the unusual and expensive solvents and harsh processing conditions [2] that are often required to solubilize it. Chemically, cellulose is a homopolysaccharide composed of β-D-glucopyranose units that are linked together by (1→4)-glucosidic bonds. Strong intramolecular and intermolecular hydrogen bonding is a major contributing factor to the strength of wood and other plant matter based on cellulose. The same strong bonds are the root cause of its difficult solubilization. Ionic liquids may offer an attractive environmentally friendly and economical alternative for the chemical and physical processing of cellulose that could have a significant impact on utilizing its full potential as a renewable raw material in special areas such as polymer manufacturing.

Ionic liquids are often referred to as "green" solvents. Recently, a discussion has emerged among green chemists about the possible global and local interpretation of "greenness" of chemical processes in general and solvents in particular [3–4]. Nevertheless, when compared to the harsh conditions required currently to solubilize cellulose, ionic liquids still can be considered as "green" alternatives. One of the greatest advantages of ionic liquids is their remarkable versatility: with different combinations of cations and anions, approximately 10^{18} different ionic liquids are possible [5]. The great flexibility and desirably low environmental impact suggests that task-specific ionic liquids with predetermined physical and chemical properties may hold great promise for future applications. In spite of the obvious interest in this area, only a few ionic liquids have been characterized even by such basic properties as density, conductivity, melting point, and so on. *In silico* experiments with predictive power can greatly facilitate the implementation of ionic liquids in industrial applications.

Molecular modeling has been demonstrated to be a useful tool in the characterization of many different types of chemical systems, including bulk materials. Many properties can be computed with an accuracy that is comparable to experimental capabilities [6]. Simulation of properties is especially important for systems that are challenging to study experimentally due to their limited solubility in common solvents. In some other cases, the experiment itself may be difficult due either to sensitivity to sample preparation or to ambiguities in the interpretation of the results. When the system consists of a relatively small number of atoms, both traditional *ab initio* and density functional methods can be employed. In the case of sugar molecules and disaccharides, a number of studies have been carried out to develop force field methods [7] to be used in molecular dynamics simulations while studying the conformations of cellulose and its interactions with water and other molecules [8–16].

Computer simulations of ionic liquids have largely focused on the development of force field parameters specific to an ionic liquid or an ionic liquid family [17–23]. In addition to simulation of structural, dynamical, electric, and thermodynamic properties of several pure ionic liquids [24–26], the solvation of small solutes in ionic liquids has also been investigated [27–31]. Compatibility of ionic liquids and cellulose derivatives has not been explored extensively with molecular theory: the current work is intended to draw attention to this area.

4.2 METHODOLOGY

All of the calculations described in this work employed the Discover simulation engine as implemented in the Materials Studio® suite of programs by Accelrys. The COMPASS force field was used throughout for cellulose-mimicking oligomers because it has been specifically optimized to yield accurate cohesive properties for a great variety of molecules [6] and because it has been successfully applied in the computation of solubility parameters for rigid chain polymers [32]. COMPASS-assigned partial charges were used throughout because exploratory calculations on small oligomers showed only mild changes when charges derived from density functional calculations were used. To test the accuracy of the approach for cellulose, some common solvents with structural similarity to the polysaccharide repeat units were also investigated using the same force field.

Densities are well known experimentally for common solvents and many polymers, but they are largely unknown for oligomers with a small degree of polymerization. To calculate these densities, each cellulose-mimicking oligomer was built using the Polymer Builder module in Materials Studio and its structure minimized (without an extensive conformer search) using Discover. Taking the structure in the optimal geometry, at least five copies of a three-dimensional (3-D) periodic amorphous cubic box were constructed (usually in the 16 Å–18 Å size range, containing about 650–750 atoms). The 3-D periodic boxes were then equilibrated for 20 ps under NPT conditions (the numbers of atoms, pressure, and temperature were kept constant at $p = 1$ atm and $T = 298$ K using velocity scaling as thermostat and the Berendsen method as barostat). In the production run, a 100-ps NPT molecular dynamics simulation was executed while sampling the conformer space at 5000- fs intervals (using the Andersen thermostat). Atom-based summation (as opposed to cell-based methods) was used throughout for the nonbonding interactions (both Coulomb and van der Waals) with atom-based cutoff values of 9.50 Å and 0.50-Å buffer width. The resulting average density for each simulation was then recorded for statistical analysis.

The protocol to calculate the solubility parameters was somewhat similar. Using the same geometry-optimized oligomer molecule, five copies of a cubic amorphous box were created at the average density (calculated as described above). The system was equilibrated for 20 ps at 298 K in an NVT simulation (constant number of atoms, volume and temperature, and velocity scaling), followed by a 100-ps production run (Andersen thermostat). The conformer space was sampled at 5000-fs intervals. Cohesive energy densities and solubility parameters were then evaluated for these trajectories. In atomistic simulations, the cohesive energy is defined as the increase in energy per mole of a material if all intermolecular forces are eliminated, whereas the cohesive energy density corresponds to the cohesive energy per unit volume. In these calculations, solubility parameters are calculated as the square root of the cohesive energy density without any further decomposition into individual components (such as dispersion and dipole).

4.3   RESULTS AND DISCUSSION

4.3.1   VALIDATION STUDIES

To assess the accuracy of the simulations and the COMPASS force field in these calculations, two organic solvents were studied first. Glycerol was chosen because it is a polyol, in common with the building blocks of cellulose; 2-methanol-tetrahydro-pyran (THPMeOH) was chosen because its ring structure is analogous to the repeat unit of cellulose. The calculated densities and solubility parameters are presented in Table 4.1 along with the experimental values.

It can be seen from Table 4.1 that both the room temperature density and the solubility parameter of glycol are in good agreement with experimental findings. For the pyran derivative, the density is somewhat overestimated in these calculations, whereas the solubility parameter is reproduced with a good accuracy.

4.3.2   DENSITY OF CELLULOSE-MIMICKING OLIGOMERS

The cellulose derivatives chosen for these studies were cellulose itself (C), methyl cellulose (MC), hydroxypropyl cellulose (HPC), and carboxymethyl cellulose (CMC). The repeat units are shown in Figure 4.1.

To explore the dependence of the calculated density on the polymer length, a series of oligomers with increasing degree of polymerization was created for each cellulose derivative. The results for cellulose are presented in Table 4.2.

It can be seen from the table that the density increases with increasing degree of polymerization up to the tetramer, remains unchanged up to the hexamer, and starts to slowly decrease for the longer chain lengths. The small standard errors indicate that the reproducibility of the simulations is fairly good and it does not show any correlation with the oligomer length. All other cellulose derivatives (not shown here in detail) behave the same way.

In this format, it is challenging to extrapolate the behavior of very long polymer chains and, therefore, instead of the degree of polymerization, the inverse of the

---

TABLE 4.1

Calculated Room Temperature Densities and Solubility Parameters of Selected Organic Solvents with Structural Similarity to the Cellulose Repeat Unit along with Experimental Values

	Density (g/cm³)		Solubility Parameter (MPa¹ᐟ²)	
	Expt[a]	Calc	Expt[a]	Calc[b]
Glycerol	1.258	1.243	36.2	35.2
THPMeOH	1.021	1.093	22	23.5

[a]   Taken from Barton, A.F.M., *Handbook of Solubility Parameters and Other Cohesion Parameters*, Boca Raton: CRC Press, 1983.

[b]   Calculated at the experimental density.

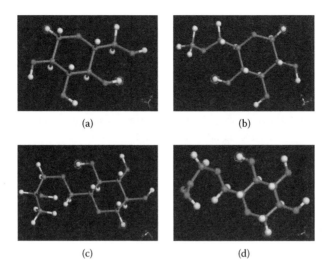

(a)   (b)

(c)   (d)

FIGURE 4.1   Repeat units of C (a), MC (b), HPC (c), and CMC (d) with the head and tail atoms marked.

molar mass was chosen as an independent variable. The results for cellulose and the three derivatives are shown in Figure 4.2.

Figure 4.2 shows that, in general, the density decreases with increasing inverse molar mass (increases with the degree of polymerization). The dependence is more or less linear with the longest oligomer (dodecamer in these calculations) lying below the straight line. Earlier work for poly(ethylene oxide) [34] found linear dependence of density on the inverse of molar mass for a series of oligomers, with a somewhat tighter fit than the current calculations. A possible explanation for this might be the length of the molecular dynamics runs: if the cellulose polymer chain is much less flexible than poly(ethylene oxide), there may not be sufficient time for the system to relax into the optimal, tight packing density from the unfavorable geometry of the initial amorphous packing. (In test runs, molecular dynamics calculations with the simulation time increased up to 500 ps failed to introduce any noticeable change in the density.)

Nevertheless, the straight line fit allows an extrapolation for the intercept with the y-axis which is the equivalent of the simulated density of polymers of very large molar masses (infinite, in principle). The extrapolated density values are listed in Table 4.3 along with the values of R^2 to show the accuracy of the fit. To check the consistency of the data sets, extrapolations were also performed with the dodecamers not included in the analysis (they are clear outliers in some cases) and those results are also shown in Table 4.3 for comparison purposes.

It can be seen in the table that omitting the dodecamer introduces only small changes into the extrapolated value of the density. The improvement in the accuracy of the fit can be partially attributed to the fact that there are fewer data points in the analysis; however, at least for microcrystalline cellulose, there is a significant improvement when the outlier dodecamer is omitted.

TABLE 4.2
Molecular Formulas, Molar Masses, and Calculated Room Temperature Densities (with Standard Error and % Standard Error) of Cellulose-Mimicking Oligomers with Increasing Degree of Polymerization

Degree of Polymerization	Cellulose-Mimicking Oligomers, C							
	1	2	3	4	5	6	8	12
Molecular formula	$C_6H_{12}O_5$	$C_{12}H_{22}O_{10}$	$C_{18}H_{32}O_{15}$	$C_{24}H_{42}O_{20}$	$C_{30}H_{52}O_{25}$	$C_{36}H_{62}O_{30}$	$C_{48}H_{82}O_{40}$	$C_{72}H_{122}O_{60}$
Molar mass (g/mol)	164	325	489	653	813	981	1299	1948
Density (g/cm³)	1.36	1.38	1.4	1.42	1.42	1.42	1.41	1.39
Standard error	0.0099	0.0131	0.0157	0.0097	0.0134	0.0104	0.00586	0.0036
% Standard error	0.73	0.95	1.12	0.68	0.95	0.73	0.42	0.26

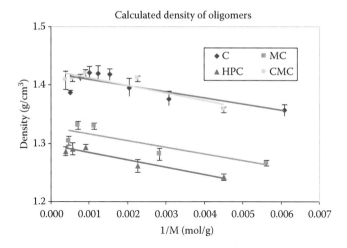

FIGURE 4.2 The density of a series (with standard errors shown as error bars) of cellulose-mimicking oligomers as a function of the inverse of the molar mass.

Experimental values for the density of cellulose are often reported as ranges rather than as single numbers. There are two reasons for this: native cellulose is a mixture of crystalline and amorphous sections and it is almost always hydrated to a certain degree. Our simulations included no water and thus represent a limiting value, and simulations were performed for the amorphous phase only. Crystalline cellulose has two polymorphs with densities of 1.582 g/cm³ and 1.599 g/cm³ [35]. Very recently, it has been shown that the true density is lower (between 1.40 g/cm³ and 1.47 g/cm³ for water concentrations of ~2%–9%) [35]. This indicates that the density of anhydrous amorphous cellulose must be even lower. Mazeau and Heux [36] used the polymer consistent force field (PCFF) force field to model both the crystalline and amorphous phases of cellulose and calculated 1.3762 g/cm³ as average density. Our value of 1.42 g/cm³ is somewhat higher than that but it is still within the range of 1.28 g/cm³–1.44 g/cm³ which was offered in Ref. [36] as expected for amorphous systems. Chen et al. [37] tested three different force fields in their simulations for the amorphous phase of cellulose and found that the PCFF

TABLE 4.3

Extrapolated Values for Room Temperature Densities to High Molar Mass Polymers for Cellulose and Some of Its Derivatives along with the Accuracy of the Straight Line Fit

Cellulose Derivative	Fitting	MCC	MC	HPC	CMC
Density (g/cm³)	a	1.42	1.33	1.3	1.42
R^2		0.6469	0.7349	0.9242	0.7867
Density (g/cm³)	b	1.43	1.34	1.3	1.43
R^2		0.8857	0.8794	0.9385	0.8532

Note: (a) All points included; (b) dodecamer omitted.

force field-yielded density of 1.385 g/cm³ was the closest to the reported literature value of 1.48 g/cm³ [38]. Choi et al. [39] studied hydroxyethyl cellulose and hydroxypropyl cellulose in the amorphous phase using the Dreiding II force field. In their paper, they refer to measured film densities of 1.10 g/cm³ for HPC (obtained from Aqualon). Our calculated value of 1.30 g/cm³ is higher than this measured value.

It is worthwhile to point out that when a nonlinear (second-order polynomial) fit to the points in Figure 4.2 is performed, the extrapolated densities are somewhat lower than those presented in Table 4.3. Densities are lowered by about 6% for cellulose itself and 2%–3% for the derivatives. Because the accuracy of the density calculations has an impact on the calculated solubility parameters (as it will be shown later), the uncertainties of densities will carry over to those calculations as well.

4.3.3 SOLUBILITY PARAMETERS FOR A SERIES OF OLIGOMERS

As described in Section 4.2, solubility parameters were calculated for each oligomer at its calculated average density. To explore the sensitivity of the results to the accuracy of the density calculations, solubility parameters were also calculated at 10% above and below the average density for the case of the octamer of the cellulose repeat unit. The results are shown in Figure 4.3.

The sensitivity analysis shows that a 10% change in density results in a significant change in the solubility parameter (around 2.5 MPa$^{1/2}$ in this case). This observation emphasizes the need for accurate values for the density of oligomers in theoretical calculations aimed to predict solubility parameters.

The dependence of the solubility parameter on the degree of polymerization was also evaluated. The effect is significantly larger than the result of changes in the density only by itself as shown in Figure 4.4.

The dependence of the solubility parameter on the inverse of molar mass is linear with a good accuracy. Longer chain lengths result in lower values and the range of

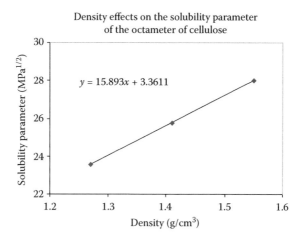

FIGURE 4.3 Solubility parameters of the eight repeat unit oligomer mimicking cellulose as obtained at the average simulated density as well as 10% higher and lower densities.

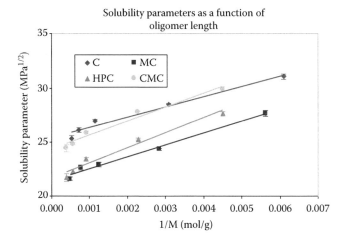

FIGURE 4.4 Solubility parameters of oligomers mimicking cellulose and its derivatives (with standard errors shown as error bars) as a function of the inverse molar mass.

change is about 6 MPa$^{1/2}$ for all four cellulose derivatives studied here. Extrapolation allows us to estimate the values for very high molar masses; results are summarized in Table 4.4 along with the accuracy of the linear fit. Again, the analysis was also performed when the dodecamers were omitted from the data set.

It can be seen from the table that the omission of the dodecamer changes the results only to a small extent. Methyl cellulose has the lowest and cellulose itself has the highest calculated solubility parameter in this series and the span is about 4 MPa$^{1/2}$. Choi and coworkers [39] recently estimated the solubility parameters of hydroxyethyl and hydroxypropyl cellulose using the Dreiding II force field. They investigated two forms of HPC: one is analogous to our system, and the other one was fully substituted. They obtained values of 23.7 MPa$^{1/2}$ and 22.1 MPa$^{1/2}$ for the two variants, respectively, which are only moderately higher than our calculated value of 21.7 MPa$^{1/2}$. Our result for HPC also agrees reasonably well with the

TABLE 4.4
Extrapolated Values for Room Temperature Solubility Parameters to High Molar Mass Polymers for Some Derivatives of Cellulose along with the Accuracy of the Straight Line Fit

Cellulose Derivative	Fitting	MCC	MC	HPC	CMC
Solubility parameter (MPa$^{1/2}$)	a	25.39	21.43	21.7	24.35
R^2		0.9687	0.9847	0.9674	0.9641
Solubility parameter (MPa$^{1/2}$)	b	25.73	21.66	21.99	24.59
R^2		0.9915	0.9916	0.9755	0.9673

Note: (a) All points included; (b) dodecamer omitted.

experimental value of 23.9 MPa$^{1/2}$ as determined by Archer [40] for fully substituted HPC. Chen and coworkers [37] simulated cellulose in the amorphous phase and predicted a solubility parameter of 19.0 MPa$^{1/2}$. Our calculated value is somewhat higher and compares less favorably with the literature values of 18.4 MPa$^{1/2}$–20.5 MPa$^{1/2}$ reported for rayon fibers [41]. At the same time, Mazeau and Heux [36] reported a calculated solubility parameter of 68 (J/cm^3)$^{1/2}$ for the amorphous phase of cellulose and also concluded that the solubility parameters of the two crystalline phases are only slightly higher. The solubility parameter of microcrystalline cellulose has been investigated by several authors using mechanical calculations [42–45], surface free energies [46–48], and inverse gas chromatography [49]. With the exception of the last method, all values are in the range of 23.6 MPa$^{1/2}$–27.7 MPa$^{1/2}$. (Inverse gas chromatography (IGC) yields a higher value, 39.3 MPa$^{1/2}$, which was attributed [42] to the sample preparation required by the method.) Roberts and Rowe [42] proposed 25.7 MPa$^{1/2}$ as indicative of the typical material used in processing. In the backdrop of these experimental values and the finding of Ref. [36] that the solubility parameter of the amorphous phase is only about 4% less than that of the crystalline phases, our calculated value of 25.4 MPa$^{1/2}$ appears to be very reasonable. The variations within the experimental and calculated values underline the difficulty of accurate determination of solubility parameters for cellulose and its derivatives both experimentally and theoretically.

4.3.4 IONIC LIQUID CALCULATIONS

To assess the compatibility of cellulose and its derivatives with selected imidazolium-based ionic liquids, we performed solubility parameter calculations for those systems as well. Force field calculations have been successfully applied in the simulation of charged systems [50–54], whereas the COMPASS force field has not been explicitly validated for cohesive energy calculations in such systems. To explore its performance for ionic liquids containing cations of the imidazolium family and some anions, we carried out density and cohesive energy calculations for 1-*R*-3-methylimidazolium (*R* = ethyl: emim, *R* = butyl: bmim, and *R* = hexyl: C$_6$mim) using chloride and trifluoroacetate as counterions. The details of these calculations are described in detail elsewhere [55]; only a brief summary is provided here to draw certain conclusions. First, it must be pointed out that COMPASS-assigned partial charges performed poorly for the calculation of the density of these ionic liquids. Instead, electrostatic potential-derived charges were used in all simulations, as determined from gradient corrected density functional calculations using the PW91 functional and double numeric polarized basis sets. In general, calculated densities for a variety of ionic liquids were found to be in agreement with experimental values. For example, the density of bmim–CF$_3$COO was calculated as 1.24 g/cm^3, whereas the experimental density is 1.22 g/cm^3 [56]. The calculated solubility parameters are summarized in Table 4.5 for compatibility considerations.

The predicted values of the solubility parameters for these ionic liquids are much higher than those of any of the cellulose derivatives in this study. It was found experimentally [2] that imidazolium-based ionic liquids are only able to dissolve cellulose after repeated microwave heating and agitation. It would require further simulation studies to determine how to mimic these experimental conditions because

TABLE 4.5
Calculated Solubility Parameters of Ionic Liquids Comprised of the 1-*R*-3-Methylimidazolium Cation (emim: *R*-Ethyl, bmim: *R* = Butyl, and C$_6$mim: *R* = Hexyl) and Two Different Anions

Cation	Anion	Solubility Parameter (MPa$^{1/2}$)
[emim]	Cl$^-$	55.86
	CF3COO$^-$	54.58
[bmim]	Cl$^-$	50.26
	CF3COO$^-$	49.79
[C$_6$mim]	Cl$^-$	46.77
	CF3COO$^-$	45.26

temperature effects have not been included so far. Solvent engineering, that is, finding a suitable ionic liquid among a very large number of candidates for dissolving a given cellulose derivative, requires exploiting variability in solubility parameters. We found only a limited variation in this small sampling of anions, whereas there was a larger sensitivity to the carbon chain length of the cations. Based on the results obtained here, it appears that the calculated solubility parameters have only limited use in the prediction of compatibility of cellulose derivatives and ionic liquids.

4.4 CONCLUSIONS

The commercially available COMPASS force field as implemented in Materials Studio by Accelrys performed well in the simulation of densities and solubility parameters of small molecules structurally related to the building blocks of some cellulose derivatives. A series of oligomers of increasing degree of polymerization mimicking the amorphous phase of cellulose and its derivatives offered some insight into the dependence of these properties on polymer chain length. Extrapolation to high molar mass polymers was performed by linear fit. Solubility parameters for longer oligomers were significantly lower than those of the monomer in all cases, whereas changes in the density exhibited only moderate dependence on the degree of polymerization. The calculated densities and solubility parameters, in most cases, agreed well with other calculated and experimental values where those were available. For imidazolium-based ionic liquids, density simulations yielded results in good agreement with experiments, but further verification of the COMPASS force field for the calculation of cohesive energies of polyelectrolytes is needed before simulations can reliably be used to assess compatibility of these solvents with organic polymers.

4.5 FUTURE WORK

One of the unexpected outcomes of the investigations was the surprisingly small change in the calculated solubility parameters upon derivatization of cellulose, which is performed precisely to increase solubility. Future studies in this area will improve the theoretical model used by including higher degrees of substitution, structural polydispersity, and the effect of moisture. Because the room temperature solubility parameter was found to be unsatisfactory to assess compatibility with ionic liquids, other means such as energy of mixing and interaction energy between layers will also be explored along with an extension of the investigations to higher temperatures. Moreover, some exploratory calculations will be performed to explore some kinetic aspects of the dissolution of cellulose in ionic liquids.

REFERENCES

1. C. M. Hansen. The three dimensional solubility parameter—Key to paint components affinities. I. Solvents, plasticizers, polymers and resins. *J. Paint Technol.* **39**, 1967, 104.
2. R. P. Swatloski, R. D. Rogers and D. Holbrey. US Patent Application Publ. No.: US 2003/0157351 A1, 2003.
3. D. Kralisch, A. Stark, S. Korsten, G. Kreisel and B. Ondruschka. Energetic, environmental and economic balances: Spice up your ionic liquid research efficiency. *Green Chem.* **7**, 2005, 301.
4. R. A. Sheldon. Green solvents for sustainable organic synthesis: State of the art. *Green Chem.* **7**, 2005, 267.
5. A. R. Katritzky, R. Jain, A. Lomaka, R. Petrukhin, M. Karelson, A. E. Visser and R. D. Rogers. Correlation of melting points of potential ionic liquids (imidazolium bromides and benzimidazolium bromides) using the CODESSA program. *J. Chem. Inf. Comput. Sci.* **42**, 2002, 225.
6. H. Sun. COMPASS: An *ab initio* force field optimized for condensed-phase applications—Overview with details on alkane and benzene compounds. *J. Phys. Chem. B* **102**, 1998, 7338.
7. A. D. French, A.-M. Kelterer, G. P. Johnson, M. K. Dowd and C. J. Cramer. HF/6-31G* energy surfaces for disaccharide analogs. *J. Comput. Chem.* **22**, 2001, 65.
8. L. M. J. Kroon-Batenburg, B. Bouma and J. Kroon. Stability of cellulose structures studied by MD simulations. Could mercerized cellulose II be parallel? *Macromolecules* **29**, 1996, 5695.
9. B. Leroux, H. Bizot, J. W. Brady and V. Tran. Water structuring around complex solutes: Theoretical modeling of α-D-glucopyranose. *Chem. Phys.* **216**, 1997, 349.
10. S. A. H. Spieser, J. A. van Kuik, L. M. J. Kroon-Batenburg and J. Kroon. Improved carbohydrate force field for GROMOS: Ring and hydroxymethyl group conformations and exo-anomeric effect. *Carbohydr. Res.* **322**, 1999, 264.
11. R. Palma, M. E. Himmel and J. W. Brady. Calculation of the potential of mean force for the binding of glucose to benzene in aqueous solution. *J. Phys. Chem. B* **104**, 2000, 7228.
12. S. Perez, K. Mazeau and C. H. du Penhoat. The three-dimensional structures of pectic polysaccharides. *Plant Physiol. Biochem.* **38**, 2000, 37.
13. K. N. Kirschner and R. J. Woods. Solvent interactions determine carbohydrate conformation. *Proc. Natl. Acad. Sci. U.S.A.* **98**, 2001, 10541.
14. C. E. Skopec, M. E. Himmel, J. F. Matthews and J. W. Brady. Energetics for displacing a single chain from the surface of microcrystalline cellulose into the active site of *Acidothermus cellulolyticus* Cel5A. *Protein Eng.* **16**, 2003, 1005.
15. S. M. Tschampel and R. J. Woods. Quantifying the role of water in protein–carbohydrate interactions. *J. Phys. Chem.* **107**, 2003, 9175.
16. F. Corzana, M. S. Motawia, C. H. du Penhoat, S. Perez, S. M. Tschampel, R. J. Woods and S. B. Engelsen. A hydration study of (1→4) and (1→6) linked α-glucans by comparative 10 ns molecular dynamics simulations and 500-MHz NMR. *J. Comput. Chem.* **25**, 2004, 573.

17. C. G. Hanke, S. L. Price and R. M. Lynden-Bell. Intermolecular potentials for simulations of liquid imidazolium salts. *Mol. Phys.* **99**, 2001, 801.

18. J. de Andrade, E. S. Boes and H. Stassen. A force field for liquid state simulations on room temperature molten salts 1-ethyl-3-methylimidazolium tetrachoroaluminate. *J. Phys. Chem. B* **106**, 2002, 3546.

19. J. de Andrade, E. S. Boes and H. Stassen. Computational study of room temperature molten salts composed by 1-alkyl-3-methylimidazolium cations—Force field proposal and validation. *J. Phys. Chem. B* **106**, 2002, 13344.

20. S. M. Urahata and M. C. C. Ribeiro. Structure of ionic liquids of 1-alkyl-3-methylimidazolium cations: A systematic computer simulation study. *J. Chem. Phys.* **120**, 2004, 1855.

21. C. Margulis, H. A. Stern and B. J. Berne. Computer simulation of green chemistry room temperature solvent. *J. Phys. Chem. B* **106**, 2002, 12017.

22. J. N. C. Lopes, J. Deschamps and A. A. H. Padua. Modeling ionic liquids using a systematic all-atom force field. *J. Phys. Chem. B* **108**, 2004, 2038.

23. T. Yan, C. J. Burnham, M. G. del Popolo and G. A. Voth. Molecular dynamics simulation of ionic liquids: The effect of electron polarizability. *J. Phys. Chem. B* **108**, 2004, 11877.

24. J. Shah, E. Brennecke and E. J. Maginn. Thermodynamic properties of the ionic liquid 1-*n*-butyl-3-methylimidazolium hexafluorophosphate from Monte Carlo simulation. *Green Chem.* **4**, 2002, 112.

25. R. M. Lynden-Bell, N. A. Atamas, A. Vasilyuk and C. G. Hanke. Chemical potentials of water and organic solutes in imidazolium ionic liquids: A simulation study. *Mol. Phys.* **10**, 2002, 3229.

26. M. G. del Popolo and G. A. Voth. On the structure and dynamics of ionic liquids. *J. Phys. Chem. B* **108**, 2004, 1744.

27. C. G. Hanke, N. A. Atamas and R. M. Lynden-Bell. Solvation of small molecules in imidazolium ionic liquids: A simulation study. *Green Chem.* **4**, 2002, 107.

28. C. G. Hanke, A. Johansson, J. B. Harper and R. M. Lynden-Bell. Why are aromatic compounds more soluble than aliphatic compounds in dimethylimidazolium ionic liquids? A simulation study. *Chem. Phys. Lett.* **374**, 2003, 85.

29. Y. Shim, J. Duan, M. Y. Choi and H. J. Kim. Solvation in molecular ionic liquids. *J. Chem. Phys.* **119**, 2003, 6411.

30. A. Cadena, J. L. Anthony, J. J. Shah, T. I. Marrow, J. F. Brennecke and E. J. Maginn. Why is CO_2 so soluble in imidazolium-based ionic liquids? *J. Am. Chem. Soc.* **126**, 2004, 5300.

31. Y. Shim, M. Y. Choi and H. J. Kim. A molecular dynamics simulation study of room temperature ionic liquids I and II. Equilibrium and nonequilibrium solvation dynamics. *J. Chem. Phys.* **122**, 2005, 044510–044511.

32. B. E. Eichinger, D. Rigby and J. Stein. Cohesive properties of Ultem and related molecules from simulations. *Polymer* **43**, 2002, 599.

33. A. F. M. Barton. *Handbook of Solubility Parameters and Other Cohesion Parameters.* Boca Raton, FL: CRC Press Inc., 1983.

34. D. Rigby, H. Sun and B. E. Eichinger. Computer simulations of poly(ethylene oxide): Force field PVT diagram and cyclization behavior. *Polym. Int.* **44**, 1997, 311.

35. C. Sun. True density of microcrystalline cellulose. *J. Pharm. Sci.* **94**, 2005, 2132.

36. K. Mazeau and L. Heux. Molecular dynamics simulations of bulk native crystalline and amorphous structures of cellulose. *J. Phys. Chem. B* **107**, 2003, 2394.

37. W. Chen, G. C. Lickfield and C. Q. Yang. Molecular modeling of cellulose in amorphous state. Part I: Model building and plastic deformation study. *Polymer* **45**, 2004, 1063.

38. H. F. Mark. *Encyclopedia of Polymer Science and Technology—Plastics, Resins, Rubbers, Fibers*, Volume 3. New York: Wiley, 1982.

39. P. Choi, T. A. Kavassalis and A. Rudin. Estimation of Hansen solubility parameters for (hydroxyethyl)- and (hydroxypropyl)cellulose through molecular simulation. *Ind. Eng. Chem. Res.* **33**, 1994, 3154.

40. W. L. Archer. Determination of Hansen solubility parameters for selected cellulose ether derivatives. *Ind. Eng. Chem. Res.* **30**, 1991, 2292.

41. J. Brandrup, E. H. Immergut and E. A. Grulke. *Polymer Handbook*, 4th edition. New York: Wiley, 1999.

42. R. J. Roberts and R. C. Rowe. The solubility parameter and fractional polarity of microcrystalline cellulose as determined by mechanical measurement. *Int. J. Pharm.* **99**, 1993, 157.

43. F. Bassam, P. York, R. C. Rowe and R. J. Roberts. Young's modulus of powders used as pharmaceutical excipients. *Int. J. Pharm.* **64**, 1990, 55.

44. R. J. Roberts and R. C. Rowe. The Young's modulus of pharmaceutical materials. *Int. J. Pharm.* **37**, 1987, 15.

45. A. B. Mashadi and J. M. Newton. The characterization of the mechanical properties of microcrystalline cellulose: A fracture mechanics approach. *J. Pharm. Pharmacol.* **39**, 1987, 961.

46. L. Zajic and G. Buckton. The use of surface energy values to predict optimum binder selection for granulations. *Int. J. Pharm.* **59**, 1990, 155.

47. S. B. Lee and P. Luner. The wetting and interfacial properties of lignin. *Tappi* **55**, 1972, 116.

48. C. M. Hancock. Material interactions and surface phenomena in size enlargement processes. PhD thesis, Bradford, 1991.

49. N. Huu-Phuoc, H. Nam-Tran, M. Buchmann and U. W. Kesselring. Experimentally optimized determination of the partial and total cohesion parameters of an insoluble polymer (microcrystalline cellulose) by gas–solid chromatography. *Int. J. Pharm.* **34**, 1987, 217.

50. J. Ennari, J. Hamara and F. Sundholm. Vibrational spectra as experimental probes for molecular models of ion-conducting polyether systems. *Polymer* **38**, 1997, 3733.

51. J. Ennari, M. Elomaa and F. Sundholm. Modeling a polyelectrolyte system in water to estimate the ion-conductivity. *Polymer* **40**, 1999, 5035.

52. J. Ennari, I. Neelov and F. Sundholm. Molecular dynamics simulation of the PEO sulfonic acid anion in water. *Comput. Theor. Polym. Sci.* **10**, 2000, 403.

53. J. Ennari, I. Neelov and F. Sundholm. Molecular dynamics simulation of the structure of PEO based solid polymer electrolytes. *Polymer* **41**, 2000, 4057.

54. J. Ennari, M. Elomaa, I. Neelov and F. Sundholm. Modeling of water-free and water containing solid polyelectrolytes. *Polymer* **41**, 2000, 985.

55. B. Derecskei and A. Derecskei-Kovacs. Exploratory molecular modeling studies of interactions between oligomers of cellulose and its derivatives with ionic liquids. *Mol. Simul.* **34** (10), 2008, 1167.

56. S. Chung, S. V. Dzyuba and R. A. Bartsch. Influence of structural variation in room-temperature ionic liquids on the selectivity and efficiency of competitive alkali metal salt extraction by a crown ether. *Anal. Chem.* **105**, 2001, 2437.

5 Molecular Modeling Simulations to Predict Density and Solubility Parameters of Ionic Liquids

Bela Derecskei and Agnes Derecskei-Kovacs

CONTENTS

5.1 INTRODUCTION

Ionic liquids are environmentally friendly solvents representing a novel medium to perform well-established reactions or to develop new ones. Their structure is composed entirely of ions and their melting points are below the boiling point of water.

Favorable properties include nonvolatility, nonflammability, chemical stability, and often thermal stability as well. One of the greatest advantages of ionic liquids is their remarkable versatility: with different combinations of cations and anions, they can generate an astronomical number of different ionic liquids [1]. In spite of the growing interest in this area, only a few ionic liquids have been characterized even by such basic properties as density, conductivity, melting point, and so on. Interactions with other substances, including inorganic, polar, and nonpolar organic and polymeric compounds, can be finely tuned by selecting the proper cation–anion pair. The potential ability to engineer task-specific ionic liquids with predetermined physical and chemical properties is an attractive prospect for facilitating the development of novel green chemical processes.

Computational chemistry has been successfully used to prescreen candidate compounds, especially in the pharmaceutical industry. *In silico* simulations of sufficient predictive accuracy could greatly aid the molecular design of task-specific ionic liquids, reducing the number of experiments to be carried out by eliminating unsuitable candidate compounds.

Mutual compatibility between a solute and a pool of candidate solvents is an important scientific and commercial issue and it is largely determined by the individual components. The basic idea of "like dissolves like" was quantified through the introduction of the Hildebrand solubility parameter and further refined by Hansen [2] who introduced the multicomponent solubility parameters. In previous work [3], we used the commercially available Materials Studio® program package to calculate the solubility parameters for some cellulose derivatives in an on-going effort to study the compatibility of some ionic liquids with these commercially important raw materials. In the current work, results of exploratory simulations are presented for the densities and two component solubility parameters of imidazolium-based ionic liquids with varying alkyl chain length and for four different anions. Density can be a technologically important property, for example, in drilling fluids or phase separation, whereas the dispersive and electrostatic contributions to the solubility parameter are useful when assessing the compatibility of solvents with different solutes.

Computer simulations of ionic liquids have been reported in increasing numbers during the past few years. Some of the modeling work has been focused on the development of force field parameters, specific to an ionic liquid or an ionic liquid family [4–10]. Structural, dynamic, electric, and thermodynamic properties of several pure ionic liquids have been simulated [10–12] using these force field tools, and the solvation of small solutes in ionic liquids has also been investigated [13–19]. More recently, an *ab initio* molecular dynamics study has also been published [20], pushing the capabilities of computational chemistry to its current limits.

5.2 COMPUTATIONAL METHODS

Because ionic liquids carry explicit charges on both the cationic and anionic components, the partial charges assigned to the individual atoms in the force field simulations strongly affect the quality of the calculations (see the Appendix for an illustration). In this work, atomic charges were determined from density functional theory (DFT) calculations to achieve good accuracy. First, all species were constructed using the Builder module in Materials Studio 4.2 by Accelrys and a +1 or −1 overall charge was assigned to the ionic species. Full geometry optimizations were performed by using the gradient-corrected density functional PW91 in conjunction with a polarized double numeric basis set in the all-electron approximation using DMol^3 in Materials Studio [21,22]. Electrostatic potential-derived charges were then calculated and assigned to all atoms. The charges were kept unchanged during the subsequent molecular dynamics simulations.

Model systems for bulk ionic liquids were constructed as amorphous three-dimensional periodic boxes using the Amorphous Cell Tool in Materials Studio. To build proper model systems, the density of the bulk phase must be specified. Because room temperature densities are not presently available for all of the ionic liquids considered here, values were determined via simulation for all of the systems and compared against the available data. The simulation cells contained 20 cations and 20 anions. They were equilibrated first in a 100-ps molecular dynamics run using the NPT ensemble (keeping the particle number, pressure, and temperature controlled while relaxing the cell parameters). This was followed by a 100-ps molecular dynamics production run for mean density using the Berendsen barostat and the Andersen thermostat. Electrostatic and van der Waals terms were determined by the Ewald summation method at the 0.001 kcal/mol accuracy using a 6-Å cutoff value and 0.5-Å buffer width. (Additional details about how the model was built are given in the Appendix.) Three to five amorphous boxes were created and simulated to obtain statistical averages and to determine overall standard deviations. All simulations were run at 298 K.

The *cohesive energy* is the amount of internal energy per mole of a substance in the condensed state arising from all of the intermolecular interactions that hold the substance together. A related quantity, the *cohesive energy density* (the cohesive energy divided by the molar volume of the material), is commonly used in the computation of the energy of mixing. The *solubility parameter* is the square root of the cohesive energy density. Because both dispersion and electrostatic forces contribute to the interaction energies, the solubility parameter will also have dispersive and electrostatic components. These (and H bonding contributions) are frequently used to quickly assess the compatibility of solvents and solutes: similar solubility parameter components imply good compatibility (for a more detailed evaluation, see, for example, Ref. [23]). In the simulation of cohesive energy densities and solubility parameters, three to five amorphous boxes were constructed at the average of the previously calculated densities. After a 100-ps-long equilibration run, 100-ps NVT molecular dynamics simulations were performed, sampling the conformer space with 1-ps frequency. The resulting trajectories were then used in the cohesive energy density and solubility parameter calculations. The calculation as performed in Materials Studio was designed for molecular solvents and the results must be corrected to reflect that ionic liquids will evaporate in the form of neutral ion pairs. The interaction energy of the gas phase ion pair had to be applied as a correction factor to the total solubility parameter as well as its components based on the Hess principle. The interaction energies were calculated at the optimal gas phase ion pair geometry (determined at the PW91/DNP level) using the same force field as the molecular dynamics simulations.

5.3 RESULTS AND DISCUSSION

5.3.1 DENSITY FUNCTIONAL LEVEL CHARACTERIZATION OF THE [RMIM] (1-*R*-3-METHYLIMIDAZOLIUM, *R* = ETHYL, BUTYL, AND HEXYL) IONS

Figure 5.1a, 5.1b, and 5.1c show the electrostatic potential field around the 1-ethyl-3-methylimidazolium [EMIM], 1-butyl-3-methylimidazolium [BMIM], and 1-hexyl-

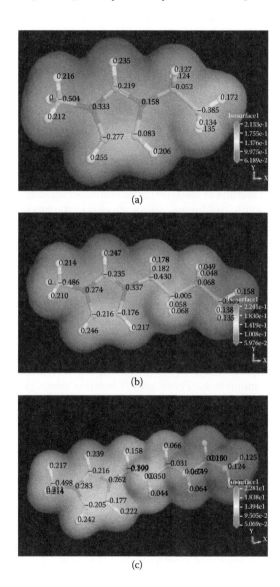

(a)

(b)

(c)

FIGURE 5.1 **(See color insert.)** Electrostatic potential field around various cations as calculated at the PW91/DNP level. (a) 1-Ethyl-3-methylimidazolium (EMIM), (b) 1-butyl-3-methylimidazolium (BMIM), and (c) 1-hexyl-3-methylimidazolium (C_6MIM).

TABLE 5.1

Calculated (D_{cal}) and Experimental D_{exp} [26] Densities of Some Organic Solvents along with the Standard Deviation (σ) of Calculations

	D_{cal} (g/cm³)	σ (g/cm³)	D_{exp} (g/cm³)
Pyridine	0.978	0.022	0.982
Carbon disulfide	1.313	0.035	1.313
Benzyl alcohol	1.037	0.019	1.05
Formamide	1.166	0.029	1.134
Methylimidazole	1.031	0.024	1.036

3-methylimidazolium [C_6MIM] cations as a result of density functional calculations described above.

It can be seen from the sequence of structures that the charge density above the five-membered ring decreases as the chain length of the alkyl substituent increases. This change in the electrostatic properties coupled with the change in the steric nature of the cation will alter many of the physicochemical properties of related ionic liquids, including viscosity, melting point, density, conductivity, and so on [24]. Point charges derived from the electrostatic potential depicted above were assigned to the cations. The anions were treated in a similar manner and these charges were used for the subsequent molecular dynamics simulations.

5.3.2 DENSITIES

5.3.2.1 Method Validation Using Traditional Solvents

To assess the applicability of the modules in Materials Studio 4.2 that are relevant for this application, the COMPASS force field [25] was chosen as currently implemented in the Forcite molecular simulation engine. To test the accuracy of densities calculated in these atomistic simulations, benchmark calculations were performed for some experimentally well-characterized traditional solvents either with some structural similarity to the imidazolium cation (pyridine, methylimidazole) or for some specifically selected value of the solubility parameter components as will be shown later (CS_2, formamide). The results of the simulations (which used 20 solvent molecules and followed the protocol described above) are presented in Table 5.1.

The very good agreement between the computed and experimental density values shown in the table demonstrates that this simple model can predict densities of traditional solvents with good accuracy (<3%) using a relatively small number of molecules. As shown in the Appendix, using a larger number of molecules in the simulation reduces the standard deviation of the calculations but does not affect the average calculated density.

5.3.2.2 Calculated Densities of Ionic Liquids

Results for ionic liquids containing the 1-*R*,3-methylimidazole cation (*R* = ethyl, butyl, and hexyl) and some common anions such as chloride, trifluoroacetate,

TABLE 5.2

Calculated Densities and Standard Deviation of Densities of Ionic Liquids Composed of the 1-R-3-Methylimidazolium Cations (R = Ethyl, Butyl, and Hexyl) and the Chloride, Trifluoroacetate, Dicyanamide, and Bis(trifluoromethylsulfonyl)imide Anions along with Available Experimental and Other Calculated Values

Cation	Anion	D^a (g/cm^3)	σ^a (g/cm^3)	D (g/cm^3)
[EMIM]	Cl⁻	0.961	0.012	1.12[b]
	[CF$_3$CO$_2$]⁻	1.336	0.017	–
	[(CN)$_2$N]⁻	1.095	0.010	–
	[Tf$_2$N]⁻	1.520	0.022	1.519[c]
[BMIM]	Cl⁻	0.919	0.012	1.05[b], 1.08[d]
	[CF$_3$CO$_2$]⁻	1.233	0.015	1.22[e]
	[(CN)$_2$N]⁻	1.051	0.010	1.058[f]
	[Tf$_2$N]⁻	1.436	0.0185	1.436[c]
[C$_6$MIM]	Cl⁻	0.910	0.011	1.00[b], 1.03[d]
	[CF$_3$CO$_2$]⁻	1.175	0.013	–
	[(CN)$_2$N]⁻	1.020	0.010	–
	[Tf$_2$N]⁻	1.400	0.0189	1.373[c]

[a] This work.
[b] Ref. [9] (calculated).
[c] Ref. [29] (experimental).
[d] Ref. [27] (experimental).
[e] Ref. [28] (experimental).
[f] Ref. [30] (experimental).

dicyanamide, and bis(trifluoromethylsulfonyl)imide are presented in Table 5.2 with the available experimental and other calculated values.

Table 5.2 shows that the calculated densities are in very good agreement with the experimental values where those are available with the exception of all systems containing the chloride ion. The larger errors in density (>10%) for Cl⁻ cases most likely indicate that this implementation of COMPASS is not appropriate for the simulation of this family of ionic liquids.* It is expected, however, that all other calculated values are reasonable estimates for the room temperature densities of ionic liquids that are yet-to-be characterized experimentally (or with unpublished density measurements).

---

* The underlying reason, most likely, is the inability of the force field to predict interaction energies between the cations and the chloride ion in a satisfactory fashion. In spite of seeing this discrepancy when comparing cation–anion interaction energies at the DFT and force field levels, we still attempted to carry out the simulations due to the frequent applications of chloride-containing ionic liquids. Unfortunately, after the density simulations, we had to give up this effort until some developments in further releases of the software.

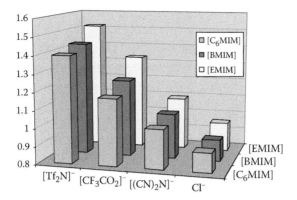

FIGURE 5.2 Cation and anion effects on the calculated room temperature density of some imidazolium-based ionic liquids.

These results are shown graphically in Figure 5.2 to observe the trends introduced by changing the anions and systematically increasing the side chain length in the [RMIM] cation.

Figure 5.2 shows that the density decreases with increasing alkyl chain length, and the extent of the change increases with the size of the anion. Similar behavior was observed experimentally by Tokuda et al. [24] using the same family of cations. The selection of the anion appears to be even more significant: densities can be engineered to span the range of 0.9 g/cm³–1.5 g/cm³ even with this very limited selection.

5.3.3 COHESIVE ENERGY DENSITIES AND SOLUBILITY PARAMETERS OF IONIC LIQUIDS

5.3.3.1 Method Validation Using Organic Solvents

To assess the compatibility of solvent and solute, it is often necessary to examine the various contributions to the solubility parameter rather than the net value. Experimental tables often list at least three components: van der Waals (dispersive), Coulombic, and H bonding (the last two may be considered as electrostatic components). Because most commercial force fields do not contain a separate term for H bonding, calculations yield a van der Waals, δ_{vdW}, and an electrostatic, δ_{ES}, component. For the sake of direct comparison to the experiment, we calculated an equivalent electrostatic term from the Coulomb and H bonding values using the formula

$$\delta_{ES} = (\delta_C^2 + \delta_H^2)^{1/2}$$

where δ_C and δ_H are the Coulomb and H bonding components of the solubility parameter, respectively.

Table 5.3 contains our calculated values along with the experimental data for some molecular solvents (at the experimental densities).

TABLE 5.3

Calculated (Calc.) and Experimental (Exp.) [23] Values of the Different Components of the Solubility Parameter of Some Molecular Solvents

	δ_{vdW}		δ_{ES}		δ_C	δ_H
	Calc.	Exp.	Calc.	Exp.	Exp.	Exp.
Pyridine	19.3	19.0	8.9	10.6	8.8	5.9
Carbon disulfide	20.3	20.5	3.8	0.6	0.0	0.6
Benzyl alcohol	18.8	18.4	16.4	15.1	6.3	13.7
Formamide	20.9	17.2	29.3	32.4	26.2	19.0
Methylimidazole	19.3	19.7	19.3	19.2	15.6	11.2

Pyridine and methyl imidazole were chosen for the validation calculations because they exhibit some structural similarity to the imidazolium cation, present in the ionic liquids under investigation. Carbon disulfide was chosen as a limiting case because it has a negligible electrostatic component, whereas formamide was chosen because it has both large dispersive and electrostatic components. Benzyl alcohol exhibits an average behavior.

It can be seen from Table 5.3 that the dispersive components were reproduced well with these simulations. Formamide had the largest deviation from the experiment. The electrostatic components had larger deviations in general, with carbon disulfide exhibiting the worst agreement. However, the trend of the polar components was reproduced correctly. Simulated components of the solubility parameter for methyl imidazole (which is structurally the closest to the cation component of the ionic liquids of interest) were the closest to the experimental values.

TABLE 5.4

Calculated Solubility Parameters of Ionic Liquids (Components as well as Total) Composed of the 1-R-3-Methylimidazolium Cation and Three Different Anions

Cation	Anion	δ_{vdW}	δ_{ES}	δ_{total}
[EMIM]	[CF$_3$CO$_2$]$^-$	25.3	24.6	34.9
	[(CN)$_2$N]$^-$	23.7	20.8	35.1
	[Tf$_2$N]$^-$	21.9	17.8	28.9
[BMIM]	[CF$_3$CO$_2$]$^-$	23.8	21.3	31.6
	[(CN)$_2$N]$^-$	24.2	20.4	34.0
	[Tf$_2$N]$^-$	21.1	16.0	26.1
[C$_6$MIM]	[CF$_3$CO$_2$]$^-$	23.7	19.7	30.5
	[(CN)$_2$N]$^-$	23.4	18.4	32.2
	[Tf$_2$N]$^-$	21.2	12.7	25.5

Note: A: R = ethyl, B: R = butyl, and C: R = hexyl.

TABLE 5.5
Comparison of Calculated and Experimental Values of the Solubility Parameter for Some Ionic Liquids Containing the Imidazolium Cation and the [Tf$_2$N]$^-$ Anion

Cation	Anion	$\delta_{total, Calc.}$[a]	$\delta_{total, Exp.}$[b]
[EMIM]	[Tf$_2$N]$^-$	28.9	27.6
[BMIM]	[Tf$_2$N]$^-$	26.1	26.7
[C$_6$MIM]	[Tf$_2$N]$^-$	25.5	25.6

[a] This work.
[b] Ref. [29].

5.3.3.2 Solubility Parameters of Ionic Liquids

The solubility parameters (total as well as the van der Waals and electrostatic components) for the family of ionic liquids studied are summarized in Table 5.4 (chloride-containing ionic liquids were omitted due to reasons explained above).

Experimental data were found only for the total solubility parameter of the [Tf$_2$N]$^-$ series [29]. The Hildebrand solubility parameters as determined from intrinsic viscosity measurements in Ref. [29] are listed along with our calculated total solubility parameter values in Table 5.5.

Table 5.5 shows a good agreement between the calculated and experimental values. In addition to the total value determined experimentally, one can also analyze the contribution of the dispersive and electrostatic components as presented in Table 5.4. For all anions, the change in the dispersive component is moderate as the carbon chain length increases. Almost all variations in the total solubility parameter are due to changes in the polar component, which decreases with increasing carbon chain length (which is most likely related to the trends in the charge density depicted in Figure 5.1). In contrast, the nature of the anion has an impact on both the disperse and polar components. With these few anions, the disperse components can be engineered to be in between 21 and 25, and the polar components can be selected in the range of 13–25.

5.4 CONCLUSIONS

Calculated room temperature densities of the imidazolium-based ionic liquids considered here were in good agreement with experimental values for a variety of anions but were poor for ionic liquids containing the chloride ion. Both the experimental values and the general trends were correctly reproduced when cations and anions were changed systematically. Based on these results, the commercially available COMPASS force field as implemented in Materials Studio can be used in simulations to predict densities of ionic liquids. However, partial charges on atoms must be determined at a higher level of theory, such as our choice to use density functional calculations and assign electrostatic potential-derived atomic charges. It must be noted that some other anions that are present in commercially available ionic liquids (such as larger halogens and polyvalent sulfur and phosphorous-containing anions) currently may not be included in simulations with this software package due to the incompleteness of the set of the necessary force field parameters.

The COMPASS force field was found to be able to predict the components of the solubility parameter for some traditional solvents and the total solubility parameter for the few ionic liquids where they were available from experimental studies. The cation selection affected mainly the electrostatic component, whereas the anion selection had an impact on both the dispersive and electrostatic contributions. Future experimental studies are necessary to check the validity of these findings, and many more ionic liquid families will have to be studied experimentally and in simulation studies to be able to select the proper solvent purely on the basis of the components of the solubility parameter.

APPENDIX: FACTORS AFFECTING THE OUTCOMES OF THE SIMULATIONS

1) EFFECT OF THE SYSTEM SIZE

To test the sensitivity of the calculated densities to the size of the model unit cell, we created sets of five amorphous cells consisting of 10, 20, and 30 ion pairs each using [BMIM] [(CN)$_2$N] as a test case. After following the protocol described earlier, the density data were evaluated using the Minitab15 statistical program package. Results of the basic statistical analysis are shown in Figure 5.3a, 5.3b, and 5.3c.

Visual inspection of the distributions shows that the density data distribution is normal for all three cases. The standard deviation becomes smaller with increasing system size. The mean density varies only moderately with the size of the simulation box: the smallest box (10 ion pairs) yielded a statistically different mean density (1.0906 g/cm^3) than the two larger boxes (1.0951 and 1.0952 g/cm^3), which were found to be statistically the same. The box plot visualization of the data set illustrates this finding.

It must be pointed out that even the density of the smallest simulation box is well within the range of technical grade (but not analytical grade) compounds, commonly used in most industries. Increasing the system size from 20 ion pairs to 30 did not introduce a significant improvement of the results; consequently, 20 ion pair amorphous boxes were used in all simulations described in this work to save computational time during the simulations without losing computational accuracy (Figure 5.4).

2) EFFECT OF THE LENGTH OF THE SIMULATION

Starting from the same five amorphous boxes (each containing 10 ion pairs and pre-equilibrated for 100,000 fs), we executed independent production runs of 100,000 and 500,000 fs. The inspection of the box plots showed that increasing the length of simulation introduced almost no change in the mean density and only a very moderate change in the standard deviation (Figure 5.5).

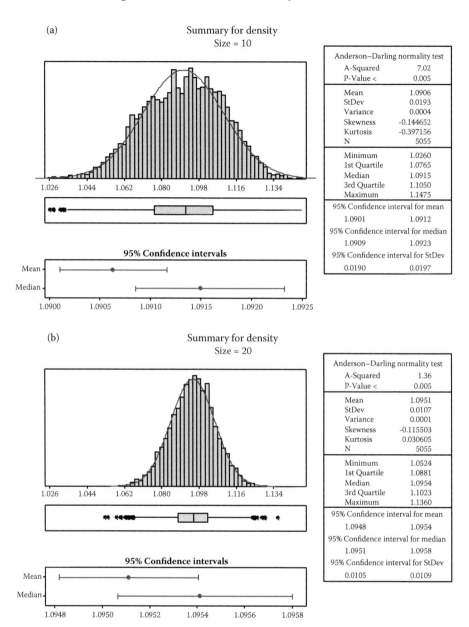

FIGURE 5.3 Results of basic statistics for the density distribution of the molecular dynamics simulation of the [BMIM][(CN)$_2$N]$^-$ ionic liquid simulation boxes containing (a) 10, (b) 20, and (c) 30 ion pairs.

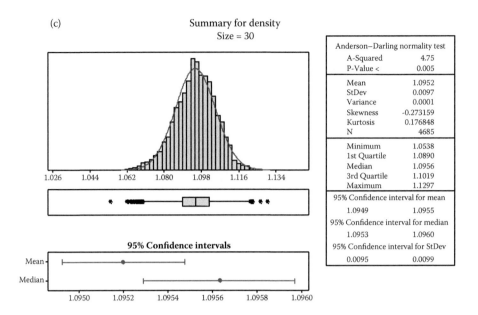

(c)

Summary for density
Size = 30

Anderson–Darling normality test	
A-Squared	4.75
P-Value <	0.005
Mean	1.0952
StDev	0.0097
Variance	0.0001
Skewness	-0.273159
Kurtosis	0.176848
N	4685
Minimum	1.0538
1st Quartile	1.0890
Median	1.0956
3rd Quartile	1.1019
Maximum	1.1297
95% Confidence interval for mean	
1.0949	1.0955
95% Confidence interval for median	
1.0953	1.0960
95% Confidence interval for StDev	
0.0095	0.0099

FIGURE 5.3 (continued)

Because there was no additional advantage to be gained for the longer production runs, we used 100,000 fs in all simulations (following 100,000-fs preequilibration molecular dynamics runs).

3) EFFECT OF THE CHARGE ASSIGNMENT

To investigate the role of the charge assignment, we performed benchmark studies on formamide, which is a molecular solvent with large dispersive and electrostatic

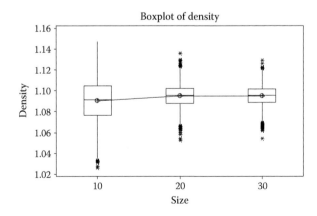

FIGURE 5.4 Box plot representation of the density distribution of the molecular dynamics simulation of the [BMIM][(CN)$_2$N]$^-$ ionic liquid simulation boxes containing 10, 20, and 30 ion pairs showing the means and the middle quartiles.

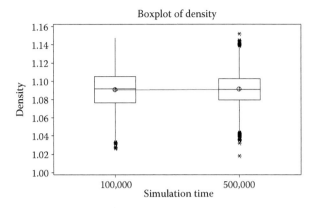

FIGURE 5.5 Box plot representation of the density distribution of two molecular dynamics simulations of different lengths for a [BMIM][(CN)$_2$N]$^-$ ionic liquid simulation box containing 10 ion pairs showing the means and the middle quartiles.

components of the solubility parameter. In two sets of calculations, we used atomic charges as assigned by COMPASS and as calculated from the electrostatic potential at the density functional level. The results along with the experimental values are shown in Table 5.6.

These data show that while the density changes only slightly when the force field assigned charges are replaced by those determined at the density functional level, the components of the solubility parameters show some improvement. Similar studies on ionic liquids (not shown here) exhibit even larger differences. Analogous observations were made earlier [3] for oligomers mimicking some cellulose derivatives in the amorphous phase.

4) EFFECT OF FORCE FIELD SELECTION

Because there are several force fields available for use in the Forcite engine each having certain advantages and disadvantages, we also carried out a set of simulations using the Dreiding force field, which is somewhat less sophisticated than COMPASS. To decouple the effect of the force field from the effect of the charge

TABLE 5.6

Two Components of the Solubility Parameter of Formamide as Determined in Molecular Dynamics Simulations Using Charges Assigned by the COMPASS Force Field and Electrostatic Potential-Derived Charges along with the Experimental Values

	D (g/cm^3)	δ_{vdW}	δ_{ES}
COMPASS	1.160	21.5	28.2
DFT	1.166	20.9	29.3
Exp.	1.134	17.2	32.4

TABLE 5.7
Two Components of the Solubility Parameter of Formamide as Determined in Molecular Dynamics Simulations Using the Dreiding Force Field with Charges Obtained by Charge Equilibration (Charge Eq.) and Determined by DFT Calculations along with the Experimental Values

	D (g/cm³)	δ_{vdW}	δ_{ES}
Dreiding, Charge Eq.	1.102	8.9	31.6
Dreiding, DFT	1.021	7.7	29.7
Exp.	1.134	17.2	32.4

assignment, we performed simulations both with using charges as determined from charge equilibration and at the DFT level. The results are shown in Table 5.7 for formamide along with experimental values.

Clearly, changing the force field from COMPASS to Dreiding had a large negative impact both on the calculated density and the dispersive component of the solubility parameter. Somewhat surprisingly, the electrostatic component of the solubility parameter showed some moderate improvement.

5) EFFECT OF DENSITY INACCURACIES ON THE SOLUBILITY PARAMETERS

Based on the data in Tables 5.6 and 5.7, the densities determined from molecular simulations have some inaccuracies associated with the selection of the force field, assignment of charges, and other details of the calculations. Because the calculated values of the solubility parameter also have the same computational inaccuracies, it seems to be worthwhile to investigate the extent of error introduced by a known error in the density calculation. For these studies, we built three copies of amorphous

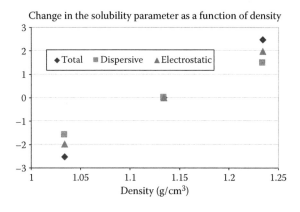

FIGURE 5.6 Changes in the calculated components of the solubility parameter of formamide as a function of density of the simulation box using the values at the experimental density as reference.

boxes containing 20 formamide molecules at the experimental density and at 0.1 g/cm^3 higher and lower densities (about 9% change). The solubility parameters for all systems were then determined as usual. In this range, the components of the solubility parameters changed linearly with the density of the system and the differences relative to the value at the experimental density are shown in Figure 5.6.

It can be seen in Figure 5.6 that even an almost 10% inaccuracy in the density introduces only a <2 unit (about 7%) change in the solubility parameters. Formamide was chosen for this test because it has both large van der Waals and electrostatic contributions and it is expected that the other molecular solvents are even less sensitive to changes in the density.

ACKNOWLEDGMENTS

The authors would like to thank the Cristal Global management team led by Bob Daniels (VP of R&D) and Rob McIntyre (Director of Performance Chemicals) for the time, hardware, and software resources provided for this work. They would also like to thank Professor Tom Welton at Imperial College London, U.K., for the valuable discussions on the subject of ionic liquids.

REFERENCES

1. A. R. Katritzky, A. Lomaka, R. Petrukhin, M. Karelson, A. E. Visser and B. D. Rogers. Correlation of the melting points of potential ionic liquids (imidazolium bromides and benzimidazolium bromides) using the CODESSA program. *J. Chem. Inf. Comput. Sci.* **42**, 2002, 225.
2. C. M. Hansen. The three dimensional solubility parameter—key to paint components affinities. I. Solvents, plasticizers, polymers and resins. *J. Paint Technol.* **39**, 1967, 104.
3. B. Derecskei and A. Derecskei-Kovacs. Molecular dynamic studies of the compatibility of some cellulose derivatives with selected ionic liquids. *Mol. Simul.* **32**, 2006, 109.
4. C. G. Hanke, S. L. Price and R. M. Lynden-Bell. A simulation study of water–dialkylimidazolium ionic liquid mixtures. *Mol. Phys.* **99**, 2001, 801.
5. J. de Andrade, E. Boes and H. Stassen. A force field for liquid state simulations on room temperature molten salts: 1-Ethyl-3-methylimidazolium tetrachloroaluminate. *J. Phys. Chem. B* **106**, 2002, 3546.
6. J. de Andrade, E. Boes and H. Stassen. Computational study of room temperature molten salts composed by 1-alkyl-3-methylimidazolium cations force-field proposal and validation. *J. Phys. Chem. B* **106**, 2002, 13344.
7. S. M. Urahata and M. C. C. Ribeiro. Structure of ionic liquids of 1-alkyl-3-methylimidazolium cations: A systematic computer simulation study. *J. Chem. Phys.* **120**, 2004, 1855.
8. C. Margulis, H. Stern and B. Berne. Computer simulation of a "green chemistry" room-temperature ionic solvent. *J. Phys. Chem. B* **106**, 2002, 12017.
9. J. N. C. Lopes, J. Deschamps and A. A. H. Padua. Modeling ionic liquids using a systematic all-atom force field. *J. Phys. Chem. B* **108**, 2004, 2038.
10. T. Yan, C. J. Burnham, M. G. del Popolo and G. A. Voth. Molecular dynamics simulation of ionic liquids: The effect of electronic polarizability. *J. Phys. Chem. B* **108**, 2004, 11877.
11. J. Shah, J. Brennecke and E. Maginn. Thermodynamic properties of the ionic liquid 1-n-butyl-3-methylimidazolium hexafluorophosphate from Monte Carlo simulations. *Green Chem.* **4**, 2002, 112.

12. R. M. Lynden-Bell, N. A. Atamas, A. Vasilyuk and C. G. Hanke. Chemical potentials of water and organic solutes in imidazolium ionic liquids: A simulation study. *Mol. Phys.* **100**, 2002, 3229.

13. M. G. del Popolo and G. A. Voth. On the structure and dynamics of ionic liquids. *J. Phys. Chem. B* **108**, 2004, 1744.

14. C. G. Hanke, N. A. Atamas and R. M. Lynden-Bell. Solvation of small molecules in imidazolium ionic liquids: A simulation study. *Green Chem.* **4**, 2002, 107.

15. C. G. Hanke, A. Johansson, J. B. Harper and R. M. Lynden-Bell. Why are aromatic compounds more soluble than aliphatic compounds in dimethylimidazolium ionic liquids? A simulation study. *Chem. Phys. Lett.* **374**, 2003, 85.

16. Y. Shim, J. Duan and H. J. Kim. Solvation in molecular ionic liquids. *J. Chem. Phys.* **119**, 2003, 6411.

17. C. Cadena, J. L. Anthony, J. K. Shah, T. I. Marrow, J. F. Brennecke and E. J. Maginn. Why is CO_2 so soluble in imidazolium-based ionic liquids? *J. Am. Chem. Soc.* **126**, 2004, 5300.

18. Y. Shim, M. Y. Choi and H. J. Kim. A molecular dynamics computer simulation study of room-temperature ionic liquids. I. Equilibrium solvation structure and free energetics. *J. Chem. Phys.* **122**, 2005, 044510.

19. Ibid., A molecular dynamics computer simulation study of room-temperature ionic liquids. II. Equilibrium and nonequilibrium solvation dynamics. 044511.

20. M. G. del Popolo, R. M. Lynden-Bell and J. Kohanoff. *Ab initio* molecular dynamics simulation of a room temperature ionic liquid. *J. Phys. Chem.* **109**, 2005, 5895.

21. B. Delley. An all-electron numerical method for solving the local density functional for polyatomic molecules. *J. Chem. Phys.* **92**, 1990, 508.

22. Ibid. From molecules to solids with the DMol³ approach. **113**, 2000, 7756.

23. C. M. Hansen. *Hansen Solubility Parameters. A User's Handbook.* Boca Raton: CRC Press LLC, 2000.

24. H. Tokuda, K. Hayamizu, K. Ishii, Md. A. B. H. Susan and M. Watanabe. Physicochemical properties and structures of room temperature ionic liquids. 2. Variation of alkyl chain length in imidazolium cation. *J. Phys. Chem. B* **109**, 2005, 6103.

25. H. Sun. COMPASS: An *ab initio* force field optimized for condensed-phase applications—Overview with details on alkane and benzene compounds. *J. Phys. Chem. B* **102**, 1998, 7338.

26. C. D. Hodgman. *Handbook of Chemistry and Physics*, 42nd edition. Cleveland, OH: The Chemical Rubber Publ. Co., 1961.

27. J. G. Huddleston, A. E. Visser, W. Reichert, M. Wilauer, D. Heather, G. A. Brocker and R. D. Rogers. Characterization and comparison of hydrophilic and hydrophobic room temperature ionic liquids incorporating the imidazolium cation. *Green Chem.* **3**, 2001, 156.

28. http://ildb.merck.de/ionicliquids/ASP/IonicLiquidsRead_PrintDetails.asp.

29. S. H. Lee and S. B. Lee. The Hildebrand solubility parameters, cohesive energy densities and internal energies of 1-alkyl-3-methylimidazolium-based room temperature ionic liquids. *Chem. Commun.* **27**, 2005, 3469.

30. C. P. Fredlake, J. M. Crosthwaite, G. D. Hert, S. N. V. K. Aki and J. F. Brennecke. Thermophysical properties of imidazolium-based ionic liquids. *J. Chem. Eng. Data* **49**, 2000, 954.

6 Semiempirical Molecular Orbital Study of Freestanding and Fullerene-Encapsulated Mo Nanoclusters

James A. Elliott and Yasushi Shibuta

CONTENTS

6.1   INTRODUCTION

The development of large-scale, high-purity production methods for carbon nanotubes (CNTs) is highly desirable for further increasing the practical applications of these fascinating materials. In addition to the previously known laser furnace [1] and arc discharge techniques [2], catalytic chemical vapor deposition (CCVD) methods [3–10] have been widely pursued as a potential low-cost, large-scale production method. In earlier versions of this technique, such as the high-pressure carbon monoxide method [7], floating catalysts were used during the CVD process. However, more recently, a high-purity technique for production of vertically aligned single-walled CNTs on a quartz substrate has been developed using a supported catalyst [8], which was followed by a later refinement by the addition of water, resulting in very long (millimeter-thick) nanotube forests [9], and later termed "super growth" [11]. Koziol et al. [12] have reported mechanical property data from high-performance CNT fibers, in which the addition of sulfur was critical for promoting the formation of thin-walled nanotubes on the catalyst surface [5]. More recently, the same authors have also demonstrated the role of nitrogen in stabilizing the structure of iron carbide catalyst particles [13], thus providing some degree of control over the chiral angle of CNTs synthesized.

To achieve higher efficiencies for the CCVD method, and a greater degree of control over the type and structure of CNTs produced, the role of catalyst metals has been widely studied [10,14]. It is well known that the catalytic metals are essential for synthesis of single-wall CNTs (SWCNTs). The mechanism by which this occurs is commonly thought to be according to the vapor–liquid–solid (VLS) model [15,16], which involves the following steps: adsorption and decomposition of the carbon source molecules at the surface of the catalytic metal, followed by diffusion of the carbon atoms within the particle, and their segregation and graphitization once the metal cluster is saturated with carbon. There exist an abundance of transmission electron microscope images of the catalytic metals attached to the tip or root of SWCNTs [17,18], which support such a mechanism. Furthermore, studies of the formation process by molecular dynamics (MD) simulation [19,20] are also consistent with the VLS model. It is therefore plausible that the action of additives such as oxygen and sulfur, which are a prerequisite for super growth [11] or rapid production of fibers [5], is to affect one or more stages of carbon incorporation into CNTs according to the VLS model.

At present, it is not yet clear which metal is most suitable as a catalyst of CNT growth, although various transition metals (including iron, cobalt, nickel, and molybdenum) and their alloys are commonly used. For example, Herrera et al. found that the selectivity of Co–Mo catalysts toward formation of SWCNTs depended on stabilization of Co species in a nonmetallic state, resulting from an interaction with Mo that is a function of the Co/Mo ratio [21]. On the other hand, Lan et al. claimed that Mo may not in fact be necessary for the formation of SWNCTs, which they produced using Co alone [22]. The catalytic activities of various transition metals in the formation of SWCNTs by a laser-furnace technique were compared [23], resulting in three criteria that could be used to determine a "better" catalyst: graphitization ability [24], low solubility in carbon, and stable crystallographic orientation on graphite. Flahaut et al. found that a variation in the proportion of Mo with respect to Co in a mixed (Mg, Co, and Mo) oxide catalyst had an influence on the yield, structure, and purity of tubes produced [25]. Hu et al. studied the morphology and chemical state of Co–Mo catalysts using transmission electron microscopy and x-ray photoelectron spectroscopy, and found that particles composed of Co molybdates and metallic Co are important for the promotion of SWCNT growth [26]. Recently, a combinatorial method for examining the best concentration ratio of Mo–Co binary metal nanoparticles [27] or Fe/Al_2O_3 catalysts [28] was investigated, and millimeter-thick forests of nanotubes were produced [28].

In parallel to the experimental approaches, numerical simulations have contributed to the interpretation of the role of the catalytic metals during CNT growth. The CNT formation process has been investigated by various types of calculation, including classical MD [19,20,29,30], *ab initio* MD [31,32], *ab initio* molecular orbital (MO) methods [33], and tight-binding Monte Carlo [32,34]. In general, each level of the calculation has its own advantages and disadvantages: *ab initio* methods have high chemical precision and a few free parameters, but the time and size scales of such simulations are extremely restricted by their computational complexity [35]. Hence, an appropriate method should be selected for the purpose of the calculation in every case. One way of overcoming the computational limit is to develop a multi-scale modeling approach for understanding the formation process of CNTs [36,37]. However, it is difficult to describe the entire process of CNT growth, from carbon source molecules to a forest of cylindrical tubes, although recent simulation work using a reactive force field has captured the step-by-step atomistic nucleation of a CNT with definable chirality [38].

Due to their important role in CNT growth, among other reasons, the thermodynamic stability of different types of catalytic metal nanoclusters, including the noble metals Au [39,40] and Ag [41], and various transition metals [42–45], has been studied using an atomistic modeling approach. The general issues of how structural properties of nanoclusters depend on their energetics, thermodynamics, and kinetic effects have been recently reviewed by Baletto and Ferrando [46], and, for transition metal clusters, in particular, Alonso has summarized their electronic, magnetic, and structural properties [47]. Therefore, we shall not give a detailed discussion here; however, it suffices to say that even using very simple empirical pair potentials (e.g., Lennard–Jones [48]) or embedded atom models (e.g., Finnis–Sinclair [49], Sutton–Chen [50], and Gupta [51]), finding the global minimum energy configuration for a pure metal nanocluster of more than 100 atoms is an extremely challenging problem [52]. For *ab initio* methods, and alloyed systems, an exhaustive search is therefore out of the question. Nevertheless, there are a range of general structural motifs, some with only short-range order (e.g., icosahedral or decahedral clusters) and some with long-range order (e.g., closed-packed clusters) that occur regularly in a wide range of systems due to stabilization either by geometric or electronic effects [46], leading to certain "magic sizes" predominating in distributions of clusters. In the case of clusters of transition metals typically used for catalyzed CNT growth, in which the d electrons in unfilled shells are localized, geometric stabilization is the dominant effect [47].

The properties of the Mo and Mo–S nanoclusters have been examined previously using *ab initio* calculations, and their structural stability is discussed in Refs. [53–60]. Most studies used density functional theory (DFT) with a generalized gradient approximation for the exchange-correlation functional and focused on the role of MoS_x clusters in heterogeneous catalysis. The main conclusion of relevance to current work is that the lowest energy structures tend to have a metal core with sulfur atoms at the surface. However, in general, the size of the clusters was too small (typically from 1 to 10 Mo atoms, although Li and Galli have recently presented results on triangular $(MoS_2)_n$ platelets with n up to 78 [60]) to discuss the role of the Mo during CNT growth due to the computational demands of DFT. For this reason, a semiempirical MO (SEMO) calculation using the AM1* Hamiltonian was used in this study to investigate the stability of freestanding and fullerene-encapsulated Mo clusters. A more detailed justification for the choice of SEMO methods is given in Section 6.2, and the rest of the paper is structured as follows. First, the energy and structure of the freestanding clusters of Mo_n and $Mo_{(n-x)}X_x$ (with X: O, S, N) are examined. Then, the energy and structure of fullerene-encapsulated $Mo_n@C_{180}$

clusters are investigated, focusing on the defect formation and breakup of fullerene. Finally, we conclude with a discussion of the effect of sulfur addition in $Mo_{(52-x)}S_x@$ C_{180} encapsulated molybdenum sulfide clusters.

6.2 COMPUTATIONAL METHODOLOGY

SEMO methods based on the neglect of diatomic differential overlap (NDDO) approximation have been widely used on small- and medium-sized organic molecules after the publication in 1977 of the first NDDO-type method: MNDO (which stands for modified neglect of diatomic overlap) by Dewar and Thiel [61] for molecules containing hydrogen, carbon, nitrogen, and oxygen [62]. Since then, more sophisticated Hamiltonians based on NDDO, such as AM1 and PM3, have extended coverage of SEMO methods to all main group elements, and the variety of applications and successes of these techniques is summarized in the following review [63]. More recently, with the publication by Stewart [64] of NDDO-type parameter sets including Co, Ni, and Fe as part of the PM6 Hamiltonian, their application has increased to more than 70 elements in the periodic table. However, it is important to realize that the degree of empiricism inherent in SEMO methods, which are based on Hartree–Fock (HF) theory, is quite different to that of the so-called "empirical" force field models, such as those for metal clusters mentioned in the preceding section. NDDO neglects only differential overlap between atomic orbitals on different atoms and retains all two-electron, two-center integrals on the same atoms [35]. To compensate for this neglect, MNDO introduces parametric expressions for the two-center integrals, which are derived from charge distributions around the interacting atoms. The parameters in these expressions are determined by fitting to heats of formation, molecular geometries, ionization potentials, and dipole moments of a wide range of molecules. In this way, it is possible to compensate for deficiencies such as neglect of electronic correlation and the simplifying assumptions of NDDO. SEMO models are therefore usually transferable between different types of molecule, provided that representative structures have been included in the training data set. By contrast, empirical force field models are usually based on much more severe semiclassical approximations (such as bead-spring or mean-field representations) and must be carefully refitted for each new type of system or change of atomic composition. SEMO methods therefore have the advantage of having fewer variable parameters and can yield results with precision comparable to some post-HF *ab initio* methods with a much lower computational cost [65]. In Section 6.3.1, we compare SEMO results for a Mo cluster (based on AM1*, described below) directly with DFT using a hybrid exchange-correlation functional (B3LYP) [66,67] with a minimal basis set.

Until recently, NDDO-type SEMO calculations based on MNDO were restricted to main group elements containing only s and p electrons, although older types of SEMO methods had been used for calculating spectroscopic properties [68] and geometries [69] of transition metal compounds. Voityuk and Rösch were the first to describe an extension of AM1 to d orbitals, which they called AM1/d, and have reported parameters for Mo [70]. Their approach was based on an extended multipole–multipole interaction scheme [69] and the introduction of two bond-specific parameters for Mo in the core–core repulsion term [70]. The AM1/d scheme has the advantage that results obtained for nontransition-metal atoms are identical to those for the original AM1 method, whereas the former is additionally able to reproduce chemical properties of complex organometallic and bioinorganic compounds of Mo with high precision [70]. The Mo parameters in AM1/d were later incorporated in a slightly modified form by the Clark group into their AM1* Hamiltonian [71,72], which uses a distance-dependent core–core repulsion parameter for some interactions. The same group then later reported AM1* parameters for Al, Si, Ti, and Zr [73]. The AM1* Hamiltonian is implemented in the SEMO software package VAMP, now part of Materials Studio® (MS) from Accelrys, which, at the time of writing (MS version 4.2), contains parameters for the following elements: H, C, N, O, F, Al*, Si*, P*, S*, Cl*, Ti*, Cu*, Zn*, Zr*, and Mo*, with the asterisked elements containing parameter sets that include d orbitals. In this way, it has been possible to use AM1* to study Mo clusters in contact with C, O, S, and N, which are of relevance to CNT production via a CCVD process, without further parameterization. Optimized geometries and single point energies were computed using the standard eigenvector following (EF) and self-consistent field (SCF) convergence algorithms in VAMP [74]. Since this work was carried out, further recent progress has been made using PM6 to extend this work to comparative studies of other transition metal elements typically used in the CCVD processes described earlier [52]. The results from PM6 are broadly consistent with AM1* in the case of molybdenum, but cover a wider range of transition metals including iron, nickel, and cobalt.

Before describing the results obtained with AM1* on Mo-based clusters, we will briefly remark on one further simplification made in this study: the use of spin-restricted (RHF) methods for all calculations. Because the ground state electronic configuration of Mo is $[Kr]5s^14d^5$, it may generally be expected that the unpaired d electrons will give rise to a net magnetic moment for a Mo cluster. For small clusters, interactions between the electronic spins of adjacent atoms can be very significant, although the magnetic moments for typical transition metal clusters (such as Fe, Co, and Ni) are known to converge to their bulk values for sizes larger than a few hundred atoms [47]; this effect can be understood in terms of the local under-coordination of surface atoms in the smaller clusters. Although it is straightforward to express HF theory, and therefore SEMO methods, in a spin-polarized framework, it was found at a very early stage of this work that spin-unrestricted (UHF) calculations on pure Mo clusters resulted in very high values of the total electronic spin. To avoid the effects of spin contamination, in which contributions from excited states are mixed with the ground state wavefunction, it is conventional to require that the expectation value of total spin operator $\langle S^2 \rangle$ lies within 10% of $S(S + 1)$, where S is the total electronic spin [75]. Because the spin state of the Mo clusters is, in general, unknown *a priori*, it is difficult to eliminate the effects of spin contamination of the ground state wavefunction and, hence, to avoid any systematic errors due to mixing with excited states using UHF. Therefore, all optimizations were carried out using RHF under the assumption that the electron spins are paired. This is possible, in general, for all Mo_n and $Mo_{(n-x)}X_x$ clusters, where $X = \{O, N\}$, because each atom has an even number of valence electrons, but where $X = \{N\}$, nitrogen atoms were added in pairs to maintain an even total number of electrons. Under this assumption, the effects of magnetism are thus ignored as a first approximation. However, previous DFT calculations by Murugan et al. on small Mo–S clusters have shown that structural changes due to spin polarization are small [59].

6.3   RESULTS

6.3.1   FREESTANDING Mo_n CLUSTERS

The first attempts to build stable Mo_n cluster were based on an initial configuration consisting of a $2 \times 2 \times 2$ supercell for bulk crystalline Mo (taken from the Materials Studio structural database, space group Im–3m, lattice parameter $a = 3.1469$ Å), as shown in Figure 6.1a. The periodic boundary conditions were then removed from the supercell to create a freestanding b.c.c. cluster containing 35 atoms. Figure 6.1b shows the resulting structure after geometry optimization at the RHF/AM1* level using VAMP [with SCF tolerance 5×10^{-7} eV atom^{-1} and root mean square (rms) force convergence 0.1 kcal mol^{-1} Å^{-1}], with atoms color-coded according to their Mulliken charges. The total decrease in energy was -123.4 eV, corresponding to -3.53 eV atom^{-1}. It is clear that the most stable structure of the Mo_{35} cluster is amorphous, in contrast to the bulk crystalline form. Of course, there may well be other structures close in energy to the one shown, and we do not claim that it is the global minimum in energy for RHF/AM1* potential energy surface.

The Mo–Mo pair distribution functions for the crystalline and amorphous clusters of Mo_{35} given in Figure 6.1 are shown in Figure 6.2 (normalized to bulk coordination number for bulk crystal of 8). They confirm that the structural order in the amorphous particle (Figure 6.1b) is only very short-ranged and also shows that there is a small decrease in the nearest neighbor separation due to a decrease in local coordination of Mo atoms at the edge of the cluster as compared to the bulk. This is a well-known effect in simulations for metal clusters using many-body potentials [29].

To check the reliability of the AM1* method for calculating the geometry and energies of Mo clusters in this study, the RHF/AM1* cluster geometry shown in Figure 6.1b was compared against DFT results using the B3LYP hybrid functional calculated with Gaussian03 [76]. Due to the computational demands of the DFT calculation, which took nearly two weeks of CPU time on a quad-processor Itanium2

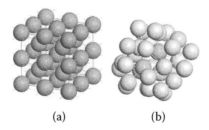

(a)   (b)

FIGURE 6.1   **(See color insert.)** Mo_{35} nanocluster. (a) Initial configuration generated from $2 \times 2 \times 2$ b.c.c. supercell, and (b) the resulting structure optimized at the RHF/AM1* level without periodic boundaries or symmetry constraints. Mo atoms in (b) are colored according to their Mulliken charges: red, negative; blue, positive.

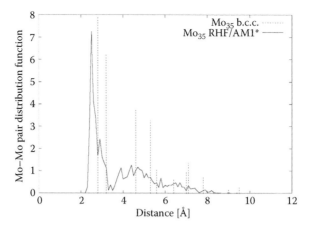

FIGURE 6.2   Mo–Mo pair distribution functions for Mo_{35} structures shown in Figure 6.1. Dashed line, Figure 6.1a; solid line, Figure 6.1b.

machine, it was only possible to reoptimize the structure using a minimal basis set (STO-3G). A comparison of the AM1* and B3LYP/STO-3G optimized clusters is shown in Figure 6.3. The change in the structure after reoptimization with DFT is only very slight, with an rms displacement of 0.22 Å atom^{-1}, and the difference in energy between the structures is $\Delta E = 205$ meV atom^{-1} (evaluated at the B3LYP/STO-3G level). Although this is by no means a conclusive test of the reliability of the AM1* method for our given application, it nonetheless adds confidence to the validation work carried out previously for organometallic compounds [70,71]. Furthermore, because the CPU time required to carry out the SEMO calculation was a factor of approximately 10^3 lower than required to achieve the DFT result, the latter technique is still impractical for a systematic study such as undertaken here.

Attempts to build larger Mo_n particles, with $n > 35$, using a similar method as described above for Mo_{35} proved difficult due to convergence failure of the SCF calculation. Instead, an alternative method was used in which individual Mo atoms were added sequentially to the previously optimized clusters before reoptimizing

FIGURE 6.3   Aligned structures of Mo_{35} nanocluster optimized with AM1* (dark, solid) and B3LYP/STO-3G (lightly shaded, transparent). The rms displacement between the two structures is 0.22 Å/atom (evaluated with Kabsch method).

at the RHF/AM1* level. In this way, clusters of arbitrary size can be generated, although the process is rather slow, and the question naturally arises as to how close these structures are to the global energy minimum for clusters of that size. More recent work using a combination of basin hopping Monte Carlo (BHMC) with classical potentials followed by refinement with SEMO theory [77] has shown that, compared to classical global minima, structures obtained via SEMO optimization tend to be distorted due to Jahn–Teller effect from incompletely occupied d orbitals. However, it must be emphasized that the aim in this study was not to carry out an exhaustive search for the global minimum. Nevertheless, as a further test of the confidence of the RHF/AM1* method for calculating viable cluster structures, and to give some indication of how close these structures might be to the global energy minimum, we compared optimized structures calculated for Mo_{38} and Mo_{55} clusters generated by sequential addition with those relaxed from coordinates of the global optima, taken from the Cambridge Cluster Database [78], and computed using the Gupta potential for Co [51] (with interatomic distances rescaled to the bulk value for Mo). The reason for choosing $n = 38$ and $n = 55$ is that these are geometrically magic numbers corresponding to the truncated octahedron and Mackay icosahedron structures, respectively. Similar structures were also predicted for both 12–6 Lennard–Jones [48] and Ni (modeled by 9–6 Sutton–Chen) [50] clusters of equivalent size, so it appears that they are global minima across a wide range of systems. Hence, there is a good chance that the global energy minimum at the RHF/AM1* level, or a structure very close to it, can be found by the straightforward EF algorithm using VAMP from these initial coordinates.

Figure 6.4 shows a comparison of RHF/AM1* optimized Mo_{38} clusters resulting from sequential addition (Figure 6.4a) and from the Gupta global minimum (Figure 6.4b). The overall shape of the Mo_{38} clusters is qualitatively similar, although the

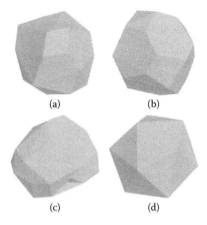

(a)   (b)

(c)   (d)

FIGURE 6.4 Comparison of optimized Mo_n cluster structures generated by RHF/AM1* optimization after sequential addition: (a) Mo_{35} and (c) Mo_{55}, and starting at global minimum for Gupta potential: (b) Mo_{35} and (d) Mo_{55}. Representation is polyhedral, with Mo atoms (not shown) at the vertices.

surface facets in Figure 6.4a are not as well defined for the cluster made by sequential addition, and there is some distortion preferred in Figure 6.4b even when compared to a perfect truncated octahedron. The overall energy difference is $\Delta E = 1.860$ eV, corresponding to 48.9 meV atom^{-1} in favor of the structure in Figure 6.4b, which is approximately twice $k_B T$ at 300 K. Therefore, it is reasonable to suppose that the structure in Figure 6.4a may be one of a family of structures that are all thermally indistinguishable from the global minimum. Figure 6.4c and 6.4d show equivalent structures for Mo$_{55}$ made by sequential addition and relaxation from Gupta global minimum, respectively. In this case, the structures differ qualitatively in appearance, with the cluster made by sequential addition, shown in Figure 6.4c, being less spherical. However, the difference in energies is still only $\Delta E = 5.412$ eV, corresponding to 98.4 meV atom^{-1} in favor of the structure in Figure 6.4d. We conclude that although optimized structures formed by sequential addition are not in general global minima, they lie very close in energy to more regular structures that could be global minima, even for clusters that differ markedly in shape. Of course, only an exhaustive search would establish with more certainty the global minimum structure, and this has since been attempted elsewhere [77].

6.3.2 FREESTANDING MO–X CLUSTERS

Having established a suitable methodology for generating freestanding Mo$_n$ clusters, it was subsequently possible to investigate the effects of addition by random substitution of oxygen and sulfur atoms, and nitrogen dimers, to Mo clusters with a fixed total number of atoms. We adopt the nomenclature Mo$_{(35-x)}X_x$, where $X = \{S, O, N\}$ and x is the number of nonmetal atoms. Each nonmetal atom substitution was made by transforming one Mo atom chosen at random and then reoptimizing the cluster geometry. As in the case of pure Mo clusters studied in Section 6.3.1, no attempt was made to make an exhaustive search for global minimum energy, but it was presumed that the resulting structures would be representative of realistic cluster geometries. Figure 6.5 shows a series of optimized structures for Mo$_{(35-x)}S_x$ for $0 \le x \le 19$, and similar series were generated for O and N substitutions. Although it is difficult to draw any definite conclusions from just a qualitative inspection of the structures in Figure 6.5, an interesting event occurs in going from $x = 7$ to $x = 8$: a tetrahedral cluster of four S atoms in the center of $x = 7$ cluster is disrupted by the addition of a further S atom, resulting in the majority of S atoms moving to the surface of the $x = 8$ cluster with a corresponding drop in the heat of formation. This behavior is similar to that found for small Mo–S particles by Murugan et al. [59], who reported that lowest energy structures tended to have a metal core with sulfur atoms at the surface.

To assess quantitatively the stability of Mo$_{(35-x)}X_x$ clusters as a function of their stoichiometry, we define a binding energy per atom, B.E.^{free}, according to the following equation:

$$\text{B.E.}^{\text{free}} \equiv \frac{xE^{\text{AM1}*}(X) + yE^{\text{AM1}*}(\text{Mo}) - E^{\text{AM1}*}(X_x, \text{Mo}_y)}{x + y}, \tag{6.1}$$

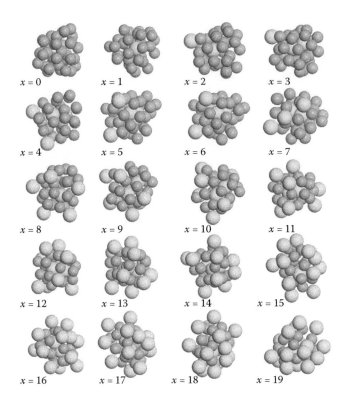

$x = 0$ $x = 1$ $x = 2$ $x = 3$

$x = 4$ $x = 5$ $x = 6$ $x = 7$

$x = 8$ $x = 9$ $x = 10$ $x = 11$

$x = 12$ $x = 13$ $x = 14$ $x = 15$

$x = 16$ $x = 17$ $x = 18$ $x = 19$

FIGURE 6.5 $Mo_{(35 - x)}S_x$ nanoclusters optimized at the RHF/AM1* level. Mo atoms are shown as smaller and darker than the larger and lightly colored sulfur atoms.

where x and y are the number of X and Mo atoms, respectively (with total number of atoms $n = x + y$), $E^{AM1*}(X)$ and $E^{AM1*}(Mo)$ are the energies of isolated atoms of X and Mo, respectively, and $E^{AM1*}(X_x,Mo_y)$ is the total energy of RHF/AM1* geometry-optimized cluster.

The behavior of B.E.^{free} for clusters of $n = 35$ atoms with $X = \{S, O, N\}$ is shown as a function of x in Figure 6.6. In all cases, the B.E.^{free} of the Mo_{35} cluster is reduced from an initial value of 15.658 eV atom^{-1} (or 5.062 eV bond^{-1} by dividing by the number of bonds according to Mayer bond orders calculated from the electron density and overlap matrices) by the substitution of nonmetal atoms. This is expected due to the disruption of metallic bonding between the Mo atoms. In the case of N, the rate of decrease is highest and monotonic with an increasing number of atoms substituted. However, for S and O substitution, the rate of decrease is less rapid and nonmonotonic. In particular, for S substitution, the expulsion of S atoms to the surface of the $x = 8$ cluster (as shown in Figure 6.5) corresponds to a small local rise in the cluster stability. A similar transition occurs for O substitution when $x = 7$. These may be interpreted as magic compositions for Mo–X clusters, which are stabilized by surface addition of a nonmetal additive, although an exhaustive search of cluster geometries for each composition would be required to establish

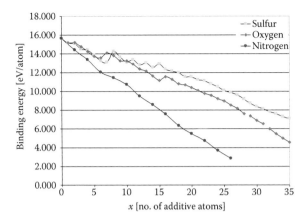

FIGURE 6.6 Behavior of binding energy for $Mo_{(35-x)}X_x$ clusters as a function of nonmetal additive and number of additive atoms.

this definitively. There are smaller subsidiary fluctuations in the B.E.^{free} curve for both S- and O-substituted clusters between $9 \leq x \leq 16$, but for $x > 16$, both curves become monotonically decreasing as the nonmetal atoms dominate the behavior of the cluster. For a cluster with a larger total number of atoms, it may be possible to observe magic behavior for a greater number of substituted atoms. In summary, calculations on freestanding Mo–X clusters show preferential segregation of S and O to the surface of Mo clusters for certain compositions, which is consistent with previous DFT studies on small clusters [59] and experimental studies of Fe-catalyzed CNT growth in which a sulfur-rich layer was seen to form at the surface of the catalyst particle [79]. Furthermore, recent unpublished simulation work by present authors using BHMC with classical potentials, followed by refinement with SEMO theory, suggests that surface segregation is not an artifact of the initial configuration. We next describe the behavior of fullerene-encapsulated clusters.

6.3.3 ENDOHEDRAL $Mo_n@C_{180}$

Even using SEMO techniques, the size of the system required for simulating a metal cluster interacting with a graphene sheet presents a formidable challenge, and therefore a more simplified model is required. Instead, the freestanding Mo_n clusters studied in Section 6.3.1 were encapsulated by a spherical fullerene molecule, which has the advantage of being highly symmetrical (i.e., no periodic boundaries or dangling bonds) and whose properties are already well known. While buckminsterfullerene, C_{60}, itself is too small to encapsulate anything larger than isolated atoms [80–85], there are other related structures sufficiently large to completely contain a Mo_{35} cluster; we chose the I_h isomer of C_{180} that satisfies the isolated pentagon rule. Of course, due to its closed π-system, C_{180} is more stable than a graphene sheet of equivalent size, but by changing the size and composition of the encapsulated cluster, we were able to observe the effects of perturbing the C_{180} ground state and thereby

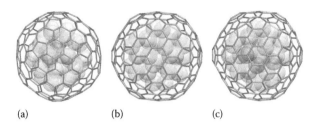

(a)   (b)   (c)

FIGURE 6.7   (See color insert.) Endohedral fullerene Mo_n clusters: (a) $Mo_{35}@C_{180}$, (b) $Mo_{36}@$ C_{180}, and (c) $Mo_{37}@C_{180}$. Mo atoms are represented by semitransparent spheres, colored according to their Mulliken charges (red, negative; blue, positive) with ranges set the same for each image.

draw inferences about the nature of interaction between the cluster and fullerene. We adopt the standard nomenclature for endohedral fullerenes of $X@C_{180}$, where X denotes the chemical formula of the encapsulated cluster.

Figure 6.7 shows a set of visualizations, aligned and rendered using the VMD package [86], for RHF/AM1* optimized geometries of $Mo_n@C_{180}$ nanoclusters, where $n = 35–37$. The Mo atoms are represented by semitransparent spheres, colored according to their Mulliken charges (with red being negative and blue being positive) and the color ranges are identical for each image. It is possible to discern a general trend of increasing polarization in the Mo_n cluster, which coincides with a slight distortion of the C_{180} structure and a decrease in the calculated C_{180} band gap, indicating that there is increasing charge transfer from the cluster to the C_{180} π-system as the cluster becomes larger. In Figure 6.8, the RDFs show a slight expansion of Mo–Mo distance in $Mo_{35}@C_{180}$ compared to a freestanding Mo_{35} cluster, again consistent with there being charge transfer from the Mo_n cluster to the fullerene molecule.

As the size of the Mo cluster was increased beyond $n = 37$ by sequential addition, as described in Section 6.3.1, the cluster polarization and strain in the C_{180} molecule

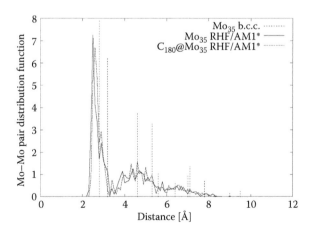

FIGURE 6.8   Mo–Mo pair distribution functions for the freestanding Mo_{35} cluster (shown in Figure 6.1a) and the $Mo_{35}@C_{180}$ endohedral structure (shown in Figure 6.7a). The distribution for bulk b.c.c. Mo is also included for comparison.

were further increased, resulting in the formation of defects in the fullerene structure for $n \geq 44$. Figure 6.9 shows a set of visualizations, aligned and rendered using the VMD package [86], for RHF/AM1* optimized geometries of $Mo_n@C_{180}$ nanoclusters, where $n = 44$–53, showing formation of first fullerene defect for $Mo_{44}@C_{180}$, a second defect for $Mo_{51}@C_{180}$, and finally rupture of fullerene cage structure for $Mo_{53}@C_{180}$. The defects observed in Figure 6.9a and 6.9b are both of the 6–6 type, denoting the breakage of bonds between adjacent hexagons in the fullerene cage. This is consistent with the general observation that they are more reactive in chemical processes than 5–6 bonds between pentagons and hexagons, for example, in the oxidative etching of CNTs [87]. It is interesting to note that the rupture appears to originate at the second defect site (Figure 6.9b, yellow), whereas the original defect site (Figure 6.9a and 6.9b, green) heals after the rupture has occurred (Figure 6.9c). It was also found that reducing the size of the Mo cluster after the formation of defects caused them to heal.

To study the interaction of Mo nanocluster and fullerene more quantitatively, we define two types of binding energy: encapsulated cluster binding energy, $B.E._{endo}$ (Equation 6.2), taking into account atomic decomposition of metal cluster only, and the total encapsulated cluster binding energy, $B.E._{endo}'$ (Equation 6.3), taking into account total atomic decomposition of all components. The analytical expressions are given by

$$B.E._{endo} \equiv \frac{E^{AM1*}(C_{180}) + nE^{AM1*}(Mo) - E^{AM1*}(Mo_n@C_{180})}{n} \tag{6.2}$$

$$B.E._{endo}' \equiv \frac{E^{AM1*}(C_{180}') + nE^{AM1*}(Mo) - E^{AM1*}(Mo_n@C_{180})}{n}, \tag{6.3}$$

where n is the number of Mo atoms in the encapsulated cluster, $E^{AM1*}(C_{180})$ and $E^{AM1*}(Mo)$ are the energies of an isolated C_{180} molecule and Mo atom, respectively, and $E^{AM1*}(Mo_n@C_{180})$ is the total energy of RHF/AM1* geometry-optimized

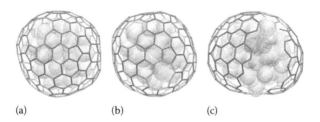

(a)   (b)   (c)

FIGURE 6.9 (See color insert.) RHF/AM1* optimized geometries of endohedral fullerene Mo_n clusters: (a) $Mo_{44}@C_{180}$, (b) $Mo_{51}@C_{180}$, and (c) $Mo_{53}@C_{180}$. Mo atoms are represented by semitransparent spheres, colored according to their Mulliken charges (red, negative; blue, positive) with ranges set the same for each image. Fullerene defects in (a) and (b) are highlighted in green and yellow, and full opening of the fullerene cage is observed in (c), with bonding to Mo atoms (which terminate the dangling bonds) not shown for clarity.

endohedral cluster. $E^{AM1*}(C'_{180})$ is the energy of the strained C_{180} molecule calculated in isolation from the encapsulated cluster, but without further optimization.

The difference between the two quantities given by Equations 6.2 and 6.3 is that B.E.$_{\cdot endo}$ does not take into account the strain energy induced in C_{180} due to expansion of the cluster, whereas B.E.$'_{\cdot endo}$ does. Figure 6.10 shows the behavior of B.E.$_{\cdot endo}$ and B.E.$'_{\cdot endo}$, calculated as a function of the number of cluster atoms. Both quantities show an increase of approximately 0.75 eV atom^{-1} at the point of fullerene breakage ($n = 53$), but are relatively unchanged by formation of isolated 6–6 defects (at $n = 44$, $n = 51$). The general trend is for B.E.$_{\cdot endo}$ to decrease between $n = 39$ until the point of breakage ($n = 53$), which, by comparison with the curve for B.E.$'_{\cdot endo}$ that increases monotonically, can be seen to be due to a buildup of strain energy in the fullerene cage. After opening of the fullerene, the binding energies rise steadily, showing the stabilizing influence of the d electrons in the Mo cluster on the broken π-system of the fullerene.

To separate the effects of cluster encapsulation from changes in binding energy, we define cluster encapsulation energy, E.E. (Equation 6.4), which only involves dissociation of the strained C_{180} and Mo$_n$ cluster. As shown in Figure 6.11, the E.E. changes only very slightly for $n > 35$, except for a small discrete jump around the breakage point ($n = 53$). Nevertheless, in all cases, encapsulation of Mo nanoclusters by C_{180} lowered the energy by around 1.75 eV–2.5 eV atom^{-1} in the cluster. The encapsulation energy is given by

$$\text{E.E.}' \equiv \frac{E^{AM1*}(C'_{180}) + E^{AM1*}(Mo_n) - E^{AM1*}(Mo_n @ C_{180})}{n}, \qquad (6.4)$$

where $E^{AM1*}(Mo_n)$ is the energy of the isolated Mo$_n$ cluster (taken from endohedral geometry without further optimization) and all other terms have their previously defined meanings.

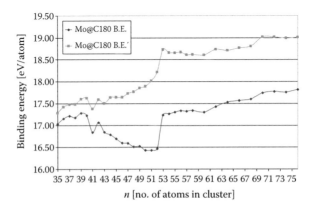

FIGURE 6.10 Binding energies of Mo$_n$@C$_{180}$ endohedral clusters, B.E.$_{\cdot endo}$ and B.E.$'_{\cdot endo}$, defined by Equations 6.2 and 6.3, as a function of the number of cluster atoms.

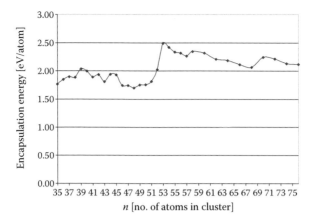

FIGURE 6.11 Encapsulation energies of $Mo_n@C_{180}$ endohedral clusters, E.E., defined by Equation 6.4, as a function of the number of cluster atoms.

In terms of structural changes induced by encapsulation, a correlation was found between average carbon–carbon bond strain and the total energy of isolated strained C_{180} molecule computed at the RHF/AM1* level. The maximum average C–C bond strain was found to be around 5.5% before breakage occurs (corresponding to cluster size $n = 53$). Furthermore, Mo–Mo RDFs for $Mo_{35}@C_{180}$ (relatively unstrained) and $Mo_{51}@C_{180}$ (just before breakage) showed only a small difference; there was sharpening of nearest-neighbor peak due to confinement, but no increase in overall crystalline order.

6.3.4 ENDOHEDRAL $Mo_{(52-x)}S_x@C_{180}$

To investigate the effect of sulfur addition on the stability of endohedral fullerene structures, preliminary results have been obtained from C_{180}-encapsulated molybdenum sulfide clusters. Because the parameter space of possible simulations for the binary $Mo_{(n-x)}S_x@C_{180}$ system is very large, it was decided to limit the study to $Mo_{(52-x)}S_x@C_{180}$ for $0 \leq x \leq 10$ to investigate specifically whether there was any systematic change in ordering of Mo atoms with substitution of sulfur atoms, and what effect this may have on the breakage of the fullerene π-system. As described in Section 6.3.3, for $Mo_{53}@C_{180}$ shown in Figure 6.9c, the fullerene cage is partially ruptured compared with the presence of two smaller isolated 6–6 defects in the $Mo_{51}@C_{180}$ cluster (Figure 6.9b). The structure of $Mo_{52}@C_{180}$ cluster, shown in Figure 6.12a, displays a rupture of intermediate size, but when a single Mo atom is transformed to S, the rupture is immediately enlarged (Figure 6.12b). Because the degree of polarity of the metal atoms between Figure 6.12a and 6.12b is relatively unchanged, it is likely that this effect is simply due to the increased size of the S atom relative to a Mo atom. As in the case of a freestanding cluster of $Mo_{51}S$, the S atom is preferentially contained in the interior of the Mo cluster rather than close to the surface. As successively more Mo atoms are transformed to S, the size of the fullerene rupture increases slightly but the distribution of S atoms tends to remain closer to the interior

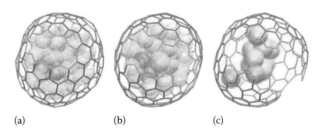

(a)   (b)   (c)

FIGURE 6.12   (See color insert.) RHF/AM1* optimized geometries of endohedral fullerene $Mo_{(52-x)}S_x@C_{180}$ clusters: (a) $Mo_{52}@C_{180}$, (b) $Mo_{51}S@C_{180}$, and (c) $Mo_{42}S_{10}@C_{180}$. In (a) and (b), Mo atoms are represented by semitransparent spheres, colored according to their Mulliken charges (red, negative; blue, positive) with ranges set the same for each image. In (b) and (c), sulfur atoms are highlighted by overlaying a semitransparent yellow sphere. The fullerene defects in all three structures are highlighted in green, with full opening of the fullerene cage observed in (c). As in Figure 6.9, bonding to Mo atoms (which terminate the dangling bonds) is not shown for clarity, and Mo atoms are also hidden in (c).

of the encapsulated particle (see Figure 6.12c, where Mo atoms are hidden for clarity) rather than near the site of the rupture. From these albeit limited results, we can nevertheless conclude that the metal atoms play the dominant role in stabilizing the edges of a broken fullerene structure, with substituted nonmetal atoms influencing the structure via a secondary steric affect. However, a more conclusive investigation would require the study of much larger catalyst particles, and further studies are underway.

6.4   CONCLUSIONS

SEMO calculations using the AM1* Hamiltonian have been used to study the structure and energetics of freestanding and fullerene-encapsulated Mo nanoclusters. It was observed that freestanding nanoclusters of Mo_n and $Mo_{(n-x)}X_x$ are always amorphous for $x < 35$, where $X = \{O, N, S\}$. In general, the incorporation of chalcogens lowers the binding energy of cluster, although some small increases are seen for smaller x when chalcogen atoms move to the surface. There is reasonable quantitative agreement between AM1* geometries and energies when compared with B3LYP using a minimal basis set. In all cases, encapsulation of Mo_n or $Mo_{(n-x)}X_x$, where $X = \{O, N, S\}$, nanoclusters by a C_{180} molecule lowers energy by around 1.75 eV to 2.5 eV atom^{-1} in cluster. This is due to stabilization by interaction between d-electrons of Mo and the C_{180} π-system, leading to formation of defects by attack at 6–6 bonds. Eventually, the C_{180} cage opens, with a corresponding increase in binding energy due to release of strain in fullerene. No systematic change in ordering of Mo atoms was seen on substitution of S species, but further investigations using larger nanoparticles are underway. More recently, other transition metal catalysts (such as Ni, Co, and Fe) have been studied, and effects of magnetism need to be considered carefully in future studies.

ACKNOWLEDGMENTS

J. A. Elliott acknowledges the award of an Invitation Fellowship from the Japan Society for Promotion of Science no. L08536. Part of this work was financially supported by Grand-in-Aid for Young Scientist (a) (no. 18686017) from MEXT, Japan.

REFERENCES

1. A. Thess, R. Lee, P. Nikolaev, H. J. Dai, P. Petit, J. Robert, C. H. Xu, et al. Crystalline ropes of metallic carbon nanotubes. *Science* **273**, 1996, 483–487.
2. C. Journet, W. K. Maser, P. Bernier, A. Loiseau, M. L. de la Chapelle, S. Lefrant, P. Deniard, et al. Large-scale production of single-walled carbon nanotubes by the electric-arc technique. *Nature* **388**, 1997, 756–758.
3. Y.-L. Li, I. A. Kinloch and A. H. Windle. Direct spinning of carbon nanotube fibers from chemical vapor deposition synthesis. *Science* **304**, 2004, 276–278.
4. S. Maruyama, R. Kojima, Y. Miyauchi, S. Chiashi and M. Kohno. Low-temperature synthesis of high-purity single-walled carbon nanotubes from alcohol. *Chem. Phys. Lett.* **360**, 2002, 229–234.
5. M. Motta, A. Moisala, I. A. Kinloch and A. H. Windle. High performance fibres from 'Dog bone' carbon nanotubes. *Adv. Mater.* **19**, 2007, 3721–3726.
6. L. Ci, Z. Rao, Z. Zhou, D. Tang, X. Yan, Y. Liang, D. Liu, et al. Double wall carbon nanotubes promoted by sulfur in a floating iron catalyst CVD system. *Chem. Phys. Lett.* **359**, 2002, 63–67.
7. P. Nikolaev, M. J. Bronikowski, R. K. Bradley, F. Rohmund, D. T. Colbert, K. A. Smith and R. E. Smalley. Gas-phase catalytic growth of single-walled carbon nanotubes from carbon monoxide. *Chem. Phys. Lett.* **313**, 1999, 91–97.
8. Y. Murakami, S. Chiashi, Y. Miyauchi, M. H. Hu, M. Ogura, T. Okubo and S. Maruyama. Growth of vertically aligned single-walled carbon nanotube films on quartz substrates and their optical anisotropy. *Chem. Phys. Lett.* **385**, 2004, 298–303.
9. K. Hata, D. N. Futaba, K. Mizuno, T. Namai, M. Yumura and S. Iijima. Water-assisted highly efficient synthesis of impurity-free single-waited carbon nanotubes. *Science* **306**, 2004, 1362–1364.
10. A. Loiseau, X. Blasé, J.-C. Charlier, P. Gadelle, C. Journet, C. Laurent and A. Peigney. Synthesis methods and growth mechanisms. *Lect. Notes Phys.* **677**, 2006, 49–130.
11. D. N. Futaba, K. Hata, T. Namai, T. Yamada, K. Mizuno, Y. Hayamizu, M. Yumura and S. Iijima. 84% Catalyst activity of water-assisted growth of single walled carbon nanotube forest characterization by a statistical and macroscopic approach. *J. Phys. Chem. B* **110**, 2006, 8035–8038.
12. K. Koziol, J. Vilatela, A. Moisala, M. Motta, P. Cunniff, M. Sennett and A. Windle. High-performance carbon nanotube fiber. *Science* **318**, 2007, 1892–1895.
13. K. K. K. Koziol, C. Ducati and A. H. Windle. Carbon nanotubes with catalyst controlled chiral angle. *Chem. Mater.* **22**, 2010, 4904–4911.
14. G. Lolli, L. A. Zhang, L. Balzano, N. Sakulchaicharoen. Y. Q. Tan and D. E. Resasco, Tailoring (n,m) structure of single-walled carbon nanotubes by modifying reaction conditions and the nature of the support of CoMo catalysts. *J. Phys. Chem. B* **110**, 2006, 2108–2115.
15. R. S. Wagner and W. C. Ellis. Vapor–liquid–solid mechanism of single crystal growth. *Appl. Phys. Lett.* **4**, 1964, 89–90.
16. R. T. K. Baker. Catalytic growth of carbon filaments. *Carbon* **27**, 1989, 315–323.

17. H. Dai, A. G. Rinzler. P. Nikolaev, A. Thess, D. T. Colbert and R. E. Smalley. Single-wall carbon nanotubes produced by metal-catalyzed disproportionation of carbon monoxide. *Chem. Phys. Lett.* **260**, 1996, 471–475.

18. Y. Zhang, Y. Li, W. Kim, D. Wang and H. Dai. Imaging as-grown single-walled carbon nanotubes originated from isolated catalytic nanoparticles. *Appl. Phys. Mater. Sci. Process.* **74**, 2002, 325–328.

19. Y. Shibuta and S. Maruyama. Molecular dynamics simulation of formation process of single-walled carbon nanotubes by CCVD method. *Chem. Phys. Lett.* **382**, 2003, 381–386.

20. Y. Shibuta and S. Maruyama. A molecular dynamics study of the effect of a substrate on catalytic metal clusters in nucleation process of single-walled carbon nanotubes. *Chem. Phys. Lett.* **437**, 2007, 218–223.

21. J. E. Herrera, L. Balzano, A. Borgna, W. E. Alvarez and D. E. Resasco. Relationship between the structure/composition of Co–Mo catalysts and their ability to produce single-walled carbon nanotubes by CO disproportionation. *J. Catal.* **204**, 2001, 129–145.

22. A. D. Lan, Y. Zhang, X. Y. Zhang, Z. Iqbal and H. Grebel. Is molybdenum necessary for the growth of single-wall carbon nanotubes from CO? *Chem. Phys. Lett.* **379**, 2003, 395–400.

23. M. Yudasaka, Y. Kasuya, F. Kokai, K. Takahashi, M. Takizawa, S. Bandow and S. Iijima. Causes of different catalytic activities of metals in formation of single-wall carbon nanotubes. *Appl. Phys. Mater. Sci. Process.* **74**, 2002, 377–385.

24. Y. Shibuta and J. A. Elliott. A molecular dynamics study of the graphitization ability of transition metals for catalysis of carbon nanotube growth via chemical vapor deposition. *Chem. Phys. Lett.* **472**, 2009, 200–206.

25. E. Flahaut, A. Peigney, W. S. Bacsa, R. R. Bacsa and C. Laurent. CCVD synthesis of carbon nanotubes from (Mg,Co,Mo)O catalysts: Influence of the proportions of cobalt and molybdenum. *J. Mater. Chem.* **14**, 2004, 646–653.

26. M. H. Hu, Y. Murakami, M. Ogura, S. Maruyama and T. Okubo. Morphology and chemical state of Co–Mo catalysts for growth of single-walled carbon nanotubes vertically aligned on quartz substrates. *J. Catal.* **225**, 2004, 230–239.

27. S. Noda, H. Sugime, T. Osawa, Y. Tsuji, S. Chiashi, Y. Murakami and S. Maruyama. A simple combinatorial method to discover Co–Mo binary catalysts that grow vertically aligned single-walled carbon nanotubes. *Carbon* **44**, 2006, 1414–1419.

28. S. Noda, K. Hasegawa, H. Sugime, K. Kakehi, Z. Zhang, S. Maruyama and Y. Yamaguchi. Millimeter-thick single-walled carbon nanotube forests: Hidden role of catalyst support. *Jpn. J. Appl. Phys.* **46**, 2007, L399–L401.

29. Y. Shibuta and S. Maruyama. Bond-order potential for transition metal carbide cluster for the growth simulation of a single-walled carbon nanotube. *Comput. Mater. Sci.* **39**, 2007, 842–848.

30. F. Ding, A. Rosen and K. Bolton. Molecular dynamics study of the catalyst particle size dependence on carbon nanotube growth. *J. Chem. Phys.* **121**, 2004, 2775–2779.

31. J. Y. Raty, F. Gygi and G. Galli. Growth of carbon nanotubes on metal nanoparticles: A microscopic mechanism from ab initio molecular dynamics simulations. *Phys. Rev. Lett.* **95**, 2005, 096103.

32. J. C. Charlier, H. Amara and P. Lambin. Catalytically assisted tip growth mechanism for single-wall carbon nanotubes. *ACS Nano* **1**, 2007, 202–207.

33. J. Gavillet, A. Loiseau, C. Journet, F. Willaime, F. Ducastelle and J. C. Charlier. Root-growth mechanism for single-wall carbon nanotubes. *Phys. Rev. Lett.* **8727**, 2001, 275504.

34. H. Amara, C. Bichara and F. Ducastelle. Formation of carbon nanostructures on nickel surfaces: A tight-binding grand canonical Monte Carlo study. *Phys. Rev. B* **73**, 2006, 113404.

35. A. R. Leach. *Molecular Modelling: Principles and Applications*. Harlow: Prentice-Hall, 2001.

36. Y. Shibuta and J. A. Elliott. A molecular dynamics study of the carbon–catalyst interaction energy for multi-scale modelling of single wall carbon nanotube growth. *Chem. Phys. Lett.* **427**, 2006, 365–370.

37. J. A. Elliott, M. Hamm and Y. Shibuta. A multiscale approach for modeling the early stage growth of single and multiwall carbon nanotubes produced by a metal-catalyzed synthesis process. *J. Chem. Phys.* **130**, 2009, 034704.

38. E. C. Neyts, Y. Shibuta, A. C. T. van Duin and A. Bogaerts. Catalyzed growth of carbon nanotube with definable chirality by hybrid molecular dynamics-force biased Monte Carlo simulations. *ACS Nano* **4**, 2010, 6665–6672.

39. C. L. Cleveland, W. D. Luedtke and U. Landman. Melting of gold clusters. *Phys. Rev. B* **60**, 1999, 5065–5077.

40. Y. Dong and M. Springborg. Unbiased determination of structural and electronic properties of gold clusters with up to 58 atoms. *J. Phys. Chem. C* **111**, 2007, 12528–12535.

41. D. Alamanova, V. G. Grigoryan and M. Springborg. Theoretical study of the structure and energetics of silver clusters. *J. Phys. Chem. C* **111**, 2007, 12577–12587.

42. F. Ding, A. Rosen, S. Curtarolo and K. Bolton. Modeling the melting of supported clusters. *Appl. Phys. Lett.* **88**, 2006, 133110.

43. H. K. Kim, S. H. Huh, J. W. Park, J. W. Jeong and G. H. Lee. The cluster size dependence of thermal stabilities of both molybdenum and tungsten nanoclusters. *Chem. Phys. Lett.* **354**, 2002, 165–172.

44. S. H. Huh, H. K. Kim, J. W. Park and G. H. Lee. Critical cluster size of metallic Cr and Mo nanoclusters. *Phys. Rev. B* **62**, 2000, 2937–2943.

45. Y. Shibuta and T. Suzuki. Melting and nucleation of iron nanoparticles: A molecular dynamics study. *Chem. Phys. Lett.* **445**, 2007, 265–270.

46. F. Baletto and R. Ferrando. Structural properties of nanoclusters: Energetic, thermodynamic, and kinetic effects. *Rev. Mod. Phys.* **77**, 2005, 371–423.

47. J. A. Alonso. Electronic and atomic structure, and magnetism of transition-metal clusters. *Chem. Rev.* **100**, 2000, 637–677.

48. D. J. Wales and J. P. K. Doye. Global optimization by basin-hopping and the lowest energy structures of Lennard–Jones clusters containing up to 110 atoms. *J. Phys. Chem. A* **101**, 1997, 5111–5116.

49. M. W. Finnis and J. E. Sinclair. A simple N-body empirical potential for transition metals. *Philos. Mag. A* **50**, 1984, 45–55.

50. J. P. K. Doye and D. J. Wales. Global minima for transition metal clusters described by Sutton–Chen potentials. *New J. Chem.* **22**, 1998, 733–744.

51. L. X. Zhan, J. Z. Y. Chen, W. K. Liu and S. K. Lai. Asynchronous multicanonical basin hopping method and its application to cobalt nanoclusters. *J. Chem. Phys.* **122**, 2005, 1–9.

52. J. A. Elliott, Y. Shibuta and D. J. Wales. Global minima of transition metal clusters described by Finnis–Sinclair potentials: A comparison with semi-empirical molecular orbital theory. *Philos. Mag.* **89**, 2009, 3311–3332.

53. M. V. Bollinger, K. W. Jacobsen and J. K. Norskov. Atomic and electronic structure of MoS_2 nanoparticles. *Phys. Rev. B* **67**, 2003, 085410.

54. G. Seifert, J. Tamuliene and S. Gemming. MonS2$n+x$ clusters—Magic numbers and platelets. *Comput. Mater. Sci.* **35**, 2006, 316–320.

55. P. Murugan, V. Kumar, Y. Kawazoe and N. Ota. Bonding nature and magnetism in small Mo$X2$ (X = O and S) clusters—A comparative study by first principles calculations. *Chem. Phys. Lett.* **423**, 2006, 202–207.

56. S. Gemming, J. Tamuliene, G. Seifert, N. Bertram, Y. D. Kim and G. Gantefor. Electronic and geometric structures of MoxSy and WxSy (x = 1, 2, 4; y = 1–12) clusters. *Appl. Phys. Mater. Sci. Process.* **82**, 2006, 161–166.

57. J. V. Lauritsen, J. Kibsgaard, S. Helveg, H. Topsoe, B. S. Clausen, E. Laegsgaard and F. Besenbacher. Size-dependent structure of MoS_2 nanocrystals. *Nature Nanotech.* **2**, 2007, 53–58.

58. P. Murugan, V. Kumar, Y. Kawazoe and N. Ota. Understanding the structural stability of compound Mo–S clusters at sub-nanometer level. *Mater. Trans.* **48**, 2007, 658–661.

59. P. Murugan, V. Kumar, Y. Kawazoe and N. Ota. Ab initio study of structural stability of Mo–S clusters and size specific stoichiometries of magic clusters. *J. Phys. Chem. A* **111**, 2007, 2778–2782.

60. T. S. Li and G. L. Galli. Electronic properties of MOS2 nanoparticles. *J. Phys. Chem. C* **111**, 2007, 16192–16196.

61. M. J. S. Dewar and W. Thiel. Ground-states of molecules. 38. MNDO method—Approximations and parameters. *J. Am. Chem. Soc.* **99**, 1977, 4899–4907.

62. M. J. S. Dewar and W. Thiel. Ground-states of molecules. 39. MNDO results for molecules containing hydrogen, carbon, nitrogen, and oxygen. *J. Am. Chem. Soc.* **99**, 1977, 4907–4917.

63. T. Bredow and K. Jug. Theory and range of modern semiempirical molecular orbital methods. *Theor. Chem. Acc.* **113**, 2005, 1–14.

64. J. J. P. Stewart. Optimization of parameters for semiempirical methods. V. Modification of NDDO approximations and application to 70 elements. *J. Mol. Model.* **13**, 2007, 1173–1213.

65. J. B. Foresman and A. Frisch. *Exploring Chemistry with Electronic Structure Methods.* Pittsburgh, PA: Gaussian Inc., 1996.

66. C. T. Lee, W. T. Yang and R. G. Parr. Development of the Colle–Salvetti correlation-energy formula into a functional of the electron-density. *Phys. Rev. B* **37**, 1988, 785–789.

67. A. D. Becke. Density-functional thermochemistry. 3. The role of exact exchange. *J. Chem. Phys.* **98**, 1993, 5648–5652.

68. A. D. Bacon and M. C. Zerner. *Theor. Chim. Acta* **52**, 1979, 21–54.

69. W. Thiel and A. A. Voityuk. Extension of the MNDO formalism to d-orbitals—Integral approximations and preliminary numerical results. *Theor. Chim. Acta* **81**, 1992, 391–404.

70. A. A. Voityuk and N. Rösch. AM1/d parameters for molybdenum. *J. Phys. Chem. A* **104**, 2002, 4089–4094.

71. P. Winget, C. Selcuki, A. H. C. Horn, B. Martin and T. Clark. Towards a next generation neglect of diatomic differential overlap based semiempirical molecular orbital technique. *Theor. Chem. Acc.* **110**, 2003, 254–266.

72. P. Winget, A. H. C. Horn, C. Selcuki, B. Martin and T. Clark. AM1* parameters for phosphorus, sulfur and chlorine. *J. Mol. Model.* **9**, 2003, 408–414.

73. P. Winget and T. Clark. AM1* parameters for aluminum, silicon, titanium and zirconium. *J. Mol. Model.* **11**, 2005, 439–456.

74. *Materials Studio Modeling Environment.* v. 4.2 edition, Accelrys Inc., 2007.

75. D. Young. *Computational Chemistry: A Practical Guide for Applying Techniques to Real World Problems.* New York: Wiley Interscience, 2001.

76. M. J. Frisch, G. W. Trucks, H. B. Schlegel, G. E. Scuseria, M. A. Robb, J. R. Cheeseman, J. A. M. Jr., et al. *Gaussian 03.* Wallingford, CT: Gaussian Inc., 2004.

77. J. A. Elliott and Y. Shibuta. Energetic stability of molybdenum nanoclusters studied with basin-hopping Monte Carlo and semi-empirical quantum methods. *J. Comput. Theor. Nanosci.* **6**, 2009, 1443–1451.

78. D. J. Wales, J. P. K. Doye, A. Dullweber, M. P. Hodges, F. Y. Naumkin, F. Calvo, J. Hernández-Rojas and T. F. Middleton. *The Cambridge Cluster Database.* Available online at http://www-wales.ch.cam.ac.uk/CCD.html, 2007.

79. M. S. Motta, A. Moisala, I. A. Kinloch and A. H. Windle. The role of sulphur in the synthesis of carbon nanotubes by chemical vapour deposition at high temperatures. *J. Nanosci. Nanotechnol.* **8**, 2008, 2442–2449.

80. M. Saunders, H. A. Jimenezvazquez, R. J. Cross and R. J. Poreda. Stable compounds of helium and neon—He@C60 and Ne@C60. *Science* **259**, 1993, 1428–1430.
81. M. Saunders, H. A. Jimenezvazquez, R. J. Cross, S. Mroczkowski, M. L. Gross, D. E. Giblin and R. J. Poreda. Incorporation of helium, neon, argon, krypton and xenon into fullerenes using high-pressure. *J. Am. Chem. Soc.* **116**, 1994, 2193–2194.
82. J. Lu, L. X. Ge, X. W. Zhang and X. G. Zhao. Electronic structures of endohedral Sr@C-60, Ba@C-60, Fe@C-60 and Mn@C-60. *Mod. Phys. Lett. B* **13**, 1999, 97–101.
83. J. Lu, X. W. Zhang and X. G. Zhao. Electronic structures of endohedral Ca@C-60, SC@C-60 and Y@C-60. *Solid State Commun.* **110**, 1999, 565–568.
84. J. Lu, X. W. Zhang and X. G. Zhao. Electronic structures of endohedral N@C-60, O@C-60 and F@C-60. *Chem. Phys. Lett.* **312**, 1999, 85–90.
85. J. C. Greer. The atomic nature of endohedrally encapsulated nitrogen N@C-60 studied by density functional and Hartree–Fock methods. *Chem. Phys. Lett.* **326**, 2000, 567–572.
86. W. Humphrey, A. Dalke and K. Schulten. VMD: Visual molecular dynamics. *J. Mol. Graphics* **14**, 1996, 33–38.
87. C. Y. Moon, Y. S. Kim, E. C. Lee, Y. G. Jin and K. J. Chang. Mechanism for oxidative etching in carbon nanotubes. *Phys. Rev. B* **65**, 2002, 155401.

7 Using Fractional Charges for Computing Fukui Functions in Molecular and Periodic Systems

George Fitzgerald

CONTENTS

7.1 INTRODUCTION

The chemical potential, chemical hardness and softness, and reactivity indices have been used by a number of workers to assess *a priori* the reactivity of chemical species from their intrinsic electronic properties. Perhaps one of the most successful and best known methods is the frontier orbital theory of Fukui [1,2]. Developed further by Parr and Yang [3], the method relates the reactivity of a molecule with respect to electrophilic or nucleophilic attack to the charge density arising from the highest occupied molecular orbital or lowest unoccupied molecular orbital, respectively. Parr and coworkers [4,5] were able to use these Fukui indices to deduce the hard and soft (Lewis) acids and bases principle from theoretical principles, providing one of the first applications of electronic structure theory to explain chemical reactivity. In essentially the same form, the Fukui functions (FFs) were used to predict the molecular chemical reactivity of a number of systems including Diels–Alder condensations [6,7], monosubstituted benzenes [8], as well as a number of model compounds [9,10]. Recent applications are too numerous to catalog here but include silylenes [11], pyridinium ions [12], and indoles [13].

In applications such as the ones cited above, the magnitude of the FFs is correlated with the reactivity of various sites in a molecule. These FFs can be condensed to atomic-centered indices that can be used to predict which sites are most likely to react with electrophiles or nucleophiles. This can be used to compare the activities of sites within a molecule or can be used as a measure of how various side groups alter the reactivity of a molecule.

Naturally, the reliability of the predictions will depend on the accuracy of the method used to compute the Fukui indices. Generally, these calculations are performed by finite differences of the charge density as described in the next section. This paper will discuss aspects of this finite-difference approximation used to compute condensed FFs computed via density functional theory (DFT). DFT is particularly well suited for performing analyses of this type [14,15]. DFT calculations yield a good representation of the charge density, $\rho(r)$, even in the simplest local-density approximations, yielding reliable results for the chemical properties of molecules and solids. DFT calculations are also fairly computationally inexpensive [16], making this the method of choice for accurate calculations on large molecular and solid-state systems. Finally, DFT programs for periodic systems are widely available [17], making this the method of choice for modeling chemical reactivity on surfaces or within lattices. Such periodic calculations eliminate the uncertainties introduced by using finite-sized cluster approximations.

In particular, we will illustrate the effects of performing calculations using fractional charges, that is, noninteger orbital occupations.

This study builds on earlier work for molecules [18] and extends the results to periodic structures. As illustrations, we have opted to model zeolite structures because of their scientific and commercial significance. Zeolites are porous aluminosilicate compounds that can serve as molecular sieves and catalysts. Zeolite structures contain solid frameworks defining large ($\sim$3 Å–10 Å) internal cavities where molecules can be adsorbed. Depending on the size of the openings, they can adsorb molecules readily, slowly, or not at all, thus functioning as molecular sieves—adsorbing molecules of certain sizes while rejecting larger ones. Zeolites also act as catalysts or promoters of specific chemical reactions. Zeolite catalysis is used for processes such as hydrocarbon cracking, reforming, and alkylation. Fukui analysis has been applied to such diverse examples as toxic filtration [19], electrophilic aromatic substitution [20], methanol to olefin conversion [21], alkane aromatization [22], and dehydrogenation [23]—to name just a few of many applications.

7.2 THEORY

A number of quite thorough review papers appear in the literature (see, for example, Refs. [15], [24], and [25]); hence, this section provides only a brief overview of the computation of reactivity indices, hardness, and softness. The chemical potential, μ, may be defined as the change in total energy with respect to the number of electrons, N [26]:

$$\mu = \left(\frac{\partial E}{\partial N} \right)_v, \tag{7.1}$$

where the subscript v indicates that the external potential, that is, the potential due to the atomic nuclei, is kept fixed.

The derivative with respect to N for a molecular system is necessarily discontinuous [27]: an electron is added to or taken from different orbitals yielding different derivatives. This led to the introduction [3] of the *frontier* or *Fukui function*, f(r), which measures the sensitivity of the charge density, $\rho(r)$, to changes in the number of electrons. The FF f^- measures susceptibility to electrophilic attack (or to loss of electrons),

$$f^-(\vec{r}) = \left(\frac{\partial \rho(\vec{r})}{\partial N} \right)_v^-, \tag{7.2}$$

whereas f^+ measures susceptibility to nucleophilic attack (or to gain of electrons),

$$f^+(\vec{r}) = \left(\frac{\partial \rho(\vec{r})}{\partial N} \right)_v^+. \tag{7.3}$$

The superscript "−" refers to the derivative from below, and the superscript "+" refers to the derivative from above. These expressions are commonly evaluated by adding or removing an electron:

$$f^-(\vec{r}) = \left(\frac{\partial \rho(\vec{r})}{\partial N} \right)_v^- \cong \rho_N(\vec{r}) - \rho_{N-1}(\vec{r}), \tag{7.4}$$

and

$$f^-(\vec{r}) = \left(\frac{\partial \rho(\vec{r})}{\partial N} \right)_v^+ \cong \rho_{N+1}(\vec{r}) - \rho_N(\vec{r}). \tag{7.5}$$

In this approximation, the charge densities are converged to self-consistency for the original N-electron system, $\rho_N(\vec{r})$, for the system with one extra electron, $\rho_{N+1}(\vec{r})$, and for the electron-deficient system, $\rho_{N-1}(\vec{r})$. The FFs are computed by subtracting the results over a set of grid points and may be displayed as 3D images or 2D contour plots. It has been demonstrated how to compute the FFs analytically for both Hartree–Fock and DFT [28], but finite-difference calculations remain a fixture of the literature owing to the simplicity of their implementation.

Although considerable insight may be obtained from viewing 3D renderings of the FF, it is convenient to make quantitative comparisons among molecules using the condensed FF [15,24,28,29]. For an atom k, this is defined as

$$f_k^- = q(N)_k - q(N-1)_k \tag{7.6}$$

and

$$f_k^+ = q(N+1)_k - q(N)_k. \tag{7.7}$$

The q_k are atomic-centered charges that are computed in some reasonable manner, such as from a Mulliken population analysis [8–10,28], Stockholder analysis [30], fitting the electrostatic potential [30], or numerical integration [24,25]. The term $q(N)$ refers to the atomic charges computed for the original, N-electron system; $q(N+1)$ to charges computed for the system with an additional electron; and $q(N-1)$ to an electron deficient system.

These expressions allow one to compare candidate reactive sites within a molecule or solid as well as to compare analogous sites across a series of compounds. These condensed FFs have been used by a number of authors to compare and predict reactivity for a number of compounds including Diels–Alder condensations [9], substituted benzenes [8], and zeolites [19–23], to name just a few.

The finite-difference approach becomes a more accurate approximation to the exact derivative if one uses a smaller step size; in this case, that means using a fraction of an electron, ΔN. DFT is well suited for use with noninteger occupations. The Slater transition state formula [31], for example, uses half-integer occupations to approximate the electron affinity. Fractional occupations of orbitals are also commonly employed in the use of charge smearing to improve convergence of the self-consistent field (SCF) [32,33].

Using fractional occupations, the partial derivatives are approximated as

$$f^-(\vec{r}) = \left(\frac{\partial \rho(\vec{r})}{\partial N}\right)_v^- \cong \frac{1}{\Delta N}\left(\rho_N(\vec{r}) - \rho_{N-\Delta}(\vec{r})\right) \tag{7.8}$$

and

$$f^+(\vec{r}) = \left(\frac{\partial \rho(\vec{r})}{\partial N}\right)_v^+ \cong \frac{1}{\Delta N}\left(\rho_{N+\Delta}(\vec{r}) - \rho(\vec{r})\right). \tag{7.9}$$

The expressions for the condensed FF in Equations 7.7 and 7.8 can similarly be written as

$$f_k^- = \left[q(N)_k - q(N-\Delta N)_k\right]/\Delta N \tag{7.10}$$

$$f_k^+ = \left[q(N+\Delta N)_k - q(N)_k\right]/\Delta N. \tag{7.11}$$

Equations 7.5 through 7.7 can be viewed as special cases of these approximations with $\Delta N = 1$. Computing the finite differences with smaller values of ΔN provides more accurate approximations to the partial derivative. As will be seen below, in many cases, the difference is quite startling.

We note that recent work [34] performed a similar analysis not by finite difference of the occupation but by performing finite differences of the external potential by recasting the definition of the FF

$$f^-(\vec{r}) = \left(\frac{\partial\rho(\vec{r})}{\partial N}\right)_v^- = \left(\frac{\delta\mu}{\delta v(r)}\right)_N^-. \tag{7.12}$$

Although the method provides reliable results, the authors noted that the method has a disadvantage of a relatively high cost.

7.3   COMPUTATIONAL DETAILS

To test the impact of noninteger ΔN on FFs, DFT calculations were performed on a total of 23 molecules and 4 zeolite systems. All DFT calculations were performed using the program DMol^3 [16,17], which is already configured to perform calculations with fractional numbers of electrons. The calculations employed a double-numerical basis with d-functions on second row atoms (DND basis) [16] and a medium numerical integration grid (yielding an average of 1600 points/atom). Molecular cases used the local density approximation (LDA) exchange correlation from Vosko, Wilk, and Nusair (VWN) [36], whereas zeolite calculations used the generalized gradient approximation (GGA) from Perdew, Burke, and Ernzerhof [35], known as the PBE functional. All molecular geometries were optimized at this level of theory.

The zeolites chosen for this study were CHA, FER, MFI, and HSAPO-34. Experimental geometries were used except as described below. To become active as catalysts or separators, the zeolites require some silicon atoms to be replaced with aluminum ions together with an extra-framework cation. In each case, one Si was substituted with Al. Several symmetry-unique T sites are available for substitution in each lattice, but the results of the current research—the effect of partial charge on Fukui index—are not expected to be strongly affected by the selection. T1 was used for CHA, T2 for FER, and T8 for MFI. Note that HSAPO differs from the first three conventional zeolites in that the lattice is formed from Al and P rather than Si. In this case, a Si atom was used at the T1 site to create an active structure.

The simplest extra-framework cation to employ in zeolite simulations is, of course, H$^+$, and the placement of the ion can be problematic. In this work, we computed the electrostatic potential due to the zeolite framework and placed the H$^+$ at the location of the lowest (i.e., most negative) value. Subsequently, the position of the proton was optimized while all other atoms were kept fixed at the experimental coordinates. The resulting structures were used to evaluate the FFs.

FFs and condensed FFs were computed by finite differences of the atomic charges with $\Delta N = 0.01$, 0.1, and 1.0. For each system, the DFT energy was converged to self-consistency, and atomic point charges were computed as described below. The SCF calculation was repeated using charges of $\Delta N = \pm 0.01$, ± 0.1, and ± 1.0, the atomic point charges were again computed, and condensed FFs were evaluated via Equations 7.9 and 7.10.

Recently, Ayers et al. [24] have used sound mathematical arguments to recommend using Hirshfeld partitioning to compute the charges, and in our approach, such atomic charges were evaluated by numerical integration. This is similar to the approach used by Gilardoni et al. [25]. In the case of DMol³, the program employs partition functions to divide space into regions associated with an atomic center. Atomic charges were computed by integrating the charge density over all grid points while applying an appropriate partition function

$$q_k^H = Z_k - \sum_i \rho(\vec{r}_i) \frac{\rho_k(\vec{r}_i)}{\rho_T(\vec{r}_i)}, \tag{7.13}$$

where Z_k is the nuclear charge and $\rho_k(\vec{r}_i)$ is the charge density of the isolated atom k, and

$$\rho_T(\vec{r}_i) = \sum_j \rho_j(\vec{r}_i). \tag{7.14}$$

The sum i runs over all numerical integration points in the molecule and the sum over j over all atoms. We designate this type of atomic charge as q_k^H and will refer to this as the Hirshfeld charge throughout the paper.

7.4 RESULTS FOR f^+ AND f^-

7.4.1 RESULTS FOR MOLECULES

This section summarizes the earlier molecular results [18]. FFs f^+ and f^- were computed for 23 molecules at geometries optimized with the DND basis and VWN potential. Table 7.1 lists the RMS differences relative to the VWN/DND/$\Delta N = 0.1$ results for $\Delta N = 0.01$ and $\Delta N = 1.0$. Many of the atomic FFs are quite small, on the order of 0.01–0.05. Consequently, very small differences in the results can inflate the RMS values. For this reason, the results in Table 7.1 comprise only the "chemically significant" FFs, which are arbitrarily taken as f^+ or $f^- > 0.10$.

Decreasing the fractional charge, ΔN, results in an improved approximation to the derivatives in Equations 7.3 and 7.4, but this does not yield a significant change in f^- and f^+. Table 7.1 shows that the differences between $\Delta N = 0.1$ and $\Delta N = 0.01$ are less than 1% for f^- and f^+.

Much of the published literatures on FFs employ a charge difference of $\Delta N = 1.0$. (One notable exception is the work that compared FFs from relaxed Kohn–Sham orbitals to finite difference calculations extrapolated to $\Delta N \rightarrow 0$ [37].) As seen in Table 7.1, results using $\Delta N = 1.0$ differ from the more accurate numerical derivatives by about 4%–5%, on the average. Given the relatively qualitative nature of the FFs, this would normally be adequate for most work. Note, however, that the average RMS for all atoms, not just the chemically significant ones, is about 9% for f^- and 20% for f^+, so the use of $\Delta N = 1.0$ becomes more pronounced for smaller values. We also note that several of the systems experienced difficulty with SCF convergence, which is less likely with smaller ΔN.

TABLE 7.1

Molecular Results: Average RMS Difference (%) for Atomic FFs Compared to $\Delta N = 0.1$

Molecule	$\Delta N = 1.0$		$\Delta N = 0.01$	
	f^-	f^+	f^-	f^+
C_6H_5F	3.0	1.6	0.0	0.0
C_6H_5CN	1.7	0.8	0.0	0.4
$C_6H_5NH_2$	4.8	0.8	1.0	0.0
C_6H_5OH	2.8	1.5	0.0	0.0
$CH_2CHCHCH-CH_3$	3.7	3.6	0.4	0.0
$CH_2CHCHCH-CN$	3.4	4.4	0.5	0.6
$CH_2CHCHCH-COOH$	9.8	4.5	6.0	0.9
$CH_2CHCHCH-N(CH_3)_2$	2.5	14.4	0.5	0.8
$CH_2CHCHCH-OCH_3$	4.5	3.6	0.7	0.5
C_2H_3-CHO	7.4	3.7	0.7	0.7
C_2H_3-CN	1.7	3.2	0.4	0.5
C_2H_3-COOH	13.2	14.2	1.1	0.5
$C_2H_3-NH_2$	3.1	14.2	0.6	0.0
Pyridine	1.9	29.9	0.0	0.5
Butadiene	1.4	1.7	0.0	0.0
Ethylene	1.7	3.3	0.0	0.2
H_2CO	2.9	3.7	0.2	0.5
H_2O	4.9	0.8	0.5	0.7
Maleic anhydride	3.2	1.0	0.0	0.0
Maleimide	3.6	1.4	0.0	0.0
cis-Acrolein	7.4	3.4	0.6	0.6
trans-Acrolein	6.9	3.5	0.7	0.6
NH_2OH	9.8	11.7	1.3	5.3
Average % RMS	4.6	5.7	0.7	0.6

7.4.2 Results for Zeolites

The RMS summaries for the four zeolite cases at first appear to be insensitive to the selection of ΔN. As shown in Table 7.2, the agreement is within 1% regardless of the magnitude of the finite-difference step size. The RMS results, however, do not tell the whole story. The majority of the atoms in the zeolite cell are quite unreactive; these atoms have small Fukui indices that remain relatively constant with changes in ΔN. However, zeolite calculations tend to focus on just a few atoms such as the extra-framework cation or the most reactive oxygen atom in the lattice. The Fukui indices of these atoms turn out to vary considerably with changes in ΔN. As shown in Table 7.3, the results obtained using $\Delta N = 1.0$ can differ by as much as a factor of 8 from the more accurate values, but computing an average masks this result. Table 7.4 focuses

TABLE 7.2

Zeolite Results: Average RMS Difference (%) for Atomic FFs Compared to $\Delta N = 0.1$

Zeolite	$\Delta N = 1.0$		$\Delta N = 0.01$	
	f^-	f^+	f^-	f^+
CHA	2.1	0.6	1.4	1.1
FER	2.0	0.8	0.0	0.2
MFI	0.3	0.3	0.2	0.1
HSAPO	0.9	0.3	0.3	0.2
Average % RMS	1.3	0.5	0.5	0.4

TABLE 7.3

Zeolite Results: Maximum RMS Difference (%) for Atomic FFs Compared to $\Delta N = 0.1$

Zeolite	$\Delta N = 1.0$		$\Delta N = 0.01$	
	f^-	f^+	f^-	f^+
CHA	1.9	1.8	100	300
FER	100	100	300	800
MFI	100	500	400	330
HSAPO	20	27	22	75

TABLE 7.4

Zeolite Results: Atomic FFs for H Atom and Most Reactive Oxygen Atom

Zeolite	$\Delta N = 0.01$		$\Delta N = 0.1$		$\Delta N = 1.0$	
	$O{:}f^-$	$H{:}f^+$	$O{:}f^-$	$H{:}f^+$	$O{:}f^-$	$H{:}f^+$
CHA	0.14	0.32	0.16	0.32	0.08	0.20
FER	0.14	0.19	0.13	0.17	0.04	0.07
MFI	0.12	0.38	0.13	0.39	0.07	0.23
HSAPO	0.07	0.10	0.06	0.10	0.04	0.08
Average RMS difference from $\Delta N = 0.1$	2.9%	1.5%	–	–	12.8%	10.4%

on some atoms that would typically be of interest in a study of zeolite reactivity, namely, the extra-framework proton and the most reactive oxygen atom. On the average, $\Delta N = 1.0$ produces errors of ~10%–12%, but individual atoms can show errors of as much as 60%. Such large errors would be expected to introduce significant discrepancies into predictions of chemical reactivity and the relative importance of various active sites. In the case of HSAPO, for example, using $\Delta N = 1.0$ correctly identifies the most reactive oxygen atom, but incorrectly identifies the second-most reactive site.

7.5   SUMMARY

This work has illuminated the advantages of using fractional occupation to compute Fukui indices for molecular and periodic systems. Using a full electron, $\Delta N = 1.0$, produces average errors of only ~5% in molecular atomic Fukui indices, and even smaller ones in zeolites. The average results, however, disguise the fact that the results for the most reactive atoms can be in error by as much as 60%. These calculations demonstrate that using fractional charges offers a clear advantage in terms of the numerical precision of the calculations. It also provides a speed advantage, since the SCF calculations converge very quickly when starting from the results for the neutral molecule or crystal.

REFERENCES

1. K. Fukui. *Theory of Orientation and Stereoselection.* Berlin: Springer-Verlag, 1973.
2. K. Fukui. Role of frontier orbitals in chemical reactions. *Science* **218**, 1982, 747.
3. R. G. Parr and W. Yang. Density functional approach to the frontier-electron theory of chemical reactivity. *J. Am. Chem. Soc.* **106**, 1984, 4049.
4. R. G. Parr and R. G. Pearson. Absolute hardness: Companion parameter to absolute electronegativity. *J. Am. Chem. Soc.* **105**, 1983, 7512.
5. P. Chattraj, A. Cedillo and R. G. Parr. Variational method for determining the Fukui function and chemical hardness of an electronic system. *J. Chem. Phys.* **103**, 1995, 7645.
6. S. Damoun, G. Van de Woude, F. Méndez and P. Geerlings. Local softness as an indicator of regioselectivity in pericyclic reactions. *J. Phys. Chem. A* **101**, 1997, 886.
7. L. R. Domingo, M. Arnó, R. Contreras and P. Pérez. *J. Phys. Chem. A* **106**, 2002, 952.
8. W. Langenaeker, K. Demel and P. Geerlings. Quantum-chemical study of the Fukui function as a reactivity index. Part 2: Electrophilic substitution on mono-substituted benzenes. *J. Mol. Struct. (THEOCHEM)* **234**, 1991, 329.
9. W. Langenaeker, K. Demel and P. Geerlings. Quantum chemical study of the Fukui function as a reactivity index. Part 3. Nucleophilic addition at activated CC double bonds. *J. Mol. Struct. (THEOCHEM)* **259**, 1992, 317.
10. W. Langenaeker, M. de Decker and P. Geerlings. Quantum chemical study of the Fukui function as a reactivity index: Probing the acidity of bridging hydroxyls in zeolite type model systems. *J. Mol. Struct. (THEOCHEM)* **207**, 1990, 115.
11. J. Oláh, T. Veszprémi, F. De Proft and P. Geerlings. Silylenes: A unified picture of their stability, acid/base and spin properties, nucleophilicity, and electrophilicity via computational and conceptual density functional theory. *J. Phys. Chem. A* **111**, 2007, 10815.
12. N. Gupta, R. Garg, K. Kr. Shah, A. Tanwar and S. Pal. Deprotonation of 1,2-dialkylpyridinium ions: A DFT study of reactivity and site selectivity. *J. Chem. Phys. A* **111**, 2007, 8823.
13. N. Otero, M. Mandado and R. A. Mosquera. Nucleophilicity of indole derivatives: Activating and deactivating effects based on proton affinities and electron density properties. *J. Phys. Chem. A* **111**, 2007, 5557.

14. R. G. Parr and W. Yang. *Density Functional Theory of Atoms and Molecules*. New York: Oxford University Press, 1989.
15. H. Chermette. Chemical reactivity indexes in density functional theory. *J. Comp. Chem.* **20**, 1999, 129.
16. B. Delley. An all-electron numerical method for solving the local density functional for polyatomic molecules. *J. Chem. Phys.* **92**, 1990, 508.
17. B. Delley. From molecules to solids with the Dmol³ approach. *J. Chem. Phys.* **113**, 2000, 7756.
18. G. Fitzgerald. On the use of fractional charges for computing Fukui functions. *Mol. Simul.* **34**, 2008, 931.
19. A. Chatterjee, T. Iwasaki and T. Ebina. 2:1 Dioctahedral smectites as a selective sorbent for dioxins and furans: Reactivity index study. *J. Phys. Chem. A* **106**, 2002, 641.
20. P. Geerlings, A. M. Vos and R. A. Schoonheydt. A computational and conceptual DFT approach to the kinetics of acid zeolite catalyzed electrophilic aromatic substitution reactions. *J. Mol. Struct. (THEOCHEM)* **762**, 2006, 69.
21. C.-M. Wang, Y.-D. Wang, H.-X. Liu, Z.-K. Xie and Z.-P. Liu. Catalytic activity and selectivity of methylbenzenes in Hsapo-34 catalyst for the methanol-to-olefins conversion from first principles. *J. Catal.* **271**, 2010, 386.
22. M. N. Mikhailov, I. V. Mishin, L. M. Kustov and A. L. Lapidus. Structure and reactivity of Pt/GaZSM-5 aromatization catalyst. *Microporous Mesoporous Mater.* **104**, 2007, 145.
23. M. N. Mikhailov, I. V. Mishin, L. M. Kustov and V. Z. Mordkovich. The structure and activity of Pt6 particles in ZSM-5 type zeolites. *Catal. Today* **144**, 2009, 273.
24. P. W. Ayers, R. C. Morrison and R. K. Roy. Variational principles for describing chemical reactions: Condensed reactivity indices. *J. Chem. Phys.* **116**, 2002, 8731.
25. F. Gilardoni, J. Weber, H. Chermette and T. R. Ward. Reactivity indices in density functional theory: A new evaluation of the condensed Fukui function by numerical integration. *J. Phys. Chem. A* **102**, 1998, 3607.
26. J. F. Janak. Proof that $\partial E/\partial n_i = \varepsilon$ in density-functional theory. *Phys. Rev. B* **13**, 1978, 7165.
27. J. P. Perdew, R. G. Parr, M. Levy and J. L. Balduz. Density-functional theory for fractional particle number: Derivative discontinuities of the energy. *Phys. Rev. Lett.* **49**, 1982, 1691.
28. R. Balawender and L. Komorowski. Atomic Fukui function indices and local softness *ab initio*. *J. Chem. Phys.* **109**, 1998, 5203.
29. J. Korchowiec, H. Gerwens and K. Jug. Relaxed Fukui function indices and their application to chemical reactivity problems. *Chem. Phys. Lett.* **222**, 1994, 58.
30. J. Oláh and C. Van Alsenoy. Condensed Fukui functions derived from stockholder charges: Assessment of their performance as local reactivity descriptors. *J. Chem. Phys. A* **106**, 2002, 3885.
31. J. C. Slater. Statistical exchange-correlation in the self-consistent field. *Adv. Quantum Chem.* **6**, 1972, 1.
32. G. Kresse and J. Furthmuller. Efficiency of *ab-initio* total energy calculations for metals and semiconductors using a plane-wave basis set. *Comput. Mater. Sci.* **6**, 1996, 15.
33. A. D. Rabuck and G. E. Scuseria. Improving SCF convergence by varying occupation numbers. *J. Chem. Phys.* **110**, 1999, 695.
34. P. W. Ayers, F. De Proft, A. Borgoo and P. Geerlings. Computing Fukui functions without differentiating with respect to electron number. I. Fundamentals. *J. Chem. Phys.* **126**, 2007, 224107.
35. J. P. Perdew, K. Burke and M. Ernzerhof. Generalized gradient approximation made simple. *Phys. Rev. Lett.* **77**, 1996, 3865.
36. S. J. Vosko, L. Wilk and M. Nusair. Accurate spin-dependent electron liquid correlation energies for local spin density calculations: A critical analysis. *Can. J. Phys.* **58**, 1980, 1200.
37. A. Michalak, F. De Proft, P. Geerlings and R. F. Nalewajski. Fukui functions from the relaxed Kohn Sham orbitals. *J. Chem. Phys. A* **103**, 1999, 762.

8 Density Functional Theory Study of Urea Interaction with Potassium Chloride Surfaces

Ajeet Singh and Bishwajit Ganguly

CONTENTS

8.1 INTRODUCTION

The study and engineering of crystal faces have attracted immense interest among artists and the crystal grower community since at least the Bronze Age [1]. Although significant efforts have been made over the last few decades to precisely predict the growth morphology of crystals, it still remains a challenging task to this date. Crystal growth morphology has diverse applications ranging from drug design [2] to explosives [3] and inverse gas chromatography data [4]. Therefore, knowledge of crystal growth habits and their morphological properties is important in understanding and exploiting many of their physicochemical properties. In this regard, the influence of additives on its crystal habit has received considerable attention. It has been found that the nucleation, growth, and morphology of crystals can be significantly altered by the presence of low concentrations of impurities such as reaction byproducts, impurities present in the reactants, and additives that are purposely added to

109

alter the crystallization process [5,6]. Additives can reduce crystal growth rate and alter morphology by binding to crystal faces and interfering with propagation steps [7,8]. Rome de l'Isle showed that octahedrons are formed instead of normal cubes if rocksalt is grown in the presence of urine [9]. Many authors have since reported the cube–octahedron shape transition for various experimental conditions. Some early work reported that octahedron can also be obtained from pure water solution [10–12]. Radenović et al. have experimentally studied the habit change of sodium chloride from cubic to octahedron in the presence of smaller amides [13,14]. The observed results have been rationalized based on charge distributions and strong interactions between the more exposed carbonyl oxygen of amides such as formamide and urea with the sodium ions, which stabilize the {111} surface of sodium chloride and lead to the octahedron morphology of sodium chloride [13,14]. In earlier observations, Speidel and Bunn have also considered that the additive strongly interacts with certain crystal faces to influence the morphology of the crystal [15–17]. Bunn explained that the habit modification of NaCl by urea in aqueous solutions is due to adsorption of the impurity on certain crystal faces during crystal growth [17]. There were many proposals that have been reported to explain the habit of sodium chloride; however, studies toward the interactions at the molecular level are limited. We have recently performed a detailed computational analysis of urea interaction with the surfaces of sodium chloride [18]. The calculated results suggest that the interactions of additives with certain crystal faces are one of the important factors toward the change in morphology of alkali halides. A number of studies have been performed toward the habit of sodium chloride crystals in the presence of impurities; however, the higher homologue KCl has received little attention. Recently, there have been some efforts to crystallize KCl in carbon nanotubes [19]. The KCl crystals were grown in cubic form with predominantly stable {100} faces. It is mentioned in one of the reports that urea can act as an additive for KCl [6,20]. However, we have not come across any detailed experimental study of the effect of urea on KCl crystals. Our effort to understand the growth of alkali halide crystals prompted us to examine the effect of impurities on KCl crystals. Therefore, we have undertaken the computational approach to study the influence of urea on the morphology of KCl crystals followed by the experimental observations. The experimental results would be useful to examine the predictive nature of these computational analyses.

To examine the effective interaction of urea with specific surfaces of KCl, an approach similar to surface docking developed to predict the influence of additives on the crystal morphology has been employed [21–27]. The basis of this approach is to analyze the effect of additives on the individual crystal faces, which are cleaved from a crystal. If the additive has a preferred interaction on a special face, the growth of this face will be slower. As a result, the other fast-growing surfaces will disappear, and eventually, the slow-growing surface will control the morphology. In this way, the additive influences the morphology of crystals. For simulations of surfaces of crystalline solids, slab, and cluster models are nevertheless by far more popular because they are feasible from the computational point of view [28]. However, the cluster models came under scrutiny due to their finite size representation. Slab models rather mimic the infinite surface of solids and are considered to be a better approach than the cluster models. In this study, a conventional array of these alkali halide ions has been used in slab [constructed using periodic boundary conditions (PBCs)] using respective crystal data. The stable {100} surface of potassium chloride was modeled with alternating arrangement of K^+ and Cl^- ions. However, modeling the electrostatically polar {111} surfaces of these alkali halide crystal structures was considered a mystery in surface science because it is difficult to investigate both experimentally and theoretically [29,30]. Because the bulk structure consists of alternating cationic and anionic sheets stacked along the <111> directions, the {111} polar surfaces must have a very high divergent electrostatic energy, which makes them theoretically highly unstable [6,29–31]. It has been shown in the earlier studies that the adsorption of negatively charged site of additives would be preferred with the positive ions on top of the surface of alkali halides [13,14]. Recently, Radenović et al. have presented a surface x-ray diffraction determination of the {111} NaCl–liquid interface structure in the presence of aqueous solution and formamide [31]. It was determined that the rocksalt {111} surface is Na^+ terminated for both the environmental conditions. For KCl, {111} surfaces were modeled with the K^+ ions on top of the surface.

8.2 COMPUTATIONAL AND EXPERIMENTAL METHODS

8.2.1 COMPUTATIONAL PROCEDURE

The interaction study of urea with three-dimensional (3D) slabs of KCl has been performed using the density functional program DMol^3 in Materials Studio® (version 4.1) of Accelrys; the physical wave functions are expanded in terms of numerical basis sets [32–35]. We used a double numerical basis set with d polarization (DND), which is comparable to 6-31G* basis set. The geometry of urea and its interaction on surfaces of KCl was optimized with local spin density approximation with Perdew–Wang correlational (LDA/PWC) [32–35]. The local density approximation (LDA) is one of the earliest approximations in density functional theory (DFT). It includes correction for electron correlation effects. However, one of the most important deficiencies with the LDA exchange, the incorrect asymptotic behavior, overestimates the energies for the system. This was solved by using generalized gradient approximations (GGAs) by Perdew and Wang, PW91. The larger basis set with p polarization (DNP) compared to DND was also used to examine the influence of basis set effect. Energies were calculated for the adsorbed systems with GGA/PW91 methods. 3D slabs of KCl with [{100}, {110}, and {111}] planes were generated using unit cells of these halides with PBCs as implemented in DMol^3 [32–35]. In the 3D slab model, the slab is periodically repeated along the normal direction to the surface. The 3D slab models depend on the slab thickness and the vacuum gap separating the nearest slab images in the z-direction. We have investigated this dependence via optimizing the number of layers in 3D slabs with vacuum thickness. We have optimized the supercell lattice by varying the number of layers in each case and keeping the sufficiently large vacuum thickness fixed at 20 Å. We have considered the numbers of layers in each case when the total energy of the system becomes minimum. Furthermore, the vacuum thickness was varied from 8 Å to 20 Å in 2-Å steps with the previously optimized layers and considered the values when the total energy as a function of vacuum gap separations was minimum. For KCl, the optimized Perdew–Wang correlational (LDA/PWC) {100} slab model contains seven layers (112 ions) with a vacuum thickness of 12 Å; for {110} and {111}, the optimized slabs contain eight and seven layers with a vacuum thickness of 12 Å, respectively. Our optimized supercell lattice for KCl was found to be consistent with the results of Ermoshkint et al. that six layered slabs are large enough to reproduce the surface states and bulk states of alkali halide crystals [36]. The k-points were generated by Monkhorst and Pack [37]. This scheme produces a uniform grid of k-points along the three axes in reciprocal space. The k-points used in this study were $2 \times 2 \times 1$ with separation from origins 0.30744, 0.0355, and 0.03955/Å, respectively. The tolerances of energy, gradient, and displacement convergence were 2×10^{-5} Ha, 4×10^{-3} Ha/Å, and 5×10^{-3} Å, respectively (1 Ha = 629.5095 kcal/mol). The self-consistent field convergence criterion for all calculations was 1.0×10^{-5}. For additive and crystal surface interactions, urea was placed at the midpoint of the KCl surface in each case and optimized at the defined level of theory. The maximum gradient for most of the optimized structures was less than 2×10^{-3} Ha/Å. Interaction energies were computed by subtracting the energies of the additive molecules (formamide: $E_{additive}$) and surface (alkali halides: $E_{surface}$) from the energy of the adsorption system (alkali halides with formamide: $E_{additive/surface}$), as shown in Equation 8.1:

$$E_{int} = E_{additive/surface} - \{E_{(additive)} + E_{(surface)}\}. \tag{8.1}$$

The conductor-like screening model (COSMO) for real solvents was used for solvent calculations [38,39]. The improvement of DMol^3 version (4.1) over version 4.0 is that the continuum solvation model calculations can be performed with the slab models. Therefore, the COSMO calculations were performed with the periodic slab models used for the gas-phase calculations. This is a significant advantage over the previous version of DMol^3, where the COSMO calculations were used to mimic the periodic surfaces with cluster models [18]. Furthermore, we have demonstrated in our earlier study that the DFT models used in this case are accurate enough to predict the binding energies of additives with alkali halide surface [18]. The test case was performed with water molecule, whose experimental binding energies on NaCl {100} surfaces are known. Our DFT calculations have reproduced the water adsorption energies reasonably well with the GGA model [40,41].

8.2.2 EXPERIMENTAL PROCEDURE

8.2.2.1 Batch Crystallization

We have performed experiments for crystallization of KCl in the presence of urea as a possible habit modifier in aqueous solutions. To examine the influence of urea on the morphology of KCl crystals, three different concentrations of KCl were used, namely, 4.2, 5.2 (under saturated solution), and 5.6 M (saturated solution). Urea concentration was varied from 5% to 30% [with respect to solute (w/w)] as an additive in different batch crystallizations with a regular increment of 5 wt%. We varied the concentration of the analyte, that is, KCl, dissolved in 5 ml of distilled water taken in a 10-ml capacity beaker, and to this, urea of varying weight percent was added under ambient conditions in the laboratory (room temperature 25°C). For comparison, we repeated similar experiments with NaCl instead of KCl. Urea, KCl, and NaCl used in this study were obtained from Merck, which were 99.99% pure and were used without further purification. The morphologies obtained after crystallization were examined by scanning electron microscopy (SEM) and optical microscopy, and the observed morphologies are discussed below.

8.2.2.2 Instrumentation

The surface structure of the grown crystals was observed by SEM. The samples for the SEM studies were prepared on cleaned and mirror-polished brass stubs by spreading 100 μl of the dispersion and evaporating the solvent in air. A LEO SEM model 1430 VP was used for the purpose.

8.2.2.3 Optical Microscopy

Optical microscopic images were taken by an Olympus (SZH10) research stereo microscope. The statistical distribution of the crystals was observed in the crystallization bath (a 10-ml volumetric glass beaker). Individual crystals were examined by putting them on glass slides.

8.3 RESULTS AND DISCUSSION

Urea conformers. It has generally been assumed that urea is a planar symmetrical molecule and the crystal structure reinforced this assumption [42]. However, theoretical calculations reported on urea showed that urea is nonplanar [43–51]. We have examined the conformations of urea in the gas and aqueous phases at different DFT levels of theory (Scheme 8.1) [52–54]. In agreement with previous reports, the C_2 II conformer of urea is the most stable conformer and a true minimum (no imaginary frequencies) (Scheme 8.1) [42–45].

For the interaction study with the surface of KCl, the most stable conformer C_2 II was considered in this study. The interactions of urea II with the 3D slabs of potassium chloride are shown in Figure 8.1. The optimizations of these additives on the surface of KCl were performed at the LDA/PWC/DND level of theory. Two different sets of calculation were performed to examine the relaxation effects of KCl surfaces on the interaction energies of urea. First, the slab was kept fixed while optimizing the urea on the top of the {100}, {110}, and {111} surfaces of potassium chloride. In another set of calculations, two layers of KCl surfaces were relaxed while optimizing

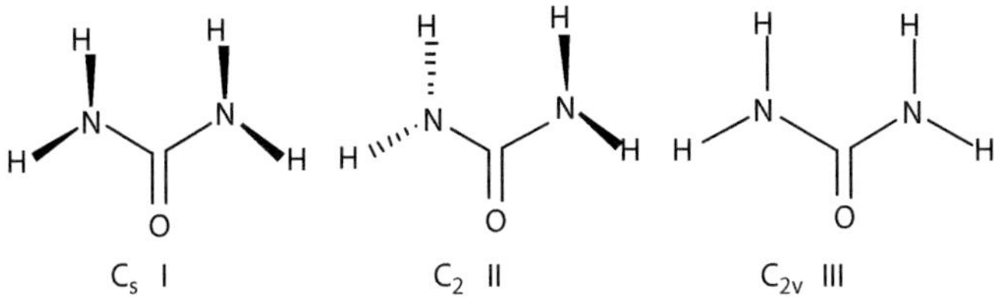

SCHEME 8.1 C_s I, C_2 II, and C_{2v} III urea conformers.

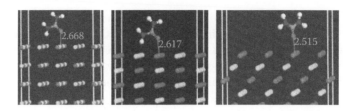

FIGURE 8.1 LDA/PWC/DND calculated geometries of urea with the fixed {100}, {110}, and {111} KCl surfaces in slab models. (From Singh, A. and Ganguly, B., *Molecular Simulation*, **34**, 973–979, 2008. With permission.)

the geometry of urea on the surfaces. In this case, the rest of the bottom layers were kept fixed in the bulk position.

The interaction energies calculated for urea with potassium chloride surfaces are shown in Table 8.1. The computed results suggest that the interactions of C_2 II urea conformer are slightly preferred for {110} surface of KCl 2–4 kcal/mol compared with {100} and {111} planes. The binding energies calculated for the urea conformer II with fixed and relaxed surface of potassium chloride are not largely different; importantly, the relative trends are the same in both cases. The calculated trend at a larger basis set DNP is similar to that obtained for DND (Table 8.1).

The mode and orientations of interaction of urea molecules with these surfaces are shown in Figure 8.1. The urea moiety prefers to interact with both the potassium and chloride ions in {100} and {110} surfaces. The carbonyl oxygen of urea interacts with the potassium ion, whereas the hydrogen of amine functionality interacts with the chloride ion. In the case of the {111} surface of KCl, the carbonyl oxygen interacts with the potassium ion.

The DFT calculated results suggest that the interaction of urea is marginally preferred with the {110} surface of KCl and such preferential interaction can have an effect to retard the growth of this face and can lead to the formation of rhombododecahedrons. However, the interaction of urea with {111} plane is comparable

TABLE 8.1

Interaction Energies Calculated at GGA/PW91/DND and GGA/PW91/DNP Levels Using LDA/PWC/DND Optimized Geometries for Urea with {100}, {110}, and {111} Surfaces of KCl Slab Models

Plane	Interaction Energies (kcal/mol)		
	{100}	{110}	{111}
GGA/PW91/DND			
(Fixed surface)	−15.8	−19.7	−17.8
(Two-layered relaxed surface)	−15.2	−20.3	−19.9
GGA/PW91/DNP	−15.9	−19.6	−17.2

Source: Singh, A. and Ganguly, B., *Molecular Simulation*, **34**, 973–979, 2008. With permission.

TABLE 8.2

Interaction Energies Calculated at GGA/PW91/DND Level Using LDA/ PWC/DND Optimized Geometries for Urea with {100}, {110}, and {111} Surfaces of KCl Slab Models in COSMO model

Plane	Interaction Energies (kcal/mol)		
	{100}	{110}	{111}
GGA/PW91/DND	−10.5	−11.9	−11.5

Source: Singh, A. and Ganguly, B., *Molecular Simulation*, **34**, 973–979, 2008. With permission.

to {110}. Therefore, a clear prediction does seem to be possible from the computed results for the formation of rhombo-dodecahedrons and or octahedrons. Nevertheless, the unstable {110} and {111} surfaces were found to be slightly energetically preferred over {100} of KCl with urea molecule. It is noteworthy that the calculations were performed in the gas phase. It has been reported that the medium can play a profound role on the crystal habit [55–59]. Therefore, the interaction of solvent on the interaction of urea with surfaces of KCl should also be examined.

It is also noteworthy that urea interacts with the alkali halide surfaces in aqueous solution; hence, the surrounding medium effect seems to be important. To examine the effect of the solvent, we performed calculations with the continuum solvation model (COSMO). The dielectrics ($\epsilon = 78.4$ for water) was used in these cases. The optimized gas-phase slabs for KCl {100}, {110}, and {111} with urea were computed with the COSMO model. The calculated interaction energies are shown in Table 8.2. Interaction energies were found to be lower in water compared to the gas-phase results. Importantly, the interaction energies for urea with all three surfaces were found to be comparable. Based on these calculated results, one could predict that the additional stabilization cannot be achieved for the unstable surfaces such as {111}

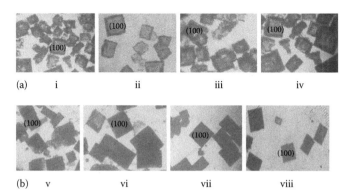

FIGURE 8.2 Optical images of KCl grown crystals obtained from (a) 4.2 M KCl with 5, 10, 15, and 20 wt% (w/w) urea, i, ii, iii, and iv, respectively, and (b) 5.5 M KCl with 5, 10, 15, and 20 wt% (w/w) urea v, vi, vii, and viii, respectively. (From Singh, A. and Ganguly, B., *Molecular Simulation*, **34**, 973–979, 2008. With permission.)

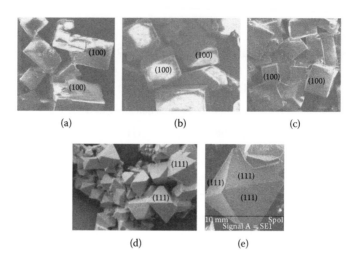

FIGURE 8.3   SEM images of KCl grown crystals obtained from (a) 4.2 M, (b) 5.5 M, and (c) 5.8 M, and NaCl crystals from (d) 4.2 M and (e) 5.5 M solutions in 25 wt% urea–water solution. (From Singh, A. and Ganguly, B., *Molecular Simulation*, **34**, 973–979, 2008. With permission.)

and {110} with urea additive, and, hence, in the aqueous solution, the stable {100} surface should grow preferentially.

To verify the computational analysis, experiments were performed for the growth of KCl crystals in the presence of urea impurity. The experiments were carried out at different saturation levels of KCl and ratios of urea impurity as mentioned in Section 8.2.2. It is noteworthy that the organic additives were required in high concentrations for habit modification of alkali halides as reported in an earlier study [60]. Initially, we have taken lower concentrations [5, 10, 15, and 20 wt% (w/w)] of an additive to examine its effects on the morphology of KCl. We have observed that there is no change in the morphology of KCl at these concentrations (Figure 8.2), and hence we used further higher concentrations of urea [25 wt% (w/w)].

Even at this concentration of impurity, the growing KCl crystals were found to be cubes (Figure 8.3 and Table 8.3). On the other hand, the growing crystals of NaCl were found to be octahedrons at 25 wt% of urea (Figure 8.3 and Table 8.3).

TABLE 8.3
Experimentally Observed Morphologies of KCl and NaCl with 25 wt% Urea

Serial No.	Conc. of the Halides	Urea	Remarks
a	4.2 M KCl	25 wt%	Cubic
b	5.5 M KCl	25 wt%	Cubic
c	5.8 M KCl	25 wt%	Cubic
d	4.2 M NaCl	25 wt%	Octahedron
e	5.5 M NaCl	25 wt%	Octahedron

Source: Singh, A. and Ganguly, B., *Molecular Simulation*, **34**, 973–979, 2008. With permission.

The experimental results corroborate the computational predictions that the growing KCl crystals in urea solution should predominantly be cubes with {100} face.

8.4 CONCLUSION

In the present study, we have examined the effect of urea on the morphology of KCl crystals. The computed results predicted that the interaction of urea is not significantly preferred to any important surfaces of KCl. Importantly, the interaction energies were found to be comparable in the solvent medium. The relaxed surfaces do not alter the interaction energies for urea with the KCl surfaces. {100} and {110} planes interact with both the carbonyl and amine functionalities of urea molecules. Experimental results reveal that the KCl crystals are cubic in the presence of urea impurity; however, the NaCl crystals are octahedrons under such conditions. The computational analyses support the model of effective interaction of additives with certain face of crystals as one of the factors responsible for the change in the morphology.

ACKNOWLEDGMENTS

This work was supported by the Department of Science and Technology, New Delhi, India. We thank Dr. P. K. Ghosh (Director), Central Salt and Marine Chemicals Research Institute, for his keen interest in this work.

REFERENCES

1. C. S. Smith. *A Search for Structure: Selected Essays on Science, Art, and History.* Cambridge, MA: MIT Press, 1981.
2. D. S. Coombes, C. R. A. Catlow, J. D. Gale, M. J., Hardy and M. R. Saunders. Theoretical and experimental investigations on the morphology of pharmaceutical crystals. *J. Pharm. Sci.* **91**, 2002, 1652.
3. J. H. ter Horst, H. J. M. Kramer, G. M. van Rosmalen and P. J. Jansens. Molecular modeling of the crystallization of polymorphs. Part I: 93. The morphology of HMX polymorphs. *J. Cryst. Growth* **237**, 2002, 2215.
4. I. M. Grimsey, J. C. Osborn, S. W. Doughty, P. York and R. C. Rowe. The application of molecular modeling to the interpretation of inverse gas chromatography data. *J. Chromatogr. A* **969**, 2002, 49.
5. D. L. Klug. *Handbook of Industrial Crystallization*, edited by A. S. Myerson. Montvale, MA: Butterworth, 1993, 65.
6. J. W. Mullin. *Crystallization*, 3rd edition. London: Butterworth, 1993, 238.
7. R. H. Doremus, B. W. Roberts and D. Turnbull. *Growth and Perfection of Crystals*. New York: Wiley, 1958, 393.
8. J. P. Van der Eerden and H. Mueller-Krumbhaar. Formation of macrosteps due to time dependent impurity adsorption. *Electrochim. Acta* **31**, 1986, 1007.
9. J. B. L. Rome de l'Isle de. Ou description des formes propres a tous les corps du regne mineral, dans l'etat de combinaison saline, pierreuse ou metallique. *Crystallographie Paris* **1**, 1783, 379.
10. R. Kern. Etude des faces de cristaux ioniques à structure simple. *Bull. Soc. Fr. Mineral. Crystallogr.* **76**, 1953, 391.

11. M. Beinfait, R. Boistelle and R. Kern. *Adsorption et croissance cristalline*, edited by R. Kern. Paris: Centre national de la recherche scientifique, **1965**, 152, 515.

12. A. Johnsen. *Wachstum and auflosing der kristallen*. Leipzig: Engelmann, 1910.

13. N. Radenović, W. van Enckevort, P. Verwer and E. Vlieg. Growth and characteristics of the {111} NaCl crystal surface grown from solution. *Surf. Sci.* **523**, 2003, 307.

14. N. Radenović, W. van Enckevort and E. Vlieg. Formamide adsorption and habit changes of alkali halide crystals grown from solutions. *J. Cryst. Growth* **263**, 2004, 544.

15. R. Speidel. *Neues Jahrbuch fur Mineralogie*. 3AD-70178 Stuttgart, Germany: E. Schweizerbart Science Publishers, Johannesstr. Part 4, 1961, 81.

16. N. Cabrera and D. A. Vermilyea. *Growth and Perfection of Crystals*. New York Chapman and Hall, London: John Wiley & Sons Inc., 1958, 393.

17. C. W. Bunn. Adsorption, oriented overgrowth and mixed crystal formation. *Proc. R. Soc. A* **141**, 1993, 567.

18. A. Singh, S. Chakraborty and B. Ganguly. Computational study of urea and its homologue glycinamide: Conformations, rotational barriers, and relative interactions with sodium chloride. *Langmuir* **23**, 2007, 5406.

19. W. K. Hsu, W. Z. Li, Y. Q. Zhu, N. Grobert, M. Terrones, H. Terrones, N. Yao, et al. KCl crystallization within the space between carbon nanotube walls. *Chem. Phys. Lett.* **317**, 2000, 77.

20. H. Hatakka, H. Alatalo and S. Palosaari. Effect of impurities and additives on crystal growth. Symposium on Crystallization and Precipitation, Lappeenranta, Finland, May 1997.

21. P. C. Coveney and W. Humphries. Molecular modeling of the mechanism of action of phosphonate retarders on hydrating cements. *J. Chem. Soc., Faraday Trans.* **92**, 1996, 831.

22. P. C. Coveney, R. Davey, J. L. W. Griffin, Y. He, J. D. Hamlin, S. Stackhouse and A. Whiting. A new design strategy for molecular recognition in heterogeneous systems: A universal crystal-face growth inhibitor for barium sulfate. *J. Am. Chem. Soc.* **122**, 2000, 11557.

23. A. Wierzbicki and H. S. Cheung. Molecular modeling of inhibition of crystals of calcium pyrophosphate dihydrate by phosphocitrate. *J. Mol. Struct. (THEOCHEM)* **454**, 1998, 287.

24. P. V. Coveney, R. J. Davey, J. L. W. Griffin and A. Whiting. Molecular design and testing of organophosphonates for inhibition of crystallisation of ettringite and cement hydration. *Chem. Commun.* 14, 1998, 1467.

25. A. Wierzbicki, C. S. Sikes, J. D. Sallis, J. D. Madura, E. D. Stevens and K. L. Martin. Scanning electron microscopy and molecular modeling of inhibition of calcium oxalate monohydrate crystal growth by citrate and phosphocitrate. *Calcif. Tissue Int.* **56**, 1995, 297.

26. A. Wierzbicki, C. S. Sikes, J. D. Madura and B. Drake. Atomic force microscopy and molecular modeling of protein and peptide binding to calcite. *Calcif. Tissue Int.* **54**, 1994, 133.

27. J. J. Lu and J. Ulrich. The influence of supersaturation on crystal morphology: Experimental and theoretical study. *Cryst. Res. Technol.* **38**, 2003, 63.

28. P. Deak. Choosing models for solids. *Phys. Status Solidi B* **217**, 2000, 9.

29. P. W. Tasker. The surface energies, surface tensions and surface structure of the alkali halide crystals. *Philos. Mag. A* **39**, 1979, 119.

30. P. W. Tasker. The stability of ionic crystal surfaces. *J. Phys. C: Solid State Phys.* **12**, 1979, 4977.

31. N. Radenović, D. Kaminski, W. van Enckevort, S. Graswinckel, I. Shah, M. In't. Veld, R. Alga and E. Vlieg. Stability of the polar {111} NaCl crystal face. *J. Chem. Phys.* **124**, 2006, 164706.

32. B. Delley. An all-electron numerical method for solving the local density functional for polyatomic molecules. *J. Chem. Phys.* **92**, 1990, 508.

33. B. Delley. Fast calculation of electrostatics in crystals and large molecules. *J. Phys. Chem.* **100**, 1996, 6107.

34. B. Delley. From molecules to solids with the *Dmol³* approach. *J. Chem. Phys.* **113**, 2000, 7756.

35. Materials Studio DMOL3, Version 4.1. Accelrys, Inc., San Diego, USA.

36. A. N. Ermoshkint, E. A. Kotomin and A. L. Shluger. The semiempirical approach to electronic structure of ionic crystal surface. *J. Phys. C: Solid State Phys.* **15**, 1982, 847.

37. H. J. Monkhorst and J. D. Pack. Special points for Brillouin-zone integrations. *Phys. Rev. B* **13**, 1976, 5188.

38. A. Klamt and G. J. Schüürmann. COSMO: A new approach to dielectric screening in solvents with explicit expressions for the screening energy and its gradient. *J. Chem. Soc., Perkin Trans.* **2**, 1993, 799.

39. J. Tomais and M. Persico. Molecular interactions in solution: An overview of methods based on continuous distributions of the solvent. *Chem. Rev.* **94**, 1994, 2027.

40. M. Foster and G. E. Ewing. Adsorption of water on the NaCl (001) surface. II. An infrared study at ambient temperatures. *J. Chem. Phys.* **112**, 2000, 6817.

41. M. Foster and G. E. Ewing. An infrared spectroscopic study of water thin films on NaCl (100). *Surf. Sci.* 427–428, 1999, 102.

42. J. E. Worsham Jr., H. A. Levy and S. W. Peterson. The positions of hydrogen atoms in urea by neutron diffraction. *Acta Crystallogr.* **10**, 1957, 319.

43. T.-K. Ha and C. Puebla. A theoretical study of conformations and vibrational frequencies in (NH2)2C=X compounds (X=O, S, and Se). *Chem. Phys.* **181**, 1994, 47.

44. D. A. Dixon and N. Matsuzawa. Density functional study of the structures and nonlinear optical properties of urea. *J. Phys. Chem.* **98**, 1994, 3967.

45. A. Masunov and J. J. Dannenberg. Theoretical study of urea. I. Monomers and dimers. *J. Phys. Chem. A* **103**, 1999, 178.

46. T. Ishida, P. J. Rossky and E. W. Castner Jr. A theoretical investigation of the shape and hydration properties of aqueous urea: Evidence for nonplanar urea geometry. *J. Phys. Chem. B* **108**, 2004, 17583.

47. P.-O. Åstrand, A. Wallqvist, G. Karlström and P. Linse. Properties of urea–water solvation calculated from a new ab initio polarizable intermolecular potential. *J. Chem. Phys.* **95**, 1991, 8419.

48. B. Mennucci, R. Cammi, M. Cossi and J. Tomasi. Solvent and vibrational effects on molecular electric properties. Static and dynamic polarizability and hyperpolarizabilities of urea in water. *J. Mol. Struct. (THEOCHEM)* **426**, 1998, 191.

49. M. Kontoyianni and P. Bowen. An ab initio and molecular mechanical investigation of ureas and amide derivatives. *J. Comput. Chem.* **13**, 1992, 657.

50. R. J. Meier and B. Coussens. The molecular structure of the urea molecule: Is the minimum energy structure planar? *J. Mol. Struct.* **253**, 1992, 25.

51. A. Gobbi and G. Frenking. Y-conjugated compounds: The equilibrium geometries and electronic structures of guanidine, guanidinium cation, urea, and 1,1-diaminoethylene. *J. Am. Chem. Soc.* **115**, 1993, 2362.

52. P. D. Godfrey, R. D. Brown and A. N. Hunter. The shape of urea. *J. Mol. Struct.* **413**, 1997, 405.

53. R. D. Brown, D. Godfrey and J. Storey. The microwave spectrum of urea. *J. Mol. Spectrosc.* **58**, 1975, 445.

54. W. Gilkerson and K. Srivastava. The dipole moment of urea. *J. Phys. Chem.* **64**, 1960, 1485.

55. R. J. Davey, J. W. Mullin and M. J. L. Whiting. Habit modification of succinic acid crystals grown from different solvents. *J. Cryst. Growth.* **58**, 1982, 304.

56. M. Lahav and L. Leiserowitz. The effect of solvent on crystal growth and morphology. *Chem. Eng. Sci.* **56**, 2001, 2245.

57. C. Stoica, P. Verwer, H. Meekes, P. J. C. M. van Hoof, F. M. Kaspersen and E. Vlieg. Understanding the effect of a solvent on the crystal habit. *Cryst. Growth Des.* **4**, 2004, 765.

58. L. A. Hurley, A. G. Jones and R. B. Hammond. Molecular packing morphological modeling, and image analysis of cyanazine crystals precipitated from aqueous ethanol solutions. *Cryst. Growth Des.* **4**, 2004, 711.

59. S. Khoshkhoo and J. Anwar. Study of the effect of solvent on the morphology of crystals using molecular simulation: Application to α-resorcinol and *N-n*-octyl-D-gluconamide. *J. Chem. Soc., Faraday Trans.* **92**, 1996, 1023.

60. C. P. Fenimore and A. Thrailkill. The mutual habit modification of sodium chloride and dipolar ions. *J. Am. Chem. Soc.* **71**, 1949, 2714.

9 Barrier Properties of Small Gas Molecules in Amorphous *cis*-1,4-Polybutadiene Estimated by Simulation

Patricia Gestoso and Marc Meunier

CONTENTS

9.1 INTRODUCTION

Many industrial applications take advantage of the diverse barrier properties of polymers. Among them, gas separation for high-purity gases production, food packaging, and the beverage industry are most common.[1] More advanced applications are concerned with the development of new polymer membranes for a higher selectivity ratio.

Modeling of gas sorption in polymers is very difficult and presents a permanent challenge to theoreticians and experimenters.[2] The gas permeation process is defined as a "solution diffusion" process, where, at first, the gas permeant is dissolved on the surface, and then the gas molecules slowly diffuse through the polymer membrane.[3] Predicting the permeability directly from a simulation is known to be quite difficult. The advantage of such a permeation model is that it allows considering each process (solubility and diffusion) separately and then combining the results to calculate the permeability

$$P = D \times S, \tag{9.1}$$

where D is the diffusion coefficient and S is the solubility of penetrants.[4]

At sufficiently low pressure, the solubility is obtained from

$$C = S \times p, \tag{9.2}$$

where C is the solubility, p is the pressure, and S is the solubility coefficient.

The prediction of physical and chemical properties by computational methods is becoming more and more common in the research area, thanks in part to the computational power available at a low cost. Various computational methods exist to model amorphous materials (e.g., polymers) that are readily available to the modeler: molecular dynamics (MD), Monte Carlo, transition state theory (TST), and mesoscale simulations, to name a few. For a complete review of these methods, see Refs. [5] and [6]. Recently, MD simulations of up to 3 ns have been performed to estimate the diffusivity of small gas molecules in amorphous cis-1,4-polybutadiene (cis-PBD).[7]

Polymers are undoubtedly widely used for "permeation materials" (e.g., in packaging and membrane for gas separation), thanks to their chemical resistance. Currently, efforts in research and development concentrate on understanding the phenomena involved during gas transport through membranes as well as synthesizing novel polymers with better separation properties. The permeability of a specific gas molecule (e.g., O_2) in different polymers varies only slightly. It has been shown that a simple relationship can be found between the ratios of the permeability constants for a series of gases through different polymers.[8,9] In fact, it has also been shown that a similar relationship can be found for the diffusion and the solubility.

The main factors affecting small penetrants' permeability in polymeric material include free volume and its distribution,[10,11] density,[12] temperature and pressure, crystallinity,[13] polymer chain length,[12] mobility[14] and packing,[10] solute size,[15] and affinity for the material. In addition, computational parameters used in the simulations such as the type of force field employed and the size of the model also affect the permeability value computed.[7] An increase in temperature generally leads to a decrease in the solubility and conversely for the diffusion. For all three physical quantities P, S, and D, the temperature dependence can be described by a Van't Hoff–Arrhenius equation.[16] In particular, for the solubility

$$S(T) = S_0 \exp(-\Delta H_S/RT), \tag{9.3}$$

where ΔH_S is the molar heat of sorption.

In this context, this work is oriented toward the assessment of atomistic simulation techniques for the calculation of the barrier properties of *cis*-PBD melts. First, two different methods, MD and TST, will be compared in terms of solubility and diffusion coefficients as well as their capabilities to reproduce their dependence on temperature. Additionally, the effect of chain length on the predictions will also be evaluated. Finally, the ability of both simulation methods to predict selectivity will be compared.

9.2 METHODOLOGY

Computational details of the simulation runs as well as a description of the atomistic models used are given in this section.

9.2.1 MODELS

9.2.1.1 Short-Chain Models

Polymer models were created using the Amorphous Cell module of the Materials Studio® suite of software[17] based on the "self-avoiding" random walk method of Theodorou and Suter[18] and on the Meirovitch scanning method.[19] Amorphous *cis*-PBD three-dimensional (3D) models consisted of 10 chains of 30-monomer oligomers and were equilibrated using a temperature cycle protocol under periodic boundary conditions (Figure 9.1).[20] For a full description of the methodology used to build the equilibrated polymer models at various temperatures, please see Ref. [7].

Polymer model validation was ensured by checking the convergence of the total energy at the end of the MD runs as well as the density value and cohesive energy density close to that of the experiment.[7] Moreover, plotting the total, intra-,

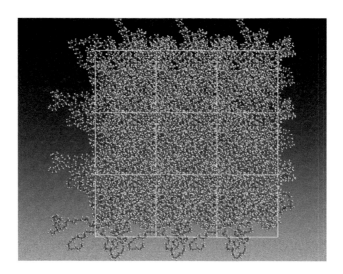

FIGURE 9.1 3D model of amorphous *cis*-PBD at 300 K with periodic boundary conditions.

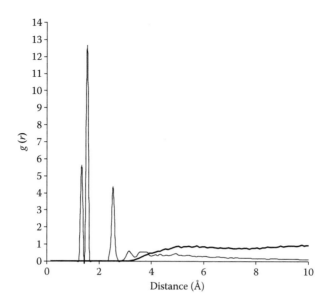

FIGURE 9.2 Carbon–carbon pair correlation functions. Thick line: intermolecular radial distribution function (RDF); thin line: intramolecular RDF.

and intercarbon–carbon pair correlation functions (Figure 9.2) shows that the intramolecular pair correlation function $g^{CC}_{intra}(r)$ has a limit value of zero, and the intermolecular pair correlation function $g^{CC}_{inter}(r)$ has a limit value of 1 at long range, thus demonstrating well-equilibrated configurations.

9.2.1.2 Long-Chain Models

Polymer models consisting of one single chain of 300 monomers were created using a Monte Carlo–based method similar to the short-chain models described above. This was done to study the effect on diffusion and solubility of the presence in these short-chain models of too many chain ends (as compared to the real material) and, therefore, of added free volume. It has been suggested that this is one of the main factors for the discrepancy between simulated and experimental values.[7] The models were validated using the similar procedure described in Section 9.2.1.1.

9.2.1.3 Sorbates

The geometry of the sorbate molecules was optimized using the density functional theory code DMol[3] using default settings.[17,21,22] Sorbates were inserted at the same time to the polymer chain using the Amorphous Cell module.[17]

9.2.2 COMPUTATIONAL METHODS

9.2.2.1 Transition State Theory

The transition state theory method for polymers was introduced by Arizzi and coworkers.[23–26] The method, as implemented in the *InsightII gsnet* and *gdiff* subroutines,[27] is initiated as a 3D fine resolution grid laid on the relaxed polymer configuration. Then, a spherical probe of radius equal to that of the gas penetrant is inserted in all grid points and the resulting nonbonded energy, $E_{ins}(x, y, z)$, is calculated between the test probe and all atoms of the polymer matrix. Following the Widom method,[28,29] the excess chemical potential, μ_{ex}, is calculated using

$$\mu_{ex} = RT \ln \left\langle \exp\left(-\frac{E_{ins}}{k_B T}\right)\right\rangle, \qquad (9.4)$$

where R is the universal gas constant, T is the temperature, k_B is the Boltzmann constant, and the brackets, $\langle \ \rangle$, denote averages over all grid points and probe insertions. Finally, the solubility coefficient is calculated from the excess chemical potential, μ_{ex}, through

$$S = \exp\left(-\frac{\mu_{ex}}{RT}\right). \qquad (9.5)$$

The identified sorption sites are separated by high-energy barrier surfaces; therefore, penetrant diffusion can be seen as a series of infrequent transitions between adjacent microstates. In TST, to calculate the diffusion coefficient, the sorbates are displaced in the coarse-grained lattice over a large number of time steps and a large population of ghost walkers through a kinetic Monte Carlo (kMC) scheme.[30,31] The diffusion coefficient is then calculated from the value of the mean square displacement (MSD) from the trajectories of all penetrant walkers

$$D = \lim_{t \to \infty} \left\{ \frac{\left\langle \left[\mathbf{r}_p(t) - \mathbf{r}_p(0)\right]^2 \right\rangle}{6t} \right\}, \qquad (9.6)$$

where the brackets, $\langle \ \rangle$, indicate average over all trajectories and all time origins. Finally, the permeability is calculated through Equation 9.1.

The thermal fluctuations of the polymer matrix are taken into account through the smearing factor "Δ^2", which is related to the MSD of the matrix segments from their equilibrium positions.

The TST method has the advantage of extending the timescale of the observation when compared to classical dynamics; however, it involves a number of assumptions. First, the polymer matrix response to the guest molecule should be elastic. This is due to the rather simplistic form of calculating the smearing factor, which limits the application of the method to the behavior of small molecules, the presence of which does not affect the polymer environment. Second, the shape of the penetrant is supposed to be isotropic. As the penetrant size increases and its shape becomes anisotropic, conventional TST fails to capture the corresponding transport behavior.

In this work, the grid size used was set to 0.3 Å, in agreement with typical values found in the literature for TST calculations.[3,32–35] The smearing factor, Δ^2, was calculated for each penetrant through a self-consistent scheme involving information about the MSD of all the polymer atoms from their respective equilibrium positions. The MSD was calculated from a 50 ps MD simulation at constant number of particles, volume, and temperature (NVT). The self-consistent scheme converged when the relative difference between two successive Δ^2 values was within 2.5%. The total duration of the kMC procedure was 10^{-4} s and the MSD was averaged over the trajectories of 1000 penetrant walkers.

All penetrant molecules were represented as single, spherical, united-atom sites whose short-range interactions with the polymer atoms are described through a 9–6 Lennard–Jones (L–J) potential. The values of collision diameters, σ, and the well depths, ε, have already been reported in the literature for the COMPASS force field.[17,36] The nonbonded interactions were truncated at 9.5 Å. All polymer atoms were represented explicitly and the potential function and atomic parameters can be found elsewhere.[36] The reported values correspond to the sampling over five structures and the error bars to standard deviations of the distributions of the calculated values.

9.2.2.2 Molecular Dynamics

Molecular dynamics simulations were used to estimate the diffusion coefficients of the sorbates in both short- and long-chain models. Four geometry-optimized sorbates were randomly inserted in the model cells and long NVT simulations were performed. The complete methodology has been described elsewhere and is not detailed here.[7]

9.3 RESULTS AND DISCUSSION

In this part, we report the results of the computation of the solubility and diffusivity values using Molecular Dynamics and TST methods. A comparison between the two methods is given as well as a comparison with experimental data when available.

9.3.1 GENERAL

We report first the variation of the total and free volume of the models. For the short-chain models, the total and free volumes vary linearly with the temperature in the temperature range 250 K–400 K, as shown in Figures 9.3 and 9.4. The free volume represents 38.6% of the total volume at 250 K, 40.2% at 300 K, and 43.7% at 400 K. For the long-chain models, the total free volume represents 39.2% at 300 K which is, as expected, slightly less than that of the short-chain models due to a reduction in the number of chain ends at equal density.

9.3.2 SOLUBILITY COEFFICIENTS

The comparison between solubility coefficient results obtained from TST calculations at $T = 300$ K for short- and long-chain models and experimental results is

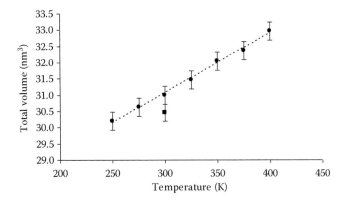

FIGURE 9.3 Comparison of the total cell volume of the models. Circle: short-chain model (average of five frames); square: long-chain model (average over five frames).

shown in Table 9.1. As can be seen, the simulation results follow the experimental tendency, that is, the solubility parameter increases with the size of the molecule. This behavior has been well documented in the literature. Additionally, it can be seen that there is a very good quantitative agreement between experiments and TST, especially for long chains, with the exception of CO_2. Failure to capture the CO_2 transport behavior is inherent to the restrictions of the current implementation of TST where all molecules are considered spherical. Because CO_2 has an anisotropic shape, conventional TST fails to capture the corresponding transport behavior.[37]

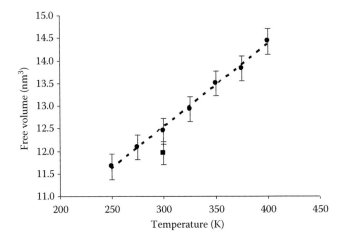

FIGURE 9.4 Comparison of the polymer free volume of the models. Circle: short-chain model (average of five frames); square: long-chain model (average over five frames).

TABLE 9.1

Solubility Coefficients ($\times 10^{-6}$ Pa^{-1}) for Various Sorbates at $T = 300$ K (Comparison of Simulation Methods and Experimental Data)

Penetrants	Exp. 1[a]	Exp. 2[b]	TST Short Chain	TST Long Chain
N_2	0.45	0.45	0.74	0.45
Ar	0.76	–	1.13	0.63
O_2	0.96	0.94	1.58	1.18
CH_4	–	–	2.69	1.33
CO_2	9.87	9.7	8.39	3.95
RMSD	–	–	0.418	1.481
RMSD (no CO_2)			0.259	0.085

[a] Ref. [13].
[b] Ref. [38].

Figure 9.5 shows the dependence of the solubility coefficient on temperature in the range 250 K $\leq T \leq$ 400 K. As expected, solubility decreases with temperature; that is, as the temperature increases, the gas molecules in the experiment become more difficult to condense. This behavior is in agreement with the experiments (CO_2 in PET[39]) and simulations [CO_2 and He in PE[20]; CO_2 and CH_4 in polyetherimide[41]; CH_4 and CO_2 in HDPE[42]; n-alkanes in poly(dimethylsilamethylene)[43]; and O_2, N_2, and CH_4 in PE for long and short chains[37]].

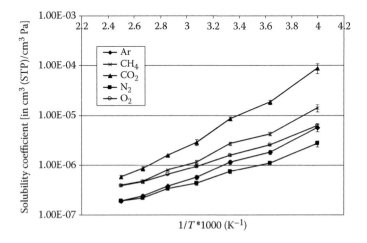

FIGURE 9.5 Logarithm of solubility coefficient [in cm^3 (STP)/cm^3 Pa] as a function of reciprocal temperature for *cis*-PBD short-chain models. STP—standard temperature and pressure.

From Figure 9.5, it is also possible to calculate the heat of the solution. The values obtained were 27.69 kJ/mol for CO_2, 19.83 kJ/mol for CH_4, 18.53 kJ/mol for Ar, 15.32 kJ/mol for O_2, and 14.54 kJ/mol for N_2. For all penetrants, it was found that the dependence with temperature followed an Arrhenius behavior (Equation 9.3). The fact that the heat of solution values is, in general, more positive as the solubility of the permeant increases is in contradiction with experimental evidence,[38,44] pointing out the limitations of TST to capture the dependence of solubility with temperature.

9.3.3 DIFFUSION COEFFICIENTS

Table 9.2 presents the comparison between the diffusion coefficients from MD calculations, TST, and experimental results at 300 K for short- and long-chain models. As can be seen, the predicted values from simulations and TST are in good agreement with experimental data, with low values of the root mean square deviation (RMSD). Deviations are of the same order as in experimental reported data. The experimental reported values of the penetrants in Ref. [13] come from two different publications, Refs. [45] and [46], thus making the comparison difficult. Although the ranking of the penetrants between experiments should be similar, errors while combining data might alter this (as it is the case here for O_2). Overall, theoretical values are systematically overestimating the experimental data, a behavior that has been already observed and reported in the literature.[7] Long-chain models perform better than short-chain ones, as expected from Figure 9.3, which shows the comparison of free volume in the models. This can be explained as experimental values were measured for very long-chain samples, which have smaller free volume and higher density when compared with short chains. Free volume and density play a fundamental role in diffusion, as has been demonstrated elsewhere;[37] therefore, it is to be expected

TABLE 9.2

Diffusion Coefficients for Various Sorbates ($\times 10^{-6}$ cm^2 s^{-1}) at T = 300 K (Comparison of Simulation Methods and Experimental Data)

		MD	TST	MD	TST
Penetrants	Exp.[a]	Long Chain	Long Chain	Short Chain[b]	Short Chain
N_2	1.1–2.96	6.11	5.69	8.8	6.08
O_2	1.5	5.07	8.37	9.5	11.19
Ar	4.06	5.06	3.51	7.03	3.84
CH_4	2.25[c]	2.96	3.22	7.5	3.47
CO_2	1.05	4.06	1.55	5.3	1.58
RMSD		2.58	3.35	5.52	4.59

[a] Ref. [13].
[b] Ref. [7].
[c] Ref. [10].

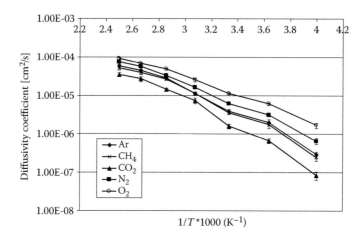

FIGURE 9.6 Logarithm of diffusivity coefficient (in cm²/s) as a function of reciprocal temperature for *cis*-PBD short-chain models.

that the diffusion coefficient values obtained for the long-chain models are closer to experimental values than those from short-chain models. Therefore, although the making of such long-chain models is computationally more expensive than the short-chain ones, it results in an improvement of the predictability of the methods.

The dependence of diffusion coefficients with the inverse of temperature is plotted in Figure 9.6 using TST. As can be seen, the diffusivity increases with the temperature, a behavior that is in perfect agreement with experimental and simulation results in the literature. The diffusion activation energies were calculated as

$$D = D_0 \exp(-E_D/RT). \qquad (9.7)$$

The values obtained for E_D are 29.70 kJ/mol for CH_4, 29.21 kJ/mol for Ar, 22.11 kJ/mol for O_2, and 26.18 kJ/mol for N_2, which compare well with experimental activation energies in the range of 20 kJ/mol–30 kJ/mol.[38,44] This is an improvement compared

TABLE 9.3
O_2 Selectivity at 300 K

Penetrants	Exp.[a]	MD Long Chain	TST Long Chain	MD Short Chain[b]	TST Short Chain
N_2	0.67	0.46	0.26	0.43	0.25
Ar	2.15	0.53	0.22	0.53	0.25
CO_2	7.27	2.68	0.62	2.96	0.75

[a] Ref. [13]; average of both data for N_2.
[b] Ref. [7].

TABLE 9.4

N_2 Selectivity at 300 K

		MD	TST	MD	TST
Penetrants	Exp.[a]	Long Chain	Long Chain	Short Chain[b]	Short Chain
Ar	3.20	1.16	0.86	1.22	0.96
CO_2	10.81	5.83	2.39	6.83	2.95

[a] Ref. [13]; average of both data for N_2.
[b] Ref. [7].

with the results obtained using MD techniques,[7] in which the values for the same penetrants were lower than 10 kJ/mol. For CO_2, the activation energy decreases with temperature as already reported for methane in polybutadiene.[7,10]

9.3.4 SELECTIVITY

A key property for the design of novel material for gas separation is its selectivity. In this study, we investigate modeling methods to predict the selectivity of small gas molecules in *cis*-PBD. The selectivity is defined as the ratio of the permeability of two penetrants, for example

$$\text{Selectivity}(O_2, N_2) = \frac{P_{O_2}}{P_{N_2}} = \frac{D_{O_2} S_{O_2}}{D_{N_2} S_{N_2}}, \tag{9.8}$$

where P_{O_2} is the permeability of oxygen and P_{N_2} is the permeability of nitrogen. For the calculations, the MD selectivities were computed from MD diffusion coefficients and TST solubility coefficients for the respective penetrants, whereas for the TST selectivities, both the diffusion and solubility coefficients were extracted from the TST simulations.

Tables 9.3 and 9.4 report the selectivity data from experiment and from simulations. In general, it can be seen that the theoretical models predict correctly the trends for oxygen and nitrogen selectivity. However, in all cases, the simulation underestimates the selectivity values when compared with the experiment. It is interesting to notice that this underestimation appears to be more noticeable for TST than for MD models.

9.4 CONCLUSION

The barrier properties of the *cis*-PBD for different small gas penetrants have been estimated using MD and TST simulation techniques. A molecular model made of one single long chain was created to gain insights on known issues due to incorrect free volume distribution and size in molecular models made of many short chains. There is an overall good agreement between experiments and the results from both methods for solubility and diffusion coefficients, although, as expected, the values corresponding to the long-chain models are in better agreement than those from the short-chain models. This difference is more noticeable for MD simulations.

The heat of solution and the diffusion activation energy for the different penetrants in *cis*-PBD were derived from the TST plots of solubility and diffusion coefficient vs. temperature, respectively. The heat of solution for the different small molecules did not follow the trends reported in the literature. On the other hand, the diffusion activation energies have a very good quantitative and qualitative agreement with experimental values, which was not the case in the MD simulations.[7]

Finally, the simulations are able to predict correctly the ranking of the selectivities for oxygen and nitrogen, although the TST simulations underestimated the values to a higher degree than the MD simulations.

REFERENCES

1. J. Crank and G. S. Park. *Diffusion in Polymers*. London: Academic Press, 1968.
2. R. Patterson and Y. Yampol'skii. Gases in glassy polymers. *J. Phys. Chem. Ref. Data* **28**, 1999, 5.
3. D. Hofmann, L. Fritz, J. Ulbrich, C. Schepers and M. Bohning. Detailed-atomistic molecular modelling of small molecule diffusion and solution processes in polymeric membrane materials. *Macromol. Theory Simul.* **9**, 2000, 293.
4. Y. Tamai, H. Tanaka and K. Nakanishi. Molecular simulation of permeation of small penetrants through membranes. *Macromolecules* **27**, 1994, 4498.
5. A. R. Leach. *Molecular Modelling, Principles and Applications*. London: Pearson Education Limited, 1996.
6. D. Frenkel and B. Smit. *Understanding Molecular Simulation. From Algorithms to Applications*. Computational Science Series. Vol 1. Second Edition. London: Academic Press, 2002.
7. M. Meunier. Diffusion coefficients of small gas molecules in amorphous *cis*-1,4-polybutadiene estimated by molecular dynamics simulations. *J. Chem. Phys.* **123**, 2005, 134906.
8. V. Stannett and M. Szwarc. Rapid determination of oxygen permeability of polymer membranes. *J. Polym. Sci.* **16**, 1955, 89.
9. H. L. Frisch. Factorization of the activation energies of permeation and diffusion of gases in polymers. *Polym. Lett.* **1**, 1963, 581.
10. R. H. Boyd and P. V. K. Pant. Molecular packing and diffusion in polyisobutylene. *Macromolecules* **24**, 1991, 6325.
11. H. Takeuchi, R. J. Roe and J. E. Mark. Molecular dynamics simulation of diffusion of small molecules in polymers. II. Effect of free volume distribution. *J. Chem. Phys.* **93**, 1993, 9042.
12. H. Takeuchi. Molecular dynamics simulations of diffusion of small molecules in polymers: Effect of chain length. *J. Chem. Phys.* **93**, 1990, 4490.
13. S. Pauly. In *Polymers Handbook*, 3rd edition, edited by J. Brandrup and E. H. Immergut. New York: Wiley, 1989.
14. H. Takeuchi and K. Okasaki. Molecular dynamics simulation of diffusion of simple gas molecules in a short chain polymer. *J. Chem. Phys.* **92**, 1990, 5643.
15. S. Trohalaki, A. Kloczkowski, J. E. Mark, D. Rigby and R. J. Roe. *Computer Simulation of Polymers*, edited by R. J. Roe. Englewood Cliffs, NJ: Prentice-Hall, 1991, 220.

16. D. W. Van Krevelen. *Properties of Polymers*, 3rd edition. Amsterdam: Elsevier, 2003.
17. Accelrys, Inc., San Diego, 2007.
18. D. N. Theodorou and U. W. Suter. Detailed molecular structure of a vinyl polymer glass. *Macromolecules* **18**, 1985, 1467.
19. H. J. Meirovitch. Computer simulation of self-avoiding walks: Testing the scanning method. *J. Chem. Phys.* **79**, 1983, 502.
20. M. P. Allen and D. J. Tidelsey. *Computer Simulation of Liquids*. Oxford: Oxford University Press, 1987.
21. B. Delley. An all-electron numerical method for solving the local density functional for polyatomic molecules. *J. Chem. Phys.* **92**, 1990, 508.
22. B. Delley. From molecules to solids with the DMol3 approach. *J. Chem. Phys.* **113**, 2000, 7756.
23. S. Arizzi. Diffusion of small molecules in polymeric glasses: A modelling approach. PhD Thesis, Massachusetts Institute of Technology, Boston, 1990.
24. A. A. Gusev, S. Arizzi, U. W. Suter and D. J. Moll. Dynamics of light gases in rigid matrices of dense polymers. *J. Chem. Phys.* **99**, 1993, 2221.
25. A. A. Gusev and U. W. Suter. Dynamics of small molecules in dense polymers subject to thermal motion. *J. Chem. Phys.* **99**, 1993, 2228.
26. A. A. Gusev, F. Müller-Plathe, W. F. van Gunsteren and U. W. Suter. Dynamics of small molecules in bulk polymers. *Adv. Polym. Sci.* **116**, 1994, 207.
27. Accelrys. InsightII. San Diego: Accelrys Software, Inc., 2007.
28. B. Widom. Some topics in the theory of fluids. *J. Chem. Phys.* **39**, 1963, 2808.
29. B. Widom. Potential-distribution theory and the statistical mechanics of fluids. *J. Phys. Chem.* **86**, 1982, 869.
30. R. L. June, A. T. Bell and D. N. Theodorou. Transition-state studies of xenon and sulphur hexafluoride diffusion in silicalite. *J. Phys. Chem.* **95**, 1991, 8866.
31. N. C. Karayiannis, V. G. Mavrantzas and D. N. Theodorou. Diffusion of small molecules in disordered media: Study of the effect of kinetic and spatial heterogeneities. *Chem. Eng. Sci.* **56**, 2001, 2789.
32. D. Hofmann, J. Ulbrich, D. Fritsch and D. Paul. Molecular dynamics simulations of the transport of water–ethanol mixture through polydimethylsiloxane membranes. *Polymer* **38**, 1997, 1035.
33. D. Hofmann, L. Fritz, J. Ulbrich and D. Paul. Molecular modelling of amorphous membrane polymers. *Polymer* **38**, 1997, 6145.
34. E. Kucukpinar and P. Doruker. Molecular simulations of small gas diffusion and solubility in copolymers of styrene. *Polymer* **44**, 2003, 3607.
35. N. C. Karayiannis, V. G. Mavrantzas and D. N. Theodorou. Detailed atomistic simulation of the segmental dynamics and barrier properties of amorphous poly(ethylene terephthalate) and poly(ethylene isophthalate). *Macromolecules* **37**, 2004, 2978.
36. H. Sun. COMPASS: An *ab initio* force-field optimized for condensed-phase applications—Overview with details on alkane and benzene compounds. *J. Phys. Chem. B* **102**, 1998, 7338.
37. P. Gestoso and N. Ch. Karayiannis. Molecular simulation of the effect of temperature and architecture on polyethylene barrier properties. *J. Polym. Phys. B* **112**, 2008, 5646.
38. G. J. van Amerongen. The permeability of different rubbers to gases and its relation to diffusivity and solubility. *J. Appl. Phys.* **17**, 1946, 972.
39. Y. Mi, X. Lu and J. Zhou. Gas diffusion in glassy polymers by a chain relaxation approach. *Macromolecules* **36**, 2003, 6898.
40. N. F. A. van der Vegt. Temperature dependence of gas transport in polymer melts: Molecular dynamics simulations of CO2 in polyethylene. *Macromolecules* **33**, 2000, 3153.
41. S. Y. Lim, T. T. Tsotsis and M. Sahimi. Molecular simulation of diffusion and sorption of gases in an amorphous polymer. *J. Chem. Phys.* **119**, 2003, 496.

42. N. von Solms, J. K. Nielsen, O. Hassager, A. Rubin, A. Y. Dandekar, S. I. Andersen and E. H. Stenby. Direct measurement of gas solubilities in polymers with a high-pressure microbalance. *J. Appl. Polym. Sci.* **91**, 2004, 1476.

43. V. E. Raptis, I. E. Economou, D. N. Theodorou, J. Petrou and J. H. Petropoulos. Methods for investigation of the free volume in polymers. *Macromolecules* **37**, 2004, 1102.

44. R. Cowling and G. S. Park. Gas permeability of polydimethylsiltrimethylene above and below the melting temperature of the crystalline phase. *J. Membrane Sci.* **5**, 1979, 199.

45. E. A. Hegazy, T. Seguchi and S. Machi. Radiation-induced oxidative degradation of poly(vinyl chloride). *J. Appl. Polym. Sci.* **26**, 1981, 2947.

46. E. Hegazy, T. Seguchi, K. Arakawa and S. Machi. Radiation-induced oxidative degradation of isotactic polypropylene. *J. Appl. Polym. Sci.* **26**, 1981, 1361.

10 On the Negative Poisson's Ratios and Thermal Expansion in Natrolite

Joseph N. Grima, Ruben Gatt, Victor Zammit, Reuben Cauchi, and Daphne Attard

CONTENTS

10.1 INTRODUCTION

Throughout history, civilizations have been distinguished according to their ability to produce and work with superior materials that perform better than the previously available ones. In fact, the different eras in our prehistory are defined by the materials available to our ancestors: stone, bronze, and then iron, where clearly the newer materials proved to be far better than the previous ones in terms of mechanical strength, durability, and workability. The search for new and superior materials has always been a crucial matter in the advancements of humanity and is still in fact an ongoing process. Since the beginning of the twentieth century, research in materials science resulted in massive developments that made possible the big advances in the communications and transport industries, among others. In particular, the last three decades have seen an increased interest in designing, discovering, and synthesizing new materials that exhibit highly unusual properties. For example, although one normally expects that a lateral contraction should occur when a material is uniaxially stretched, it has been shown that not all materials behave like this and some materials can actually get fatter when uniaxially stretched (auxetic) (Lakes 1987; Evans et al. 1991; Wojciechowski 1989). Similarly, although one normally expects a material to expand when heated, some materials actually shrink in size, that is, they undergo negative thermal expansion (NTE). As discussed below, such properties usually arise as a result of the manner in which particular features in the nano- or microstructure of the material deform upon stretching or heating and result in the materials having superior qualities when compared with their conventional counterparts.

This chapter reviews some of these properties and shows how molecular modeling simulations can aid and supplement the experimental data in the elucidation of the mechanisms that result in these observed anomalous properties.

10.1.1 Materials Exhibiting Negative Poisson's Ratios (Auxetic)

The extent of change in a lateral dimension when a material is uniaxially stretched or compressed is measured through the Poisson's ratio. In particular, for a material being loaded in the Ox_i direction, the Poisson's ratio ν_{ij} in the Ox_i–Ox_j plane (where Ox_j is perpendicular to Ox_i) is defined as

$$\nu_{ij} = -\frac{\varepsilon_j}{\varepsilon_i}, \tag{10.1}$$

where ε_i and ε_j are the applied and resulting strain in the Ox_i and Ox_j directions, respectively. For a conventional material, a positive strain ε_i (stretching in Ox_i) is accompanied by a negative strain ε_j (material gets thinner), resulting in an overall positive Poisson's ratio. However, in auxetic materials, a positive strain ε_i (stretching in Ox_i) is accompanied by a positive strain ε_j (material gets fatter), resulting in an overall negative Poisson's ratio (NPR) (Evans 1991) (Figure 10.1). In the last three decades, various molecular-level auxetics have been proposed, ranging from liquid crystalline polymers that have already been synthesized (He et al. 1998, 2005), naturally occurring cubic metals (Baughman et al. 1998) to more theoretical highly elegant systems (Wojciechowski 1989; Brańka et al. 2009). Another interesting group of materials are those with an overall zero Poisson's ratio, such as cork (Gibson et al. 1981), which show no lateral expansion or contraction when the material is stretched or compressed (Wojciechowski and Brańka 1994).

Materials with NPRs have superior properties when compared with conventional materials (Choi and Lakes 1996; Wang and Lakes 2002; Scarpa et al. 2005; Bezazi and Scarpa 2009). For example, upon impact, the material in auxetic systems flows toward the point of impact (Chan and Evans 1998) resulting in a more dense system. This is in opposition to conventional materials where the material tends to flow away from the point of impact resulting in a less dense system. This means that auxetic materials tend to show a higher impact resistance (Alderson 1999) (see Figure 10.2a). This feature is potentially applicable in personal protection equipment such as

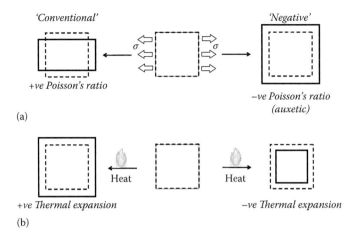

FIGURE 10.1 "Positive" and "negative" behavior toward stress and heat where (a) shows how materials behave when a uniaxial stress is applied, while (b) shows how materials act in response to a change in temperature.

bulletproof jackets, crash helmets, and fibrous seals (Stott et al. 2000). Furthermore, auxetic materials usually show an increased shear modulus (Scarpa and Tomlinson 2000) (which would result in a loss of the bulk modulus), a property that may be very useful in plates, shells, and beams used in structural parts that have to withstand high shear forces, for example, in vehicles, ships, buildings, and aircrafts (Evans 1990).

Owing to these enhanced material properties, auxetic materials can be used in a number of applications. Some proposed applications include the use of auxetic materials in rivets, gaskets, and fasteners to provide a stiffer grip upon loading (Evans 1991) and their use in fiber-reinforced composites. The strength of such composites

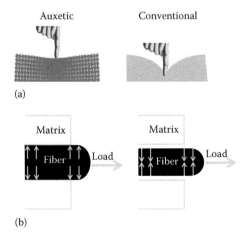

FIGURE 10.2 (a) Indentation resistance of auxetic materials compared with a conventional material. (b) Comparison between pulling an auxetic and a conventional fiber.

depends on the "pull out" strength of the fibers from the embedding matrix. Thus, if the fibers are auxetic, they would become fatter in response to an applied external stress; hence, debonding from the matrix would become much more difficult, increasing their mechanical strength (see Figure 10.2b) (Evans 1991; Stott et al. 2000; Alderson et al. 2005c).

Auxetic systems may also be used as smart filters where the pore size of the filter can be adjusted by changing the tensile stress applied to the filter. Ideal candidates for this type of filters at the molecular level include auxetic zeolites because of their open geometry (Alderson et al. 2000, 2005a,b; Grima 2000).

10.1.2 NEGATIVE THERMAL EXPANSION

One normally expects that a material should expand when heated and shrink when cooled (positive thermal expansion). Nevertheless, it is now known that not all materials behave like this, as exemplified by water, which expands when cooled near its freezing point (NTE) (Baughman and Galvao 1995; Mary et al. 1996; Evans 1999; Evans et al. 1998; Tucker et al. 2005; Miller et al. 2008, 2009; Marinkovic et al. 2009; Jakubinek et al. 2010). This property of water results in the undesirable effect of damaging plumbing systems as water expands upon freezing in extreme wintery conditions. Although NTE in water appears to have mainly unwelcome consequences, in general, this is not the case, and there are various situations where NTE is a highly desirable property; for example, in composites, materials showing NTE are used to adjust the overall thermal expansion of the composite to a specific value. A case worth noting is the development of a carbon fiber and metal composite with an extremely negative value of thermal expansion, equivalent in magnitude to the positive expansion of steel. This system also has the additional advantage of low thermal conductivity and high compressive strength (Hartwig 1995).

Materials with NTE have been proposed for use in chirped fiber gratings for the photonics sector (see Figure 10.3). By using a temperature method, the possibility of adjusting the chirp in chirped fiber gratings was proposed and experimentally demonstrated by Wei et al. (2001) by mounting the chirped fiber gratings with tapered

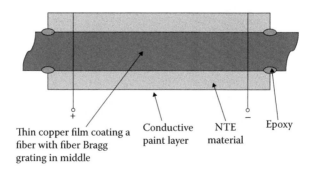

Thin copper film coating a fiber with fiber Bragg grating in middle

Conductive paint layer

NTE material

Epoxy

FIGURE 10.3 A schematic cross-sectional representation of a tunable fiber Bragg grating dispersion compensator that involves a material with an NTE encircling the coated fiber. (Adapted from Ngo, N.Q., et al., *J. Lightwave Technol.* 21, 1568–1575, 2003.)

cross-sectional areas under tension in an NTE coefficient material. Mavoori et al. (1999) and Ngo et al. (2003) have also reported similar work.

NTE materials have also been used to control surface flaws. For example, TiO_2, which has an NTE coefficient, can be deposited on the inside and outside of silica tubes used in capillary columns, so that when heated, the TiO_2 contracts, causing a compression and, therefore, reducing propagation of surface flaws (Berthou et al. 1993).

ZrW_2O_8 is used in composites to reduce their thermal expansion for application in the electronics area where substrates and heat sinks having the thermal expansion of Si are required. Cu/ZrW_2O_8 composites used in heat sinks have successfully been made to match the thermal expansion of Si over a range of several hundred degrees (Holzer and Dunand 1997).

10.1.3 Causes of Negative Poisson's Ratios and Negative Thermal Expansion

It is now known that auxeticity and NTE are phenomena that may be observed in a variety of materials. The way these properties arise in a particular material may be dependent on the way that particular features in the material's micro- or nano-structure deform when subjected to a thermal or mechanical load (the deformation mechanism). In other words, the geometry of the microstructure/nanostructure plays an important role in defining the Poisson's ratio (Alderson 1999) or the thermal expansion coefficient. In this respect, it should be noted that some of the geometry/ deformation mechanisms that lead to particular values of the Poisson's ratios or thermal expansion coefficients are independent from the scale of structure. Moreover, in the past decade, it became common practice to use macroscaled models to ease examination and explanation* of the interplaying mechanisms that lead to the deformation of the geometry at the micro- and nanoscales.

A geometry that has been studied extensively in the field of NPRs and NTE coefficients is that of rotating rigid squares (Figure 10.4a). For example, Grima and Evans (2000b) and Grima (2000d) had shown that rigid squares connected together at their vertices may exhibit NPRs of −1 as a result of a relative rotation of the rigid squares. Similarly, Giddy et al. (1993) have shown that the fully open conformation of this geometry may exhibit NTE as a result of increased vibration upon heating, which results in the squares adopting a less open average conformation (Figure 10.4b).

Recent research has shown that molecular level systems that exhibit NPRs as a result of a mechanism that in simplest of terms may be described in terms of rotating rigid squares include the zeolites in the natrolite group (Grima et al. 2000; Grima and Evans 2000a,b; Williams et al. 2007; Gatt et al. 2008), in which related work has so far not been extended to look at the thermal properties of these materials. In view of this, the current work presents new modeling studies that together with a reanalysis of existing modeling and experimental work will show that the zeolite natrolite can exhibit both NPR and NTE that can be explained through the same mechanism involving rotation of quasirigid units.

---

* Micro- and nanoscale structures are usually found in materials such as foams and metals or crystals, respectively.

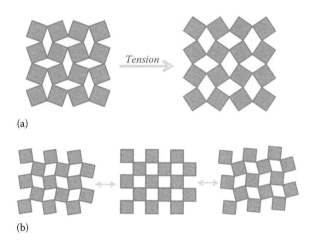

FIGURE 10.4 Idealized models of (a) rotating squares mechanism having a negative Poisson's ratio upon tension and (b) NTE as a result of vibration of hypothetical squares from the fully open conformation.

10.2 MODELING AND EXPERIMENTAL WORK ON THE THERMO-MECHANICAL PROPERTIES OF NATROLITE

10.2.1 MECHANICAL BEHAVIOR OF NAT-TYPE SYSTEMS

The potential of natrolite to exhibit auxetic behavior was first predicted in 1999 by Grima et al. (1999) and Grima and Evans (2000b) for an all-silica framework equivalent of NAT, where the aluminum atoms were replaced by silicon atoms and the extra-framework water molecules and cations that are usually present in this zeolite were removed (henceforth referred to as NAT_SI). For this silica equivalent model, it was shown that maximum auxeticity occurs at about 45° off-axis in the (001) plane, a phenomenon that was explained through a simple 2D model based on a "rigid/semirigid rotating squares" mechanism (Grima et al. 2000; Grima and Evans 2000b, 2006) (see Figure 10.5). Since then, further modeling work was performed on NAT_SI, on natrolite proper with its extra-framework cations and water molecules included (henceforth referred to NAT_CW), and also on the dehydrated form of this zeolite, henceforth referred to as NAT_C, where it was shown that all of these systems have some auxetic potential with NAT_SI being the most auxetic and NAT_CW being the least (Grima et al. 2006).

The theoretical work carried out by Grima et al. that proposed that NAT_CW was auxetic was later verified by the experimental work carried out by Sanchez-Valle et al. (2005), where the compliance matrix for NAT was measured using the Brillouin scattering techniques. Analysis of this experimental work that was carried out by Grima et al. (2007b) using standard axis transformation techniques (Nye 1957) showed that natrolite is indeed auxetic for loading in certain directions with maximum auxeticity ($\nu = -0.12$) being exhibited for loading at about 45° to the main crystallographic axis in the (001) plane as illustrated in Figure 10.6.

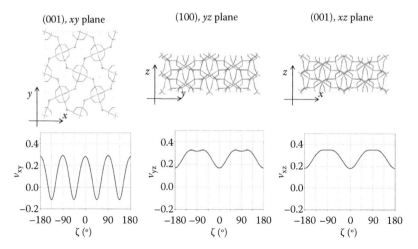

FIGURE 10.5 The off-axis plots for the Poisson's ratios in the three major planes of NAT. The data are obtained using standard axis transformation techniques (Nye 1957) of the stiffness constants obtained experimentally by Sanchez-Valle et al. (2005).

The modeling, experimental work, and its analysis through standard axis transformation techniques highlight the fact that anisotropy exhibited in the Poisson's ratios is such that (1) no auxetic characteristics can be found for loading on axis in the (001) plane or in the (100) and (010) planes, and (2) in the (001) plane, maximum auxeticity is exhibited for loading in the [110] and [1 $\bar{1}$0] directions (i.e., $v_{[1\bar{1}0][110]}$ and $v_{[110][1\bar{1}0]}$), which are directions that correspond to the lines passing through the vertices of the "projected squares" in the (001) plane. This is very significant because it

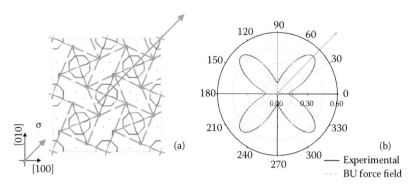

FIGURE 10.6 (a) Crystal structure of NAT with the "connected squares" model highlighted. (b) Off-axis analysis of the experimentally determined and simulated Poisson's ratios in the (001) plane of NAT. These data confirm that NAT is auxetic in the (001) plane with maximum auxeticity ($v = -0.12$) being exhibited at ±45° to the [100] and [010] crystallographic axes. The experimental data are obtained using standard axis transformation techniques (Nye 1957) of the stiffness constants obtained experimentally by Sanchez-Valle et al. (2005). Note the excellent agreement between the experimentally obtained data and the results of the simulations.

not only highlights the strong structure–property relationships in NAT but also suggests that the mechanistic explanations for the cause of auxeticity that were deduced from the modeling work performed on NAT_SI could be applied to explain the auxeticity in NAT proper. Such conclusions can, however, only be made following a proper analysis on NAT_CW.

In view of this, in an attempt to assess properly the deformation mechanism that is the underlying cause for the different values of Poisson's ratios of NAT_CW, the atomic level deformations of natrolite at various stresses σ in directions of maximum auxeticity, that is, in the (001) plane at 45° to the [100] and [010] directions, were performed through static simulations. In particular, these simulations were meant to confirm that the NPRs in NAT at ambient pressures are indeed the result of what may be described in terms of rotations of the projected squares in the (001) plane (Grima and Gatt 2010).

These simulations were performed on NAT_CW, that is, the aluminosilicate framework of NAT in the presence of cations or water molecules having a composition of $Na_2(Al_2Si_3O_{10}) \cdot 2H_2O$ using the Cerius^2 V4.1 molecular modeling package running on an SGI Octane 2 workstation. The NAT framework was arranged so as to have the [001] direction of the crystal fixed parallel to the Z-direction and the [010] direction fixed to lie in the YZ plane. The Burchart–Universal force field was used to generate the energy expressions as discussed in Gatt et al. (2008). The

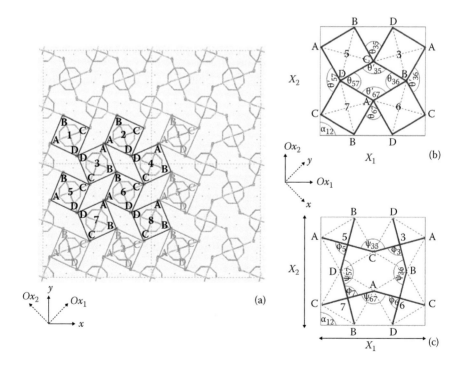

FIGURE 10.7 Connected quadrilaterals in NAT. Changes in these parameters are monitored as the temperature of the system increases.

Ewald summation technique (Ewald 1921) was used as a summation method for the nonbond terms. The SMART compound minimizer was used for the application of a number of energy minimizations, up to the default high convergence criteria of Cerius^2, whereas a series of both positive and negative external stresses of magnitude σ were being applied on the framework in the direction of maximum auxeticity. More specifically, the stresses were applied in the (001) plane at 45° to the X-direction by minimizing while subjecting the crystal to a stress

$$\sigma = \frac{1}{2}\begin{pmatrix} \sigma & -\sigma & 0 \\ -\sigma & \sigma & 0 \\ 0 & 0 & 0 \end{pmatrix}.$$

From the resulting structures at different stresses, various measurements as defined in Figure 10.7, related to the projected distances and angles that relate to the rotating squares model, were made and are summarized in Figure 10.8. The figure also shows images of the projected structure of NAT in the (100) plane at two different stress values with the projected "squares" highlighted.

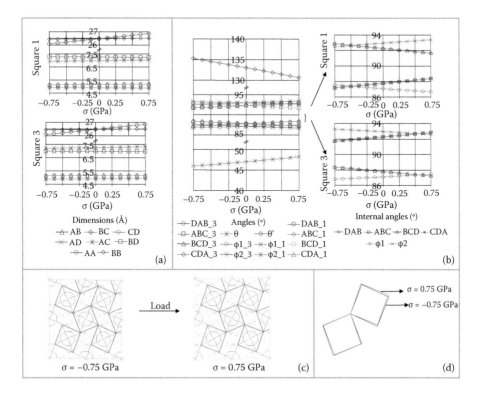

FIGURE 10.8 Absolute values of various (a) dimensions and (b) angles projected in the (001) plane for NAT as a result of changes in stress. The projected squares superimposed (c) over the structure and (d) over each other at different stresses.

Similar simulations were also performed with the aim of analyzing the molecular level deformations that occur when NAT_CW is subjected to hydrostatic pressure by performing energy minimizations on NAT_CW using the Burchart–Universal force field while subjecting it to a stress p, that is

$$\sigma = \begin{pmatrix} -p & 0 & 0 \\ 0 & -p & 0 \\ 0 & 0 & -p \end{pmatrix}.$$

Once again, from the simulations of the structures obtained at different pressures, the various measurements defined in Figure 10.7 were measured and are summarized in Figure 10.9. Also shown in Figure 10.9 are images of the projected structure of NAT in the (100) plane at two different pressure values with the projected "squares" highlighted.

10.2.2 THERMAL BEHAVIOR OF NAT-TYPE SYSTEMS

In addition to the behavior of NAT under applied uniaxial stress and hydrostatic pressure, there have been various studies in the past relating to the effect of heat

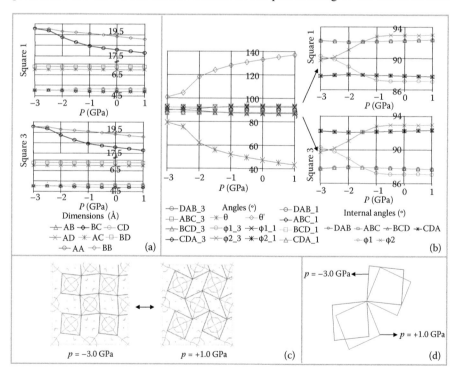

FIGURE 10.9 Absolute values of various (a) dimensions and (b) angles projected in the (001) plane for NAT as a result of changes in pressure. The projected squares superimposed (c) over the structure and (d) over each other at different pressures.

TABLE 10.1

Cell Parameters of NAT at Different Temperatures Based on Experimental Data

	Temp. (K)	a (Å)	b (Å)	c (Å)	
	298	18.43	18.71	6.52	
	361	18.43	18.71	6.52	Zero thermal
	419	18.43	18.71	6.52	expansion
Negative thermal	471	18.43	18.71	6.52	
expansion	573	16.22	17.02	6.43	
	598	16.22	17.03	6.44	Positive thermal
	723	16.58	17.43	6.46	expansion
	773	16.74	17.62	6.47	

on the lattice size, atomic positions, and nanostructure of natrolite. These existing experimental data (Peacor 1973; Baur and Joswig 1996) suggest that NAT exhibits zero thermal expansion around room temperature that becomes negative in the (001) plane when NAT is exposed to temperatures of about 300°C due to a dehydration mechanism* to form metanatrolite, a process that involves the removal of interstitial water molecules that pass through channels of the NAT framework aligned parallel to the [001]-axis.

All these are illustrated in the variations of the cell parameters with temperatures shown in Table 10.1 and Figure 10.10. These experimental data suggest that while at 25°C, the cell parameters of NAT are $a = 18.43$ Å, $b = 18.71$ Å, and $c = 6.52$ Å (Peacor 1973) those of metanatrolite measured at 325°C are given by $a = 16.223$ Å, $b = 17.029$ Å, and $c = 6.438$ Å (Baur and Joswig 1996). This suggests an 11.98% decrease in the cell parameter a, an 8.98% decrease in the cell parameter b, and a 1.25% decrease in the cell parameter c, thus indicating NTE, which is predominantly exhibited in the (001) plane.[†] In fact, a closer analysis of the data in this table suggests that available data may be divided into three regions, each corresponding to a different trend for the thermal expansion (see Figure 10.10 and Table 10.1).

1. *Region I*: From 298 K up to 471 K. This region is characterized by zero thermal expansion.
2. *Region II*: From 471 K up to 573 K. This region consists of an NTE.
3. *Region III*: From 598 K up to 773 K. This region is characterized by conventional thermal expansion.

---

* Dehydration is a well-known NTE mechanism and has been used to explain NTE in the zeolite faujasite (Wang et al. 2004).

[†] It is important to note that these results were obtained from different crystals, as the data above were obtained from different sources. Each individual crystal is different but crystals having the same chemical composition tend to have very similar geometric conformations. Any deviations that arise from minor impurities and point defects are assumed to have minimal effects on the overall crystal structure and properties.

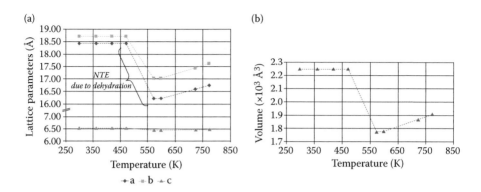

FIGURE 10.10   (a) Linear and (b) volumetric NTE in NAT based on experimental data.

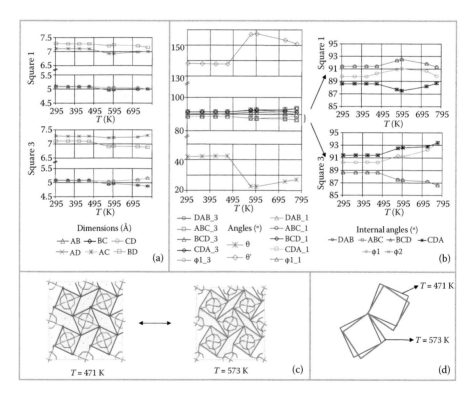

FIGURE 10.11   Absolute values of various (a) dimensions and (b) angles projected in the (001) plane for NAT as a result of changes in temperature. The projected squares superimposed (c) over the structure and (d) over each other at different temperatures. These data have been calculated from experimentally obtained x-ray data.

Apart from analyzing the variation of the lattice parameters with temperature, for the first time, an analysis of the experimentally determined crystal structure data at various temperatures is being performed in an attempt to assess whether the observed thermal expansion behavior can also be explained in terms of models involving the projected squares in the (001) plane. In fact, Figure 10.11 summarizes various measurements defined in Figure 10.7, related to the projected distances and angles that relate to the rotating squares model, and also shows images of the projected structure of NAT in the (100) plane at two different temperature values with the projected "squares" highlighted.

10.3   RESULTS AND DISCUSSION

The first important result being reported here that needs to be highlighted is the finding that NAT can exhibit both auxetic behavior and NTE, something that is very rare. In fact, although various naturally occurring or synthetic materials have been found to exhibit either NTE or NPRs (Agrawal et al. 1987; Yeganeh-Haeri et al. 1992), the identification of materials found to exhibit these two counterintuitive properties simultaneously is still in its infancy. In particular, the hypothetical (10,3)-b polyacetylene networks of Baughman and Galvão (1993) represent the only molecular level systems that have been reported so far to simultaneously exhibit NPR and NTE, although there have also been some developments on mechanical models that exhibit such properties (Grima et al. 2007a).

Second, the results of our simulations under different uniaxial stresses or pressure conditions and the analysis of the experimental data under different temperature conditions (see Figures 10.8 through 10.11) clearly suggest that, to a first approximation, NAT exhibits NPRs and NTE, which may be explained through the same mechanism, that is, a "rotating squares" type mechanism, or more accurately, "rotating quadrilaterals" because the detailed measurements suggest that the projected quadrilaterals ABCD are not perfectly square-shaped but are in fact slightly parallelogrammic. This mechanism also leads to the main deformations observed when NAT is subjected to pressure changes.

In fact, as discussed elsewhere (Grima et al. 2007c; Williams et al. 2007), the simulations on NAT_CW suggest that uniaxial loading of this system at 45° to the X-direction will result in significant changes in the angles between the quadrilaterals ABCD that are accompanied by smaller but still significant changes in the shape and size of the quadrilaterals ABCD; hence, a "semirigid rotating parallelograms" model is operating in this zeolite. All this is in line with other similar work using different force fields (Grima et al. 2007c; Williams et al. 2007) and is very significant, as it confirms the important role of this deformation mechanism.

What is also significant is the fact that in the case of the simulations of NAT_CW under hydrostatic pressure, the deformation mechanism is also describable in terms of rotation of these units. In fact, referring to Figure 10.9, the results clearly suggest that at intermediate values of applied pressure, there are significant changes in the a and b dimensions of the unit cell and a much lower change in c. This is to be expected given that the experimentally measured Young's moduli $E_x = 56.23$ GPa and $E_y = 59.35$ GPa are significantly lower than $E_z = 108.99$ GPa and can be easily explained by looking at the nanostructure of the NAT frameworks that are characterized by rigid columnar-shaped units aligned down the [001] directions (hence the high E_z), the projections of which in the (001) are the "rotating squares" units that are connected together through highly flexible X–O–X (X = Si, Al) bonds. These flexible X–O–X bonds permit significant deformations when loaded in directions orthogonal to the [001] directions (hence the low E_x and E_y) and are essential for the "rotating squares" mechanism to operate. The fact that the "rotating squares" mechanism is indeed operating when the systems are placed under moderate hydrostatic pressure is supported from measurements of various projected dimensions and angles shown in Figure 10.9. In fact, when various geometrical parameters related to the projected "rotating squares" in the (001) plane were measured, it was found that for NAT_ CW, the application of moderate amounts of pressure resulted in significant changes in the angles between the projected squares (caused by significant changes in the X–O–X hinge angles) when compared to the lengths of the sides and of the diagonals of the projected squares, the internal angles of the squares, and the angles between the diagonals of the projected squares.

At higher pressures, other deformations such as dilation and deformation of the squares start to become more dominant. In fact, at the higher values of negative hydrostatic pressures, the projected squares are nearly in their "fully open conformation" (see Figure 10.9), which means that the "rotating squares" mechanism, which at moderate pressures was giving rise to low moduli in the (001) plane, cannot operate. This permits the onset of other deformation mechanisms that result in greater deformations of the "projected squares" as opposed to changes in the angles between the squares. This is clearly indicated in the data shown in Figure 10.9.

If one now looks at the experimentally determined crystal structures at various temperatures and uses these data to examine the molecular-level deformations that occur as NAT is heated, one will also see that the deformation, in particular, the thermal shrinkage that occurs upon heating and dehydration, may also be explained through the "rotating squares" model. In fact, a more detailed analysis of the deformations of the connected quadrilateral units identified for NAT suggests that as illustrated in Figure 10.11, the net deformations of the NAT framework following heating and dehydration may be described in terms of the "rotating quadrilaterals" model, since dehydration has the net effect that the "quadrilaterals" are observed to rotate relative to each other in such a way that the acute angles between the squares become smaller, thus resulting in a less open (i.e., smaller) structure. This net mode of deformation is possible due to the same structural characteristics that result in NPR, that is, the flexibility of the Si–O–Al linking (Grima and Evans 2000a) bonds that correspond to the corners of the "connected quadrilaterals" compared to the relative rigidity of the O–Si–O and O–Al–O linking (Grima and Evans 2000a), that is, the "cage-like units" that correspond to the "rigid quadrilaterals."

All this is very significant not only because of the facts noted above, that is, that (1) not many materials exhibit both NPR and NTE, and (2) it is being demonstrated that these anomalous properties are resulting from the same structural features, but also because it highlights the importance of supplementing experimental work with the results of molecular modeling simulations, and vice versa. In fact, it should be highlighted that the experimental work aimed at determining accurately and reliably the mechanical properties of the zeolite NAT was driven by our earlier reports based on molecular modeling simulations that NAT is likely to be auxetic. Furthermore, although the experimental work has provided the proof that NAT is indeed auxetic, we still need to rely on the modeling work to understand the underlying reasons for this anomalous behavior. In this respect, given the accuracy by which the simulations are reproducing the magnitude of the Poisson's ratios, we are confident that the conclusions being drawn on the deformation mechanisms that lead to NPRs, that is, conclusions that only rely on modeling work, are likely to be reliable. Furthermore, as opposed to the experimental work, the simulations have the advantage that they can be easily fine-tuned to enable studies on other materials that are similar in form to NAT. Such simulations could make it possible to design new man-made materials that mimic the behavior of NAT but may have slightly different properties, thus possibly being more suitable for specific practical applications.

10.4   CONCLUSION

Through this work, we have shown that the zeolite NAT exhibits anomalous thermal and mechanical properties in the sense that it can get fatter rather than thinner in certain planes when stretched in particular directions (NPR, auxetic) and shrink when heated (NTE)—two properties that have very rarely been found together. Through a combined analysis of results obtained through experimental and modeling work, we have been able to show that these anomalous properties result from the same nanostructural features and, to a first approximation, may be explained through a "rotating squares" mechanism. This is very significant as it further highlights the strong relationship between the materials' nanostructure and their macroscopic thermal and mechanical properties.

Given the many advantages offered by such "negative" materials when compared with their conventional counterparts, we are confident that our findings will stimulate further modeling and experimental work that looks for other materials that also exhibit both of these properties simultaneously, possibly by mimicking the behavior of NAT.

ACKNOWLEDGMENTS

This work is supported by a grant awarded to the University of Malta by the Malta Council for Science and Technology through their Research, Technological Development and Innovation scheme. Reuben Cauchi acknowledges the support of a Strategic Educational Pathways Scholarship, financed in part by the European Union—European Social Fund under Operational Programme II—Cohesion Policy 2007–2013. Daphne Attard acknowledges the support of a Malta Government Scholarship Scheme (Grant No. ME 367/07/17) awarded by the government of Malta.

REFERENCES

Agrawal, D. K., Roy, R. and McKinstry, H. A. Ultra low thermal-expansion phases—Substituted PMN perovskites. *Mater. Res. Bull.* **22**, 1987, 83–88.

Alderson, A. A triumph of lateral thought. *Chem. Ind.* **10**, 1999, 384–391.

Alderson, A., Davies, P. J., Evans, K. E., Alderson, K. L. and Grima, J. N. Modelling of the mechanical and mass transport properties of auxetic molecular sieves: An idealised inorganic (zeolitic) host–guest system. *Mol. Simul.* **31**, 2005a, 889–896.

Alderson, A., Davies, P. J., Williams, M. R., Evans, K. E., Alderson, K. L. and Grima, J. N. Modelling of the mechanical and mass transport properties of auxetic molecular sieves: An idealised organic (polymeric honeycomb) host–guest system. *Mol. Simul.* **31**, 2005b, 897–905.

Alderson, A., Rasburn, J., Ameer-Beg, S., Mullarkey, P. G., Perrie W. and Evans, K. E. An auxetic filter: A tuneable filter displaying enhanced size selectivity or defouling properties. *Ind. Eng. Chem. Res.* **39**, 2000, 654–665.

Alderson, K. L., Webber, R. S., Kettle, A. P. and Evans, K. E. Novel fabrication route for auxetic polyethylene. Part 1. Processing and microstructure. *Polym. Eng. Sci.* **45**, 2005c, 568–578.

Baughman, R. H. and Galvão, D. S. Crystalline networks with unusual predicted mechanical and thermal properties. *Nature* **365**, 1993, 735–737.

Baughman, R. H. and Galvao, D. S. Negative volumetric thermal expansion for proposed hinged phases. *Chem. Phys. Lett.* **240**, 1995, 180–184.

Baughman, R. H., Shacklette, J. M., Zakhidov, A. A. and Stafström, S. Negative Poisson's ratios as a common feature of cubic metals. *Nature* **392**, 1998, 362–365.

Baur, W. H., Joswig, W. and Muller, G. Mechanics of the Feldspar Framework; Crystal Structure of Li-Feldspar. *Journal of Solid State Chemistry*, **121**, Number 1, January 1996, pp. 12–23(10).

Berthou, H., Neumann, V., Dan, J. P. and Hintermann, H. E. Mechanically enhanced capillary columns. *Surf. Coat. Technol.* **61**, 1993, 93–96.

Bezazi, A. and Scarpa, F. Tensile fatigue of conventional and negative Poisson's ratio open cell PU foams. *Int. J. Fatigue* **31**, 2009, 488–494.

Brańka, A. C., Heyes, D. M. and Wojciechowski, K. W. Auxeticity of cubic materials. *Phys. Status Solidi B* **246**, 2009, 2063–2071.

Chan, N. and Evans, K. E. Indentation resilience of conventional and auxetic foams. *J. Cell. Plast.* **34**, 1998, 231–260.

Choi, J. B. and Lakes, R. S. Fracture toughness of re-entrant foam materials with a negative Poisson's ratio: Experiment and analysis. *Int. J. Fract.* **80**, 1996, 73–83.

Evans, K. E. Tailoring a negative Poisson's ratio. *Chem. Ind.* **20**, 1990, 654–657.

Evans, K. E. Auxetic polymers—A new range of materials. *Endeavour* **15**, 1991, 170–174.

Evans, J. S. O., Mary, T. A. and Sleight, A. W. Negative thermal expansion in $Sc_2(WO_4)_3$. *J. Solid State Chem.* **137**, 1998, 148–160.

Evans, J. S. O. Negative thermal expansion materials, *J. Chem. Soc. Dalton Trans.* **19**, 1999, 3317.

Evans, K. E., Nkansah, M. A., Hutchinson, I. J. and Rogers, S. C. Molecular network design. *Nature* **353**, 1991, 124.

Ewald, P. P. *Ann. d. Physik*, **64**, 1921, 253.

Gatt, R., Zammit, V., Caruana, C. and Grima, J. N. On the atomic level deformations in the auxetic zeolite natrolite. *Phys. Status Solidi. B* **245**, 2008, 502–510.

Gibson, L. J., Easterling, K. E. and Ashby, M. F. The structure and mechanics of cork. *Proc. R. Soc. London, Ser. A* **377**, 1981, 99–117.

Giddy, A. P., Dove, M. T., Pawley, G. S. and Heine, V. The determination of rigid-unit modes as potential soft modes for displacive phase transitions in framework crystal-structures. *Acta Crystallogr. A* **49**, 1993, 697–703.

Grima, J. N. New auxetic materials, Ph.D. Thesis, University of Exeter, Exeter, UK, 2000.

Grima, J. N., Alderson, A. and Evans, K. E. Zeolites with negative Poisson's ratios. Paper presented at the RSC 4th International Materials Conference (MC4), Dublin, Ireland, July 1999, 81.

Grima, J. N. and Evans, K. E. Auxetic behavior from rotating squares. *J. Mater. Sci. Lett.* **19**, 2000a, 1563–1565.

Grima, J. N. and Evans, K. E. Self-expanding molecular networks. *Chem. Commun.* **16**, 2000b, 1531–1532.

Grima, J. N. and Evans, K. E. Auxetic behavior from rotating triangles. *J. Mater. Sci.* **41**, 2006, 3193–3196.

Grima, J. N., Farrugia, P. S., Gatt, R. and Zammit, V. Connected triangles exhibiting negative Poisson's ratios and negative thermal expansion. *J. Phys. Soc. Jpn.* **76**, 2007a, 025001.

Grima, J. N., Gatt, R., Zammit, V., Williams, J. J., Evans, K. E., Alderson, A. and Walton, R. I. Natrolite: A zeolite with negative Poisson's ratios. *J. Appl. Phys.* **101**, 2007b, 086102.

Grima, J. N. and Gatt, R. On the behaviour of natrolite under hydrostatic pressure. *J. Non-Cryst. Solids* **356**, 2010, 1881–1887.

Grima, J. N., Jackson, R., Alderson, A. and Evans, K. E. Do zeolites have negative Poisson's ratios? *Adv. Mater.* **12**, 2000, 1912–1918.

Grima, J. N., Zammit, V. and Gatt, R. Negative thermal expansion. *Xjenza* **11**, 2006, 17–29.

Grima, J. N., Zammit, V., Gatt, R., Alderson, A. and Evans, K. E. Auxetic behaviour from rotating semi-rigid units. *Phys. Status Solidi B* **244**, 2007c, 866–882.

Hartwig, G. Support elements with extremely negative thermal-expansion. *Cryogenics* **35**, 1995, 717–718.

He, C., Liu, P. and Griffin, A. C. Toward negative Poisson ratio polymers through molecular design. *Macromolecules* **31**, 1998, 3145–3147.

He, C. B., Liu, P. W., McMullan, P. J. and Griffin, A. C. Toward molecular auxetics: Main chain liquid crystalline polymers consisting of laterally attached *para*-quaterphenyls. *Phys. Status Solidi B* **242**, 2005, 576–584.

Holzer, H. and Dunand, D. Processing, structure and thermal expansion of metal matrix composites containing zirconium tungstate. Fourth International Conference on Composite Engineering, Hawaii, 1997.

Jakubinek, M. B., Whitman, C. A. and White, M. A. Negative thermal expansion materials. *J. Therm. Anal. Calorim.* **99**, 2010, 165–172.

Lakes, R. Foam structures with a negative Poisson's ratio. *Science* **235**, 1987, 1038–1040.

Marinkovic, B. A., Ari, M., de Avillez, R. R., Rizzo, F., Ferreira, F. F., Miller, K. J., Johnson, M. B. and White, M. A. Correlation between AO(6) polyhedral distortion and negative thermal expansion in orthorhombic Y2Mo3O12 and related materials. *Chem. Mater.* **21**, 2009, 2886–2894.

Mary, T. A., Evans, J. S. O., Vogt, T. and Sleight, A. W. Negative thermal expansion from 0.3 to 1050 Kelvin in ZrW_2O_8. *Science* **272**, 1996, 90–92.

Mavoori, H., Jin, S., Espindola, R. P. and Strasser, T. A. Enhanced thermal and magnetic actuations for broad-range tuning of fibre Bragg grating-based reconfigurable add–drop devices. *Opt. Lett.* **24**, 1999, 714.

Miller, W., Mackenzie, D. S., Smith, C. W. and Evans, K. E. A generalised scale-independent mechanism for tailoring of thermal expansivity: Positive and negative. *Mech. Mater.* **40**, 2008, 351–361.

Miller, W., Smith, C. W., Mackenzie, D. S. and Evans, K. E. Negative thermal expansion: A review. *J. Mater. Sci.* **44**, 2009, 5441–5451.

Ngo, N. Q., Li, S. Y., Zheng, R. T., Tjin, S. C. and Shum, P. Electrically tunable dispersion compensator with fixed center wavelength using fibre Bragg grating. *J. Lightwave Technol.* **21**, 2003, 1568–1575.

Nye, J. F. *Physical Properties of Crystals*. Oxford, UK: Clarendon, 1957.

Peacor, D. R. High-temperature single-crystal study of the cristobalite inversion. *Z. Kristallogr.* **138**, 1973, 274–298.

Sanchez-Valle, C., Sinogeikin, S. V., Lethbridge, Z. A. D., Walton, R. I., Smith, C. W., Evans, K. E. and Bass, J. D. Brillouin scattering study on the single-crystal elastic properties of natrolite and analcime zeolites *J. Appl. Phys.* **98**, 2005, doi:10.1063/1.2014932.

Scarpa, F., Pastorino, P., Garelli, A., Patsias, S. and Ruzzene, M. Auxetic compliant flexible PU foams: Static and dynamic properties. *Phys. Status Solidi B* **242**, 2005, 681–694.

Scarpa, F. and Tomlinson, G. Theoretical characteristics of the vibration of sandwich plates with in-plane negative Poisson's ratio values. *J. Sound Vib.* **230**, 2000, 45–67.

Stott, P. J., Mitchell, R., Alderson, K. and Alderson, A. Abstracted from Materials World, Vol. 8, pp. 12–14, 2000. A growth industry. Available online at http://www.azom.com/ details.asp?ArticleID=168. Accessed on 22nd January 2011.

Tucker, M. G., Goodwin, A. L., Dove, M. T., Keen, D. A., Wells, S. A. and Evans, J. S. O. Negative thermal expansion in ZrW_2O_8: Mechanisms, rigid unit modes, and neutron total scattering. *Phys. Rev. Lett.* **95**, 2005, 255501.

Wang, X., Hanson, J. C., Szanyi, J. and Rodriguez, J. A. Interaction of H_2O and NO_2 with BaY faujasite: Complex contraction/expansion behavior of the zeolite unit cell. *J. Phys. Chem. B* **108**, 2004, 16613–16616.

Wang, Y. and Lakes, R. Analytical parametric analysis of the contact problem of human buttocks and negative Poisson's ratio foam cushions. *Int. J. Solids Struct.* **39**, 2002, 4825–4838.

Wei, Z. X., Yu, Y. S., Xing, H., Zhou, Z. C., Wu, Y. D., Zhang, L., Zheng, W. and Zhang, Y. S. Fabrication of chirped fibre grating with adjustable chirp and fixed central wavelength. *IEEE Photonics Technol. Lett.* **13**, 2001, 821–823.

Williams, J. J., Smith, C. W., Evans, K. E., Lethbridge, Z. A. D. and Walton, R. I. An analytical model for producing negative Poisson's ratios and its application in explaining off axis elastic properties of the NAT-type zeolites. *Acta Mater.* **55**, 2007, 5697–5707.

Wojciechowski, K. W. Two-dimensional isotropic system with a negative Poisson ratio. *Phys. Lett. A* **137**, 1989, 60–64.

Wojciechowski, K. W. and Brańka, A. C. Auxetics—Materials and models with negative Poisson's ratios. *Mol. Phys. Rep.* **6**, 1994, 71–85.

Yeganeh-Haeri, Weidner, D. J. and Parise, J. B. Elasticity of alpha-cristobalite: A silicon dioxide with a negative Poisson's ratio. *Science* **257**, 1992, 650–652.

11 Structure–Property Relations between Silicon-Containing Polyimides and Their Carbon-Containing Counterparts

Silke Pelzer and Dieter Hofmann

CONTENTS

11.1 INTRODUCTION

Polyimides have been important membrane materials for gas separation for many years. Nevertheless, there is still a strong need for further improved transport properties of these polymers particularly regarding small molecule permeabilities and the selectivities for certain gas pairs. One problem is that improving the permeability goes normally along with a loss of selectivity. Numerous experiments have been performed to find out which factors might improve the permeability of a polyimide. Higher permeabilities can be, for example, reached by increasing the *ortho* alkylation in the diamine; inserting –CF_3 groups, $SiO(Me)_2$, $Si(Me)_3$, or $C(Me)_3$ groups into the dianhydride; increasing rigidity; decreasing the diamine length; or inserting bulky groups. It also turned out that producing a higher free volume leads to a higher permeability [1–18]. Kim and coworkers [19–23] designed and examined various bulky silicon-containing structures and their carbon-containing counterparts (Figure 11.1) with regard to their oxygen permeabilities (Table 11.1). The permeabilities between each pair of polyimide (e.g., Si-containing structure and C-containing

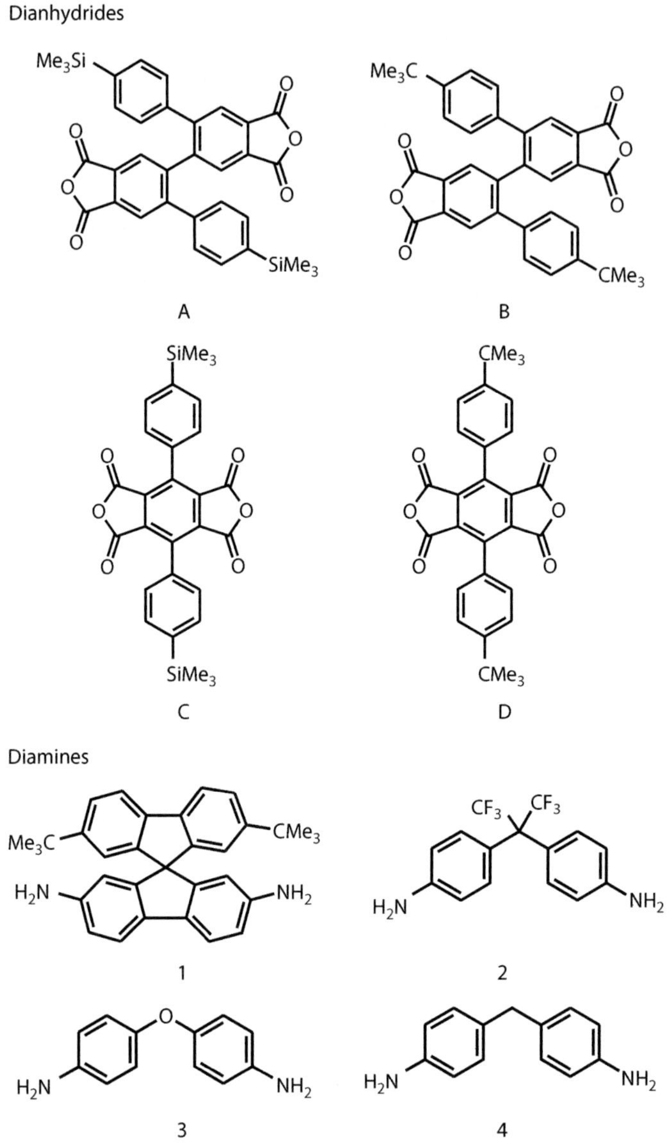

Dianhydrides

A

B

C

D

Diamines

1

2

3

4

FIGURE 11.1   Dianhydride and diamine structures of the examined polyimides.

counterpart) are, except for one case, higher for the Si than for the respective C case. With regard to the diffusion (D) and the solubility (S), no experimental data for these types of polyimides are available. To validate the chosen models, experimental wide angle x-ray diffraction (WAXD) measurements reported in the literature [21] are available for two polyimides (A-1 and B-1). The major model properties considered to assess the critical structural differences between the polymers are the fractional

TABLE 11.1

Experimental Oxygen Permeabilities of the Chosen Polyimides [19–23]

Polyimide	$P(O_2)$ [(7.5005 × 10^{-18}) m^2/sPa]	Polyimide	$P(O_2)$ [(7.5005 × 10^{-18}) m^2/sPa]
A-1	121		
A-2	105	C-2	56
A-3	61	C-3	32
A-4	52	C-4	14
B-1	52		
B-2	110	D-2	50
B-3	43	D-3	20
B-4	31	D-4	13

Note: Exp. setup: membrane thickness = 30 μm–60 μm, T = RT, p_{feed} = 0.94 MPa.

free volume distribution (FFV) and the mean squared displacement (MSD) of the respective polymer atoms. These properties should be closely related to the transport properties of interest.

11.2 COMPUTATIONAL DETAILS

The amorphous packing model construction and the atomistic simulations of the shown polymers have been investigated using Accelrys Materials Studio® (MS) Modeling 3.2 program package and the COMPASS force field [24,25] followed by extensive equilibration procedures. Initial bulk polymer packing cells were created using the amorphous cell module of Materials Studio. The polymer chains were grown at 308 K under cubic periodic boundary conditions. To avoid packing algorithm-related catenations and spearings of aromatic units, the initial packing density was very low (0.1 kg/m^3). Because no experimental densities were provided through the literature, the Synthia tool of Materials Studio was used to estimate the densities of the different polyimides, which were used to help create the final basic cell for the systems (Table 11.2).

The atomistic chain segment packing cells used in this work consist of about 4000–5000 atoms. For each of the polyimides, 10 independent packing models were obtained to increase the statistical significance of the results. The resulting initial packing models were equilibrated using force field parameter scaling [26] according to the procedure outlined in Table 11.3. The individual stages lasted for 1000 fs with a time step of 0.2 fs and were always preceded by a short energy minimization of several hundred iterations.

After equilibrating the systems, the MSD results were obtained using the corresponding routines of MS Modeling. For calculating FFV, the procedure described by Hofmann et al. [27] has been used. The van der Waals radii of the examined atoms were C: 1.55 (10^{-10} m), H: 1.10 (10^{-10} m), O: 1.73 (10^{-10} m), F: 1.30 (10^{-10} m), and Si:

TABLE 11.2
Simulation Data

Polymer	No. of Repeat Units	No. of Atoms	Density (Synthia) (kg/m³)	Average Size of the Periodic Cell (10^{-10} m)
A-1	37	4997	1.143	37.97
A-2	45	4547	1.237	37.71
A-3	45	4277	1.139	36.68
A-4	45	4187	1.154	36.56
B-1	37	4997	1.169	37.35
B-2	45	4547	1.276	37.90
B-3	45	4277	1.175	35.79
B-4	45	4187	1.191	35.66
C-2	45	4097	1.234	36.65
C-3	55	4677	1.126	38.00
C-4	55	4567	1.142	37.86
D-2	45	4097	1.198	36.51
D-3	55	4677	1.164	36.98
D-4	55	4567	1.181	36.84

2.20 (10^{-10} m). To determine FFV and the fractional accessible volume (FAV) distributions, the overlaid three-dimensional probe molecule insertion grid had a grid size δ of 0.5. An oxygen molecule [Ø: 1.73 (10^{-10} m)], a positronium [Ø: 1.1 (10^{-10} m)], and a particle with a diameter of 0.4 (10^{-10} m) were used as probe molecules. x-ray scattering calculations have been performed by using the scattering routine implemented in the forcite module of MS Modeling. The used cutoff was 30 (10^{-10} m); to avoid artifacts, a model size correction of a 1.7 (10^{-10} m) radius was applied and the scattering intensity was plotted against the 2-theta angle. Systematic conformational search procedures applying the Discover module of the MS Modeling software of Accelrys were also performed to determine the energetic potential for the rotation around the central C–C bond of some of the system.

TABLE 11.3
Scaling Factors of the Five-Step Equilibration Used for Packing Models

Stage of Equilibration	Scaling Factor for the Torsion Term	Scaling Factor for Nonbonded Interactions
1	10^{-3}	10^{-3}
2	10^{-2}	10^{-3}
3	1	10^{-3}
4	1	10^{-2}
5	1	1

11.3 RESULTS AND DISCUSSION

Experimental data available in the literature are the oxygen and nitrogen permeabilities and WAXD measurements for the polyimides A-1 and B-1. For checking the quality of the models, calculations have been performed for A-1 and B-1 WAXD, as shown in Figure 11.2.

As it can be seen, the calculated and the measured graphs are in good agreement with each other, and, therefore, the chosen equilibration procedure, giving useful results for these polyimides, will be applied to all other polyimides as well. For further validation of the models, experimental information about the density ρ, the solubility coefficient S, and the diffusion coefficient D would be helpful, as these values can also be calculated from the models. They are, however, not available in the literature. With regard to the permeability (P), experience shows that comparing calculated and measured P values is normally not sufficient. Computationally, the permeability will be calculated as the product of S and D, and the deviations of the values of S and D compared to the experimental ones, via error propagation, will lead to usually quite unsatisfying results of P. The direct model validation is, therefore, based on the already mentioned WAXD data for A-1 and B-1, indicating reasonably well-equilibrated models in these cases. Unfortunately, these are the only WAXD data being measured experimentally, but because for the more or less related other polyimides the same basic approach of model construction and equilibration was taken as for A-1 and B-1, it is assumed that the other models are also suited for evaluation. This view is further confirmed by the following approach where instead of experimental density data, the respective prediction from the Synthia module of

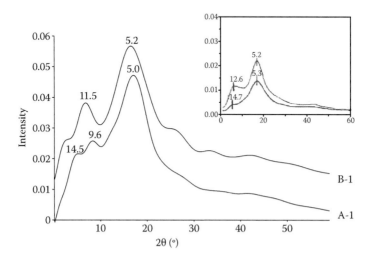

FIGURE 11.2 Comparison of the calculated (big graph) and the measured (small graph) [21] WAXD of A-1 and B-1. The given numbers in the graph show the calculated d-spacings in a 10-nm scale. The additional peak at 9.6×10 nm is probably due to limited size effects and a slightly higher order in the model than the real system.

Materials Studio was taken for comparison with the density values reached in the 1 bar (NpT) MS simulation (Table 11.4).

Synthia is a method developed by Bicerano [28], which relates the properties of a polymer to a combination of four indices considering the composition and topology of the respective polymer repeat. As Table 11.4 shows, the Synthia results are in good agreement with the MD-simulated results.

The main focus of this paper is the interpretation of the calculated free volume distribution and MSDs.

As seen in Figure 11.1, there are two different types of dianhydrides both being very extended in size. The dianhydrides A and B [19,20] have a very bulky structure, whereas the dianhydrides C and D [22,23] are rather flat as the geometry optimizations show (Figure 11.3). The geometry optimizations also show that in between the dianhydride couples (A/B and C/D), exchanging the $Si(Me)_3$ group with the $C(Me)_3$ group does not significantly change the geometry of the respective dianhydride. Except for the diamine 1 [21] being very bulky as well and structurally comparable to some diamines that Langsam examined [1], Kim and coworkers used the common diamines 6FDA (2), MDA (3), and ODA (4). The experimentally measured oxygen permeabilities (P) vary for the polyimides A and B from high values of around 121 (A-1), 110 (B-2), and 105 (A-2) Barrer to smaller ones of 43 (B-3) and 31 (B-4) Barrer, whereas the polyimides C and D have permeabilities from 56 (C-2) Barrer down to 14 (C-4) and 13 (D-4) Barrer.

It should be noticed that looking at the permeabilities pairwise, that is, the Si-containing structures and their C-containing counterparts, the P values for the

TABLE 11.4
Calculated Density Values

	Density Values (kg/m³)	
Polymer	Synthia	NpT (Average over 8–10 Packing Models)
A-1	1.1434	1.1370
A-2	1.2374	1.2375
A-3	1.1398	1.1395
A-4	1.1541	1.1512
B-1	1.1691	1.1570
B-2	1.2760	1.2746
B-3	1.1747	1.1709
B-4	1.1905	1.1902
C-2	1.2340	1.2345
C-3	1.1264	1.1250
C-4	1.1420	1.1405
D-2	1.2762	1.2985
D-3	1.1639	1.1601
D-4	1.1813	1.1772

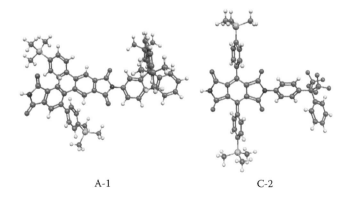

A-1   C-2

FIGURE 11.3 (**See color insert.**) Optimized geometrical structures of A-1 and C-2. A-1 has a very bulky structure, whereas C-2, especially looking at the dianhydride, is quite flat.

Si-containing polyimides are consistently higher than for the C-containing ones, except for A-2:B-2 where the C-containing structure is higher in permeability than the Si-containing one. The respective P differences are, however, not very large, and considering the error of the experimental measurements being around 10% of the P value, the data pairs A-2:B-2, C-2:D-2, and C-4:D-4 could be considered equal in their permeation values. The considerably higher difference in permeability for the A-1:B-1 pair is a bit unexpected. While all the other pairs differ by not more than 21 Barrer from each other (usually a lot less), the difference of the P values for this pair is 69 Barrer. Because in the following discussion A-1 and B-1 are also the only poly-mers extremely falling out of the recognizable correlation between atomic structure and macroscopic transport parameters, it is assumed that either A-1 or B-1 might be influenced by some systematic error in the measured permeability.

The calculated size distributions of the free volume regions accessible for an oxy-gen molecule are given for all investigated polyimides in Figure 11.4 for groups A and B and in Figure 11.5 for groups C and D. Looking at the free volume analysis, the graphs follow the same trend as the permeabilities; the Si-containing polyimides have a larger accessible free volume than their carbon counterparts (see also Table 11.5). Looking at these data, the FAV behavior follows

FAV-A-2 > FAV-A-3 > FAV-A-4 > FAV-A-1
FAV-B-2 > FAV-B-1 > FAV-B-3 = FAV-B-4
FAV-C-2 > FAV-C-3 > FAV-C-4
FAV-D-2 > FAV-D-3 > FAV-D-4,

whereas the permeabilities are

PA-1 > PA-2 > PA-3 > PA-4
PB-2 > PB-1 > PB-3 > PB-4
PC-2 > PC-3 > PC-4 >
PD-2 > PD-3 > PD-4.

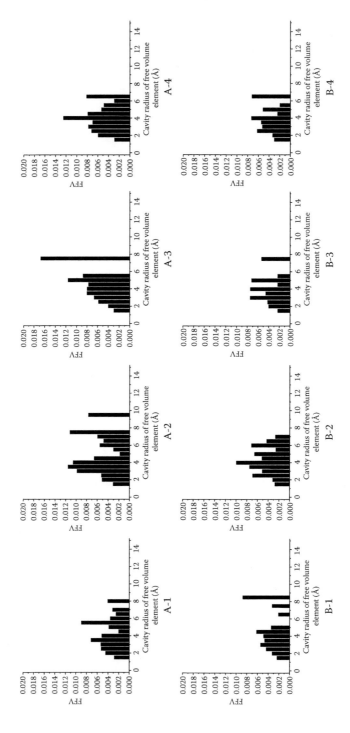

FIGURE 11.4   Bar graph size distribution of free volume elements accessible for O_2 for groups A and B.

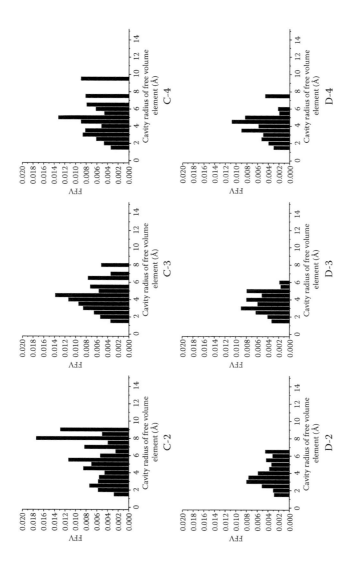

FIGURE 11.5 Bar graph size distribution of free volume elements accessible for O_2 for groups C and D.

TABLE 11.5

Calculated Accessible Free Volumes (FAV) for Oxygen, a Positron, and a Particle $r = 0.4$ (10^{-10} m) and the Estimated Free Volume (FFV)

Polyimide	FAV (O_2)	FAV (Positron)	FAV (0.4)	Estimated FFV (0.0)	$P(O_2)$ [(7.5005 × 10^{-18}) m^2/sPa]
A-1	0.058	0.211	0.310	0.330	121
A-2	0.092	0.275	0.330	0.350	105
A-3	0.080	0.196	0.328	0.405	61
A-4	0.072	0.232	0.324	0.360	52
B-1	0.049	0.261	0.298	0.305	52
B-2	0.064	0.173	0.313	0.390	110
B-3	0.047	0.191	0.306	0.340	43
B-4	0.047	0.202	0.310	0.340	31
C-2	0.114	0.302	0.341	0.345	56
C-3	0.094	0.275	0.338	0.340	32
C-4	0.088	0.264	0.334	0.340	14
D-2	0.071	0.234	0.349	0.380	50
D-3	0.053	0.225	0.319	0.330	20
D-4	0.059	0.226	0.316	0.330	13

The groups B, C, and D show the same order in the sequences of the FAV and the P values. Exceptions are B-3 and B-4 showing the same value of FAV but having different permeabilities (43 and 31 Barrer, respectively). Group A shows an inconsistency regarding the polyimide A-1. The FAV and the P value do not match in the way that A-1 shows the smallest FAV value in its group but has the highest permeability. Because in the following discussion A-1 is also the only polymer extremely falling out of the recognizable correlation between atomic structure and macroscopic transport parameters, it is assumed that B-1 might be influenced by some systematic error in the measured permeability.

All polyimides show a maximum distribution for cavities with a radius of 2–4 × 10^{-10} m. Additionally, some of the polyimides have a second maximum for cavities greater than 5 × 10^{-10} m, which might be important for the permeability as stated in the literature [29,30]. Table 11.6 shows the percentages of the FAV elements with radii greater than 5 × 10^{-10} m. It can be seen that the Si-containing structures have a higher percentage of cavities > 5 × 10^{-10} m than their counterparts containing C. In between the groups B, C, and D, the decrease in the percentage of FAV goes along with the decrease in permeability. Group A is again the exception, where A-1 is the second highest and A-2 shows the highest value.

Summarizing the main result from the FAV analysis, polyimides containing Si have a higher free volume and cavities of bigger radii than their C-containing counterparts. Another parameter to discuss is the MSD that describes the self-diffusion of the polymer chain segments. It is defined as

$$MSD(t) = \langle \Delta r_i(t)^2 \rangle = \langle (r_i(t) - r_i(0))^2 \rangle \qquad (11.1)$$

TABLE 11.6

FAV for Oxygen, Fraction of Cavities Smaller than 5×10^{-10} m and Bigger than 5×10^{-10} m, and the Percentage of the Cavities Greater than 5×10^{-10} m All Over FAV

Polyimide	$P[O_2]$ $[(7.5005 \times 10^{-18})$ m²/sPa]	FAV[O_2]	<5 (10^{-10} m)	>5 (10^{-10} m)	Percentage of FAV >5 (10^{-10} m)
A-1	121	0.058	0.036	0.022	0.381
A-2	105	0.092	0.054	0.038	0.415
A-3	61	0.080	0.055	0.025	0.313
A-4	52	0.072	0.056	0.016	0.222
B-1	52	0.049	0.035	0.014	0.286
B-2	110	0.064	0.047	0.017	0.265
B-3	43	0.047	0.039	0.008	0.170
B-4	31	0.047	0.040	0.007	0.148
C-2	56	0.114	0.047	0.067	0.585
C-3	32	0.094	0.058	0.036	0.380
C-4	14	0.088	0.064	0.023	0.268
D-2	50	0.071	0.039	0.012	0.169
D-3	20	0.053	0.050	0.003	0.066
D-4	13	0.059	0.051	0.002	0.033

with $\mathbf{r}_i(t) - \mathbf{r}_0(t)$ being the distance traveled by molecule i over the time t. The MSD is an average over all time intervals and molecules.

Figures 11.6 and 11.7 show the MSD for the whole molecule for the groups A and B, and C and D, respectively.

The MSD results for the whole molecules reveal that the Si-containing structures show an always higher self-diffusion than the respective C-containing ones.

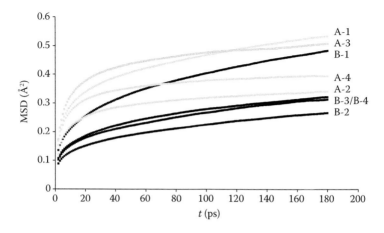

FIGURE 11.6 MSD vs. time for groups A and B.

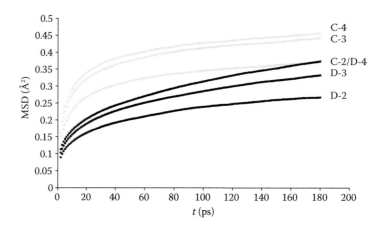

FIGURE 11.7 MSD vs. time for groups C and D.

Figures 11.8 through 11.10 contain the respective results for the backbone and side group mobilities.

Compared to the side groups, the backbones are very stiff, but in this connection, the C-containing polyimides are more flexible than their Si counterparts. This is in good agreement with the energetic potential for the dihedral rotation around the central C–C bond (Figure 11.11) of A and B, as it was obtained from a systematic conformational search procedure applying the Discover module of the MS Modeling software of Accelrys. Whereas for the case of B the rotational barrier reaches not more than 100 kJ/mol, the barrier for A is extremely high.

For the side groups, the opposite effect is observed, that is, the Si-containing ones are more flexible. The suggestion lies close, taking the results of the free volume also into account, that the size of the silicon atom is the most important cause of the

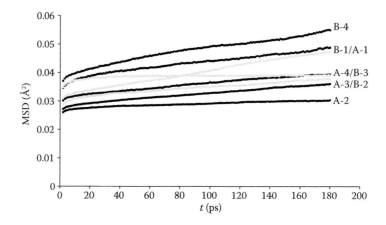

FIGURE 11.8 MSD of the backbone vs. time for groups A and B.

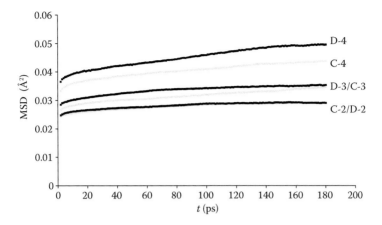

FIGURE 11.9 MSD of the backbone vs. time for groups C and D.

observed differences in free volume distribution and flexibility of the side groups. As the silicon is a lot bigger than the carbon, it needs more space and the packing is therefore looser. As the packing is looser, there is more free volume, making the side chains to be able to move more. On the other hand, the backbone for the dianhydride A due to the attached bulky $Si(Me_3)$ side groups is less flexible with regard to the rotation around the central C–C bond. The backbone flexibilities of the related pairs of groups C and D (e.g., C2:D2; C3:D3; C4:D4) are closer to each other especially for C-3:D-3 and C-2:D-2. Because of the very stiff structure of the involved dianhydrides, the main mobility contribution belongs to the respective diamines, being the same for each pair, which is why their MSD curves lie close together. The size of the Si does not influence the flexibility as much as in the cases of A and B because the imide structure in the

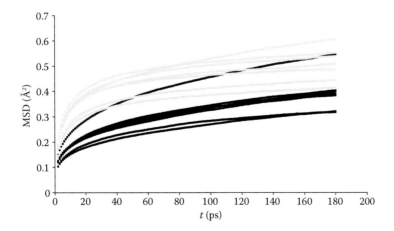

FIGURE 11.10 MSD of the side chain vs. time for all groups. Gray lines represent the Si-containing molecules, whereas the black lines represent the C-containing ones.

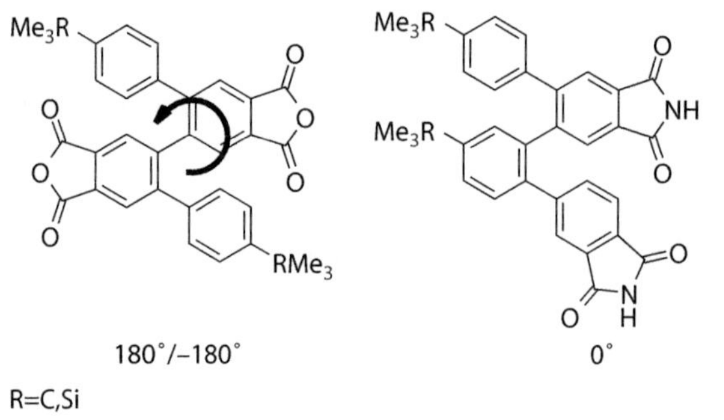

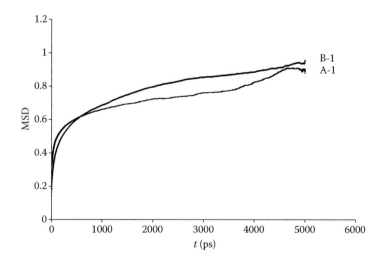

180°/–180°   0°

R=C,Si

FIGURE 11.11 Dianhydride structure showing the conformation at 0° and 180°/–180°, respectively.

center of the dianhydrides C and D (cf. Figure 11.1) is not flexible at all, so that the two Si-containing side chains of each repeat unit do not interfere with each other as much as in the case of A and B.

Taking a closer look at Figure 11.10, the gray and black lines crossing each other at 280 ps on top of the diagram present the polyimides A-1 and B-1. At the beginning, the flexibility of the side chains of B-1 is quite low, but as time passes, its side chains become as flexible as those of A-1. Whereas all the other structures are pretty parallel to each other in the MSD diagram, A-1 and B-1 cross. An MSD run over 5 ns (Figure 11.12) shows that after crossing, the two curves run parallel to each other as well. This supports again the earlier mentioned idea that the experimentally obtained P value for A-1 might be too high and should be closer to the one of B-1 as in all the other cases.

FIGURE 11.12 MSD vs. time for A-1 and B-1.

11.4 CONCLUSIONS

Parallel to the finding that the oxygen permeabilities between Si-containing polyimides and their C-containing counterparts typically are higher for the Si case, these polymers also differ in their behavior with regard to the free volume and the MSD the way that the Si-containing structures have a higher free volume and are more flexible in their side groups. This is due to the fact that the silicon atom is bigger than the carbon atom, and because of that, it needs more space. The more space there is, the more the side chains can move and the more free volume there is. The backbone of the Si-containing polyimides, however, is less flexible than the carbon ones especially in the case of groups A and B, as the big $Si(Me_3)$ group makes a rotation around the centered C–C bond energetically more demanding.

ACKNOWLEDGMENT

The authors would like to thank the European Commission 6th Framework Program Project MULTIMATDESIGN "Computer aided molecular design of multifunctional materials with controlled permeability properties" for financial support.

REFERENCES

1. M. Langsam, Polyimides for gas separation, Chapter 22, in *Fundamentals and Applications*, K.J. Mittal, ed., Marcel Dekker, New York, 1996.
2. M. Langsam and W. F. Burgoyne. Effects of diamine monomer structure on the gas permeability of polyimides. I. Bridged diamines. *J. Polym. Sci., Part A: Polym. Chem.* **31**, 1993, 909–921.
3. S. A. Stern, Y. Mi, H. Yamamoto and A. K. St. Clair. Structure/permeability relationships of polyimide membranes. Applications to the separation of gas mixtures. *J. Polym. Sci., Part B: Polym. Phys.* **27**, 1989, 1887–1909.
4. K. Tanaka, H. Kita, K. Okamoto, A. Nakamura and Y. Kusuki. Gas permeability and permselectivity in polyimides based on 3,3′,4,4′-biphenyltetracarboxylic dianhydride. *J. Membr. Sci.* **47**, 1989, 203–215.
5. M. R. Coleman and W. J. Koros. Isomeric polyimides based on fluorinated dianhydrides and diamines for gas separation applications. *J. Membr. Sci.* **50**, 1990, 285–297.
6. K. Tanaka, H. Kita, M. Okano and K. Okamoto. Permeability and permselectivity of gases in fluorinated and non-fluorinated polyimides. *Polymer* **33**, 1992, 585–592.
7. K. Tanaka, M. Okano, H. Toshino, H. Kita and K. Okamoto. Effect of methyl substituents on permeability and permselectivity of gases in polyimides prepared from methyl-substituted phenylenediamines. *J. Polym. Sci., Part B: Polym. Phys.* **30**, 1992, 907–914.
8. S. A. Stern, Y. Liu and W. A. Feld. Structure/permeability relationships of polyimides with branched or extended diamine moieties. *J. Polym. Sci., Part B: Polym. Phys.* **31**, 1993, 939–947.
9. K. Matsumoto, P. Xu and T. Nishikimi. Gas permeation of aromatic polyimides. II. Influence of chemical structure. *J. Membr. Sci.* **81**, 1993, 23–30.
10. K. C. O'Brien, W. J. Koros and G. R. Husk. Polyimide materials based on pyromellitic dianhydride for the separation of carbon dioxide and methane gas mixtures. *J. Membr. Sci.* **35**, 1988, 217–230.

11. T. H. Kim, W. J. Koros, G. R. Husk and K. C. O'Brien. Relationship between gas separation properties and chemical structure in a series of aromatic polyimides. *J. Membr. Sci.* **37**, 1988, 45–62.

12. S. L. Liu, M. L. Chng, T. S. Chung, K. Goto, S. Tamai, K. P. Pramoda and Y. J. Tong. Gas-transport properties of indan-containing polyimides. *J. Polym. Sci. B* **42**, 2004, 2769–2779.

13. C. Staudt-Bickel and W. J. Koros. Improvement of CO2/CH4 separation characteristics of polyimides by chemical crosslinking. *J. Membr. Sci.* **155**, 1999, 145–154.

14. S. L. Liu, R. Wang, T. S. Chung, M. L. Chng, Y. Liu and R. H. Vora. Effect of diamine composition on the gas transport properties in 6FDA-durene/3,3′-diaminodiphenyl sulfone copolyimides. *J. Membr. Sci.* **202**, 2002, 165–176.

15. S. L. Liu, R. Wang, Y. Liu, M. L. Chng and T. S. Chung. The physical and gas permeation properties of 6FDA-durene/2,6-diaminotoluene copolyimides. *Polymer* **42**, 2001, 8847–8855.

16. A. Shimazu, T. Miyazaki, T. Matsushita, M. Maeda and K. Ikeda. Relationships between chemical structures and solubility, diffusivity, and permselectivity of 1,3-butadiene and *n*-butane in 6FDA-based polyimides. *J. Polym. Sci. B* **37**, 1999, 2941–2949.

17. A. Shimazu, T. Miyazaki, M. Maeda and K. Ikeda. Relationships between the chemical structures and the solubility, diffusivity, and permselectivity of propylene and propane in 6FDA-based polyimides. *J. Polym. Sci. B* **38**, 2000, 2525–2536.

18. W. H. Lin, R. H. Vora and T. S. Chung. Gas transport properties of 6FDA-durene/1,4-phenylenediamine (pPDA) copolyimides. *J. Polym. Sci., B: Polym. Phys.* **38**, 2000, 2703–2713.

19. Y.-H. Kim, H.-S. Kim, S.-K. Ahn, S. O. Jung and S.-K. Kwon. Synthesis of new highly organosoluble polyimides bearing a noncoplanar twisted biphenyl unit containing *t*-butyl phenyl group. *Bull. Korean Chem. Soc.* **23**, 2002, 933–934.

20. H.-S. Kim, Y.-H. Kim, S.-K. Ahn and S.-K. Kwon. Synthesis and characterization of highly soluble and oxygen permeable new polyimides bearing a noncoplanar twisted biphenyl unit containing *tert*-butylphenyl or trimethylsilyl phenyl groups. *Macromolecules* **36**, 2003, 2327–2396.

21. Y.-H. Kim, H.-S. Kim and S.-K. Kwon. Synthesis and characterisation of highly soluble and oxygen permeable new polyimides based on twisted biphenyl dianhydride and spirofluorene diamine. *Macromolecules* **38**, 2005, 7950–7956.

22. Y. H. Kim, S.-K. Ahn, H. Sun Kim and S.-K. Kwon. Synthesis and characterization of new organosoluble and gas-permeable polyimides from bulky substituted pyromellitic dianhydrides. *J. Polym. Sci., A: Polym. Chem.* **40**, 2002, 4288–4296.

23. Y.-H. Kim, S.-K. Ahn and S.-K. Kwon. Synthesis and characterization of novel polyimides containing bulky trimethylsilylphenyl group. *Bull. Korean Chem. Soc.* **22**, 2001, 451.

24. H. Sun and D. Rigby. Polysiloxanes: Ab initio force field and structural, conformational and thermophysical properties. *Spectrochim. Acta, Part A* **53**, 1997, 1301–1323.

25. D. Rigby, H. Sun and B. E. Eichinger. Computer simulations of poly(ethylene oxide): Force field, PVT diagram and cyclization behaviour. *Polym. Int.* **44**, 1997, 311–330.

26. D. Hofmann, L. Fritz, J. Ulbrich, C. Schepers and M. Böhning. Detailed atomistic molecular modeling of small molecule diffusion and solution processes in polymeric membrane materials. *Macromol. Theory Simul.* **9**, 2000, 293–327.

27. D. Hofmann, M. Entrialgo-Castaño, A. Lerbret, M. Heuchel and Y. Yampolskii. Molecular modeling investigations of free volume distributions in stiff chain polymers with conventional and ultrahigh free volume: Comparison between molecular modeling and positron lifetime studies. *Macromolecules* **36**, 2000, 8528–8538.

28. J. Bicerano. *Prediction of Polymer Properties*. New York: Marcel Dekker, 1993.

29. M. Heuchel, D. Hofmann and P. Pullumbi. Molecular modeling of small-molecules permeation in polyimides and its correlation of free-volume distributions. *Macromolecules* **37**, 2004, 201–214.

30. B. R. Wilks, W. J. Chung, P. J. Ludovice, M. R. Rezac, P. Meakin and A. Hill. Impact of average free-volume element size on transport in stereoisomers of polynorborne. I. properties at 35°C. *J. Polym. Sci, B: Polym. Phys.* **41**, 2003, 2185–2199.

12 Density Functional Theory Computational Study of Phosphine Ligand Dissociation versus Hemilability in a Grubbs-Type Precatalyst Containing a Bidentate Ligand during Alkene Metathesis

Margaritha Jordaan and H.C. Manie Vosloo

CONTENTS

12.1   INTRODUCTION

Research in coordination and organometallic chemistry, strongly supported in the last decade by theoretical studies,[1–13] has provided much insight into the mechanism of catalytic processes involving M–C or M–H bonds. Alkene metathesis is one example of a catalytic process involving M–C bonds that has been successfully applied in both academic and industrial environments with combined experimental and theoretical support.[8,10,14] This resulted in the calculation of certain mechanistic parameters for the alkene metathesis mechanism with ruthenium carbenes.[6,7,9,10,12,15–20] However, these studies focused on the catalytic cycle and ligand dissociation of the methylidene species $RuCl_2(PR_3)_2(=CH_2)$ with model PR_3 (R = H, Me) ligands and/or ethene as a model substrate to mainly reduce the computing cost. Therefore, the steric and electronic influence of the actual ligands [PCy_3 vs. PR_3, (R = H, Me)] and substrates (1-octene vs. ethene) with the benzylidene complex (vs. methylidene complex) as a precatalyst was not taken into consideration. Nonetheless, with the use of molecular modeling, a deeper insight was gained on the mechanism of the alkene metathesis reaction.

The increasing interest in more stable and catalytically active systems for metathesis reactions has encouraged various researchers to modify the well-defined, highly active, and robust ruthenium-based Grubbs systems, $RuCl_2(PCy_3)L(=CHPh)$ [L = PCy_3 (**I**) or H_2IMes (**II**)], with the incorporation of chelating ligands. (Note: H_2IMes is an *N*-heterocyclic carbene ligand commonly abbreviated as NHC.) Hemilabile ligands, a class of chelating ligands, have the ability to increase the thermal stability of the ruthenium carbene complexes by releasing a coordination site "on demand" of *inter alia* an alkene (such as norbornene) and occupying it otherwise— thus preventing decomposition via free coordination sites (Scheme 12.1).[21–25]

In a previous study,[25] we investigated the 1-octene metathesis reaction with a hemilabile first- (**III**) and second-generation (**IV**) Grubbs-type precatalyst with the use of ^1H nuclear magnetic resonance (NMR). The results suggested that two different mechanisms might be involved for **III** and **IV**, since five carbene species were observed during the 1-octene metathesis reaction with **IV**, whereas only three carbene species were visible in the presence of **III**.[25] However, it was observed that in the presence of **IV** at 50°C in $CDCl_3$, the heptylidene is the catalytically active species that preferentially forms during the 1-octene metathesis reaction, whereas both the heptylidene and methylidene species form simultaneously during the reaction with **III**. This necessitated a computational investigation of the alkene metathesis

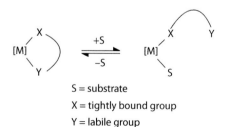

S = substrate
X = tightly bound group
Y = labile group

SCHEME 12.1   Schematic representation of the concept of hemilability.[21–25]

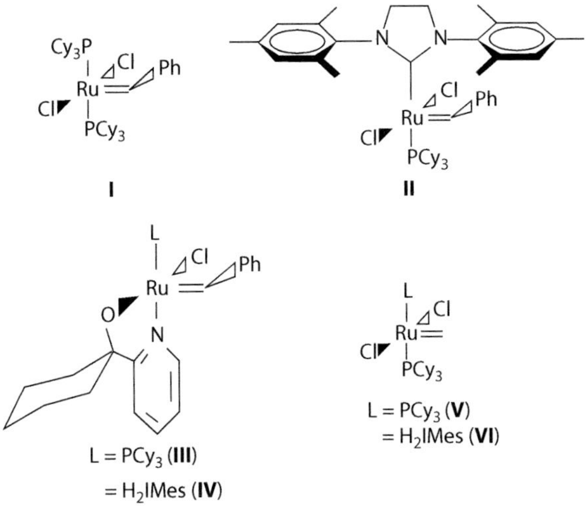

SCHEME 12.2   Grubbs-type precatalysts and catalytically active methylidene species.

reaction with a hemilabile precatalyst to gain more insight into the mechanism of this reaction. Therefore, in this study, we investigated the dissociative 1-octene metathesis mechanism with **III** and **IV** according to a conceptual model we proposed for the productive dissociative mechanism of the 1-octene metathesis reaction in the presence of **I**.[8] The computational results from this model were in agreement with the experimental results obtained with NMR and gas chromatography/mass selective detector experiments, in which it was indicated that the heptylidene species was the catalytically active species that preferentially formed in the 1-octene metathesis reaction with **I** (Scheme 12.2).[8]

12.2   EXPERIMENTAL

12.2.1   COMPUTATIONAL DETAILS

The quantum-chemical calculations were carried out by density functional theory (DFT) because it usually gives realistic geometries, relative energies, and vibrational frequencies for transition metal compounds. All calculations were performed with the DMol[3] DFT code[26–28] as implemented in Accelrys Materials Studio® 4.0[29] on either a 4-CPU (HP Proliant CP4000) or a 52-CPU cluster (HP Proliant CP4000 Linux Beowulf with Procurve Gb/E Interconnect on compute nodes). The non-local generalized gradient approximation (GGA) functional by Perdew and Wang (PW91)[30] was used for all geometry optimizations. The convergence criteria for these optimizations consisted of threshold values of 2×10^{-5} Ha, 0.004 Ha/Å, and 0.005 Å for energy, gradient, and displacement convergence, respectively, whereas a self-consistent field density convergence threshold value of 1×10^{-5} Ha was specified. DMol[3] uses a basis set of numeric atomic functions, which are exact solutions to the Kohn–Sham equations for the atom.[31] These basis sets are generally more complete than a comparable set of linearly independent Gaussian functions and have been demonstrated to have small basis set superposition errors.[31] In this study, a polarized split valence basis set, termed double numeric polarized (DNP) basis set, was used. All geometry optimizations used highly efficient delocalized internal coordinates.[32] The use of delocalized coordinates significantly reduces the number of geometry optimization iterations needed to optimize larger molecules compared to the use of traditional Cartesian coordinates. Some of the geometries optimized were also subjected to full frequency analyses at the same GGA/PW91/DNP level of theory to verify the nature of the stationary points. Equilibrium geometries were characterized by the absence of imaginary frequencies. Preliminary transition state (TS) geometries were obtained by the integrated linear synchronous transit/quadratic synchronous transit algorithm available in Materials Studio 4.0. This approach was used before in computational studies in homogeneous alkene trimerization and metathesis.[11,33] These preliminary structures were then subjected to full TS optimizations using an eigenvector following algorithm. For selected transition state geometries confirmation calculations, involving intrinsic reaction path (IRP), calculations were performed in which the path connecting reagents, TS, and products are mapped. The IRP technique used in Materials Studio 4.0 also corresponds to the intuitive minimum energy pathway connecting two structures and is

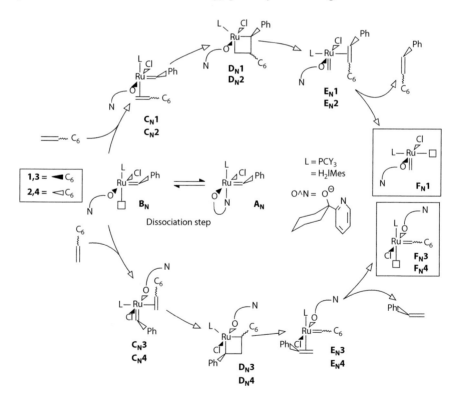

SCHEME 12.3 Dissociation (A_N to B_N) and activation (B_N to F_N) steps in the mechanism of productive 1-octene metathesis using $RuCl_2(PCy_3)L(O^\wedge N)(=CHPh)$ [L = PCy_3 or H_2IMes, $O^\wedge N$ = 1-(2′-pyridinyl)cyclohexan-1-olate].

based on the nudged elastic band algorithm of Henkelman and Jonsson.[34] The IRP calculations, performed at the same GGA/PW91/DNP level of theory, ensured the direct connection of transition states with the respective reactant and product geometries. All transition structure geometries exhibited only one imaginary frequency in the reaction coordinate. All results were mass balanced for the isolated system in the gas phase. The energy values that are given in the results are the electronic energies at 0 K, and, therefore, only the electronic effects are considered in this paper. For the purpose of this study, the minimum structure obtained from the geometry optimization (GGA/PW91/DNP) of 1-octene was used in further calculations. In this structure, the hexyl chain was "straight" and remained so in most of the optimizations of intermediates. The influence of the various confirmations of 1-octene on the energetics of the reaction pathway was not considered in this study.

12.2.2   MODEL SYSTEM AND NOTATION

Conceptually, the productive metatheses of 1-octene in the presence of the hemilabile Grubbs carbene complexes are illustrated in Schemes 12.3 and 12.4. The mechanism

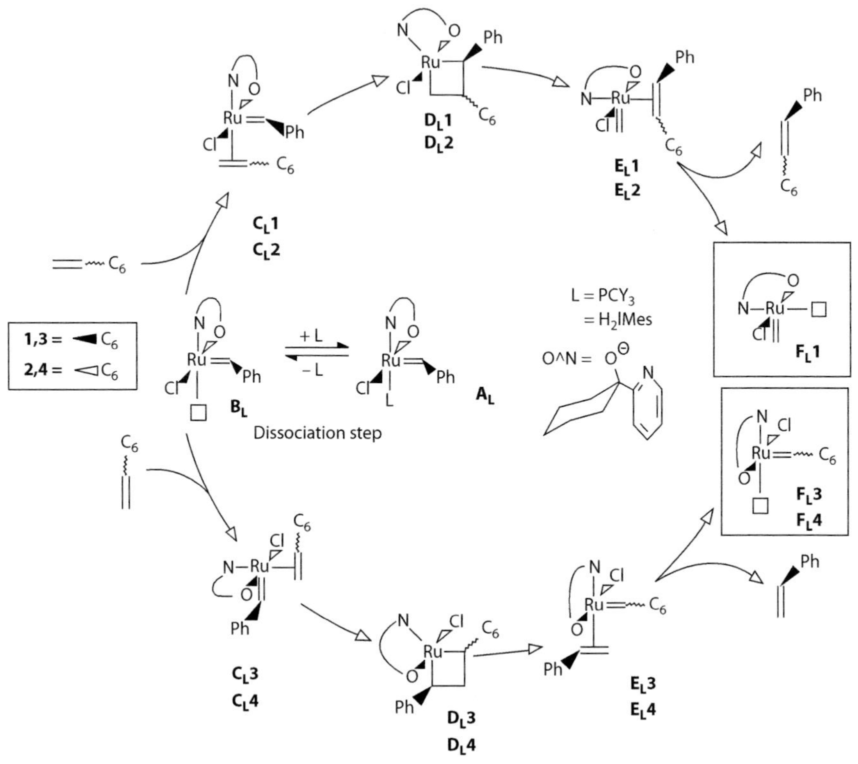

SCHEME 12.4   Dissociation (A_L to B_L) and activation (B_L to F_L) steps in the mechanism of productive 1-octene metathesis using RuCl$_2$(L)(O^N)(=CHPh) [L = PCy$_3$ or H$_2$IMes, O^N = 1-(2′-pyridinyl)cyclohexan-1-olate].

is initiated by either dissociation of the labile N-atom of the O,N-chelating ligand (Scheme 12.3) or the dissociation of the phosphine or NHC ligand from the respective first- and second-generation hemilabile carbenes (Scheme 12.4).

The generic labels **A–F** are given to the individual hemilabile ruthenium carbene and derived species involved in the conceptualized reaction mechanism. The mechanism consists of either the dissociation of the labile N-atom of the O,N-chelating ligand from **III** or **IV** (A_N) to yield $RuCl(L)(O^\wedge N\text{-open})(=CHPh)$ [L = PCy$_3$ (**III**) or H$_2$IMes (**IV**)] (**B$_N$**) or the loss of L (L = PCy$_3$ or H$_2$IMes) to yield RuCl(O^N) (=CHPh) (**B$_L$**). The different stereochemical approaches of 1-octene toward the catalytically active species **B** lead to four activation steps (**1** to **4** notations). To identify which step is under consideration, a numerical suffix (**1** to **4**) is associated with the labels **C** to **F** (e.g., **C$_N$1** to **F$_N$1** represent activation step 1 when the labile N-atom of the O,N-chelating ligand is dissociated) and the additional subscript **N** or **L** indicates the dissociated ligand under consideration, that is, **N** for the dissociation of the labile N-atom of the O,N-chelating ligand and **L** for the dissociation of the ligand L (L = PCy$_3$ for **III** and L = NHC for **IV**). Transition states are denoted analogously; for example, **C$_L$3–D$_L$3** is the transition state for the conversion of **C$_L$3** to **D$_L$3** when L is dissociated from **A$_L$**.

12.3   RESULTS AND DISCUSSION

12.3.1   Catalyst Initiation

It is generally accepted that the Ru-catalyzed alkene metathesis reaction proceeds via a dissociative mechanism, which is initiated by the dissociation of a phosphine ligand from RuX$_2$(PR$_3$)L(=CHR) to form a 14-electron species ("**B**").[6,15,35] In this sense, catalyst initiation involves the dissociation of PCy$_3$ for both **I** and **II**. However,

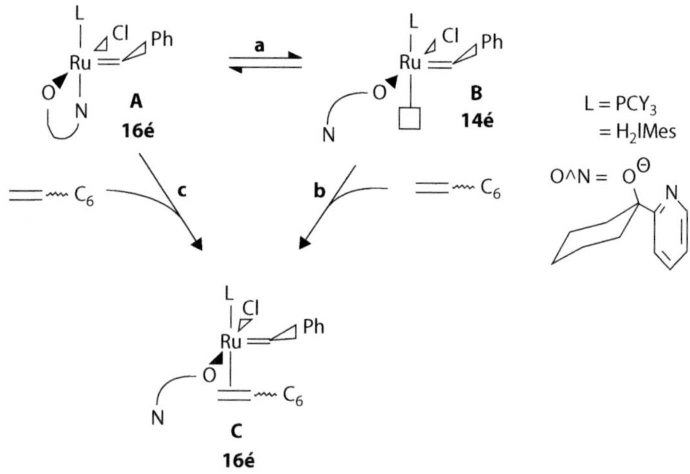

SCHEME 12.5   Illustration of the dissociative (**a–b**) and associative (**c**) catalyst initiation steps for **III** and **IV**.

catalyst initiation of hemilabile complexes is more complex because either a dissociative or associative mechanism may play a role during the initiation and activation
phases. This is due to the belief that a hemilabile ligand releases a free coordination
site "on demand" of competing substrates and occupies it otherwise.[24] As illustrated
in Scheme 12.5, the ruthenium center of the hemilabile complexes might become
coordinatively unsaturated with or without the influence of the incoming alkene.

In this study, two possible precatalyst initiations of only the dissociative mechanism were theoretically investigated, that is, (1) the dissociation of the labile N-atom
of the O^N-ligand (Schemes 12.6 and 12.7) and (2) the dissociation of ligand L
(with L = PCy$_3$ or H$_2$IMes) (Scheme 12.7). This was a result of the free PCy$_3$ ligand
being observed experimentally with gas chromatography/flame ionization detector throughout the 1-octene metathesis investigations in the presence of the first-
generation hemilabile complexes.[8]

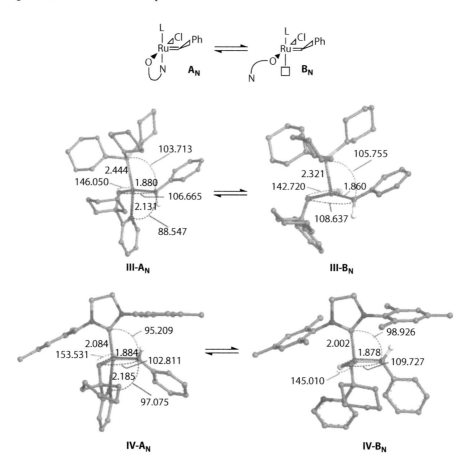

SCHEME 12.6 **(See color insert.)** Dissociation (A$_N$ to B$_N$) of labile N-atom of O,N-ligand
along with the optimized geometries for the lowest energy catalyst precursor complexes.
The hydrogen atoms on the ligands are omitted for clarity and the unit of the indicated bond
distances is angstrom (Å).

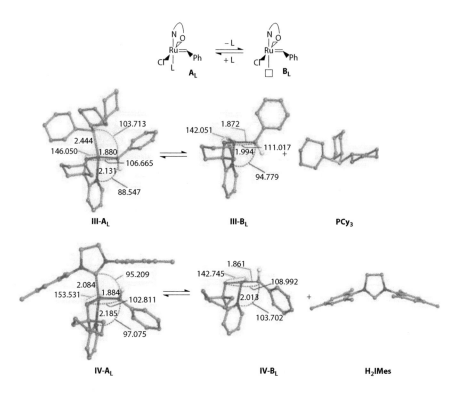

SCHEME 12.7   (See color insert.) Dissociation (A_L to B_L) of ligand L along with the optimized geometries for the lowest energy catalyst precursor complexes. The hydrogen atoms on the ligands are omitted for clarity and the unit of the indicated bond distances is angstrom (Å).

Although various orientations of the alkylidene moiety in "**B**" are possible (Figure 12.1), this was not explored in detail, since it has been shown that the **B1** orientation of the alkylidene is lower in energy for the Grubbs first- and second-generation methylidene species.[10] Therefore, complexes corresponding to "**B1**" were used as starting structures prior to optimizations. Spontaneous formation of the perpendicular alkylidene orientation ("**B2**") resulted during optimization of the unsaturated **I**, **III**, and **IV** benzylidene complexes while remaining parallel for the second-generation Grubbs system (**II–B1**). The Ph-ring of the alkylidene rotated upwards toward the PCy_3 for the first-generation systems, whereas the steric bulk on NHC caused the Ph-ring to rotate downwards into the open coordination site for **IV**.

Table 12.1 summarizes the calculated electronic (ΔE) energies in kcal/mol for the initiation phase ("**A**"–"**B**") of the 1-octene metathesis reaction in the presence of **I**, **II**, **III**, and **IV** according to the dissociation of the N-atom or L.

Dissociation of PCy_3 from the Grubbs first-generation 16-electron precatalyst complex (**I–A**) proceeds with $\Delta E = 21.89$ and 25.92 kcal/mol for the second-generation precatalyst (**II–A**). This is in agreement with the experimental kinetic studies by Grubbs[15,35] in which $\Delta H^{\ddagger}$ values of 23.6 ± 0.5 kcal/mol and 27 ± 2 kcal/mol were obtained for **I** and **II**, respectively.

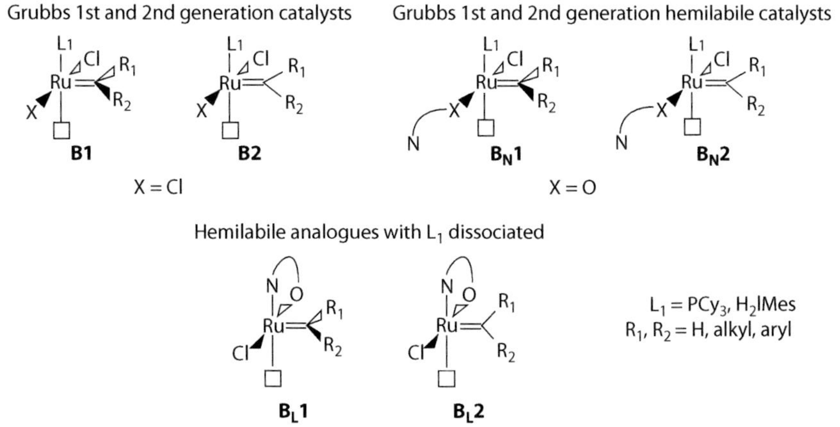

Grubbs 1st and 2nd generation catalysts   Grubbs 1st and 2nd generation hemilabile catalysts

X = Cl   X = O

Hemilabile analogues with L_1 dissociated

L_1 = PCy$_3$, H$_2$IMes
R_1, R_2 = H, alkyl, aryl

FIGURE 12.1 Illustration of the possible orientations of the alkylidene moiety for the Ru–carbene unsaturated complexes.

The dissociation of PCy_3 in the first-generation hemilabile analogue **III–A** (**III–B$_L$** ΔE = 23.26 kcal/mol) is calculated to be 3 kcal/mol less favorable than the dissociation of the labile N-atom of the O^N-bidentate ligand (**III–B$_N$**). This implies that **III** might rather initiate via the dissociation of the labile N-atom, which, as compared to **I**, is about 1.5 kcal/mol more favorable. Although **III–B$_L$** is about 3 kcal/mol less favorable than **III–B$_N$**, it is only 1.5 kcal/mol less favorable than **I** and should, therefore, not be completely discarded. Only after investigating the complete mechanism, that is, initiation, activation, and catalytic cycle, can concluding remarks be made with regard to which dissociation step **B$_N$** or **B$_L$** is favored. The dissociation of H_2IMes in the second-generation hemilabile analogue **IV–A** (**IV–B$_L$**, ΔE = 61.52 kcal/mol) is found to be more unfavorable (45 kcal/mol) than the dissociation of the labile N-atom (**IV–B$_N$**). This suggests that **IV** also initiates via the dissociation of the

TABLE 12.1
Calculated Electronic (ΔE) Energies (in kcal/mol) for the Initiation Phase Sequence "A"–"B" as Catalyzed by I, II, III, and IV

Mechanistic Sequence No.	I	II	III	IV
A	0.0	0.0	0.0	0.0
B$_L$	21.89[a]	25.92[a]	23.26[a]	61.52[b]
B$_N$	–	–	20.26	16.08

Note: All energies are reported relative to the respective precatalysts "A" and balanced with the energies of free ligand where necessary.

[a] Dissociation of L = PCy$_3$.

[b] Dissociation of L = H$_2$IMes; no PCy$_3$ ligands are present in this complex.

labile N-atom (ΔE = 16.08 kcal/mol), which as compared to **II**, is about 10 kcal/mol more favorable. However, the possibility of another mechanistic route dominating *inter alia* an associative coordination of the alkene prior to dissociation of the labile N-atom cannot be excluded entirely.

12.3.2 Activation Phase

After catalyst initiation, an alkene coordinates to the unsaturated intermediate species "**B**" to form the corresponding π-complexes "**C**." The alkene can coordinate in two discrete, perpendicular orientations *trans* to L for the various Ru=C systems (Scheme 12.8). In a recent study,[10] it was shown that the coordination of the alkene in the C_{II} mode is energetically more favored for the first- and second-generation Grubbs methylidene complexes (**V** and **VI**). However, both the C_p–C_{II}–**D** and C_{II}–**D** conversions were investigated for the hemilabile analogues to determine if the bidentate ligand will influence the stability of the various intermediates. The optimized geometries of the lowest energy equilibrium and transition state structures involved in the metathesis reaction with **III** and **IV** for activation step 4 in the C_{II} mode are given in Schemes 12.9 and 12.10 [the labile N-atom is dissociated, L = PCy_3 (Scheme 12.9) and H_2IMes (Scheme 12.10)] and Scheme 12.11 (L = PCy_3 or H_2IMes is dissociated). For the sake of simplicity, all structures in the electronic profile figures (Figures 12.2 through 12.8) are drawn in the C_{II} mode. However, in the text, the C_p–C_{II}–**D** conversions will be referred to as the C_p mode, whereas the C_{II} mode will refer to the C_{II}–**D** conversions.

12.3.2.1 When the Labile N-Atom is Dissociated from III and IV

Figures 12.2 and 12.4 illustrate the activation steps for the C_p mode of **III** and **IV**, whereas Figures 12.3 and 12.5 illustrate the steps for the C_{II} mode according to Scheme 12.1. The relative energies for the optimized structures are summarized in Table 12.2. Activation step 3 is not shown for the hemilabile systems because of its

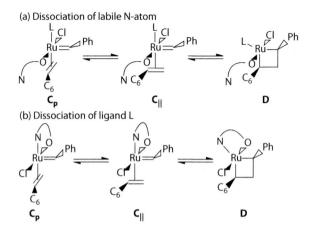

SCHEME 12.8 *Trans* alkene coordination in the dissociative pathway for the Grubbs hemilabile precatalysts (L = PCy_3 or H_2IMes).

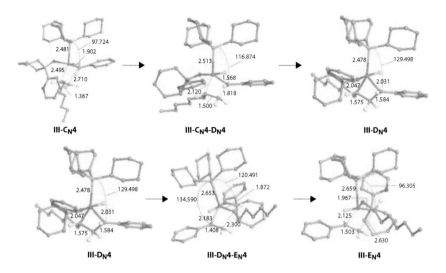

SCHEME 12.9   (See color insert.) Optimized geometries of the lowest energy equilibrium and transition state structures involved in the metathesis reaction with **III** for activation step 4 (C_N to E_N) in the C_{II} mode when the labile N-atom is dissociated. The hydrogen atoms on the ligands are omitted for clarity and the unit of the indicated bond distances is angstrom (Å).

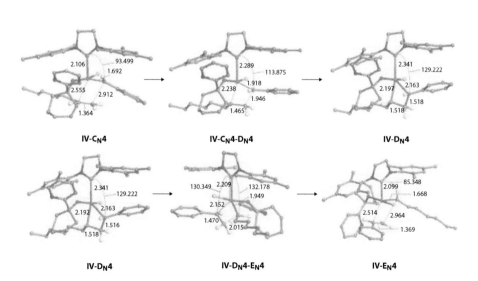

SCHEME 12.10   (See color insert.) Optimized geometries of the lowest energy equilibrium and transition state structures involved in the metathesis reaction with **IV** for activation step 4 (C_N to E_N) in the C_{II} mode when the labile N-atom is dissociated. The hydrogen atoms on the ligands are omitted for clarity and the unit of the indicated bond distances is angstrom (Å).

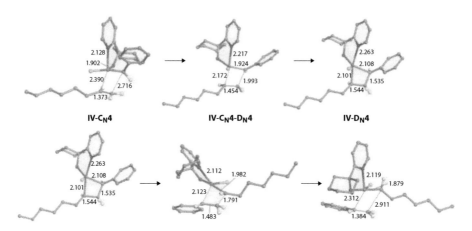

SCHEME 12.11 **(See color insert.)** Optimized geometries of the lowest energy equilibrium and transition state structures involved in the metathesis reaction with **IV** for activation step 4 (C_L to E_L) in the C_{II} mode when L is dissociated. The hydrogen atoms on the ligands are omitted for clarity and the unit of the indicated bond distances is angstrom (Å).

similarity to activation step 4. No energy barriers were calculated for the dissociation of the labile N-atom or the association of the alkene because it was assumed that these steps proceed without considerable rearrangement of the complex. This should be investigated further to take the rotation of the hemilabile ligand around the Ru–O–bond into consideration.

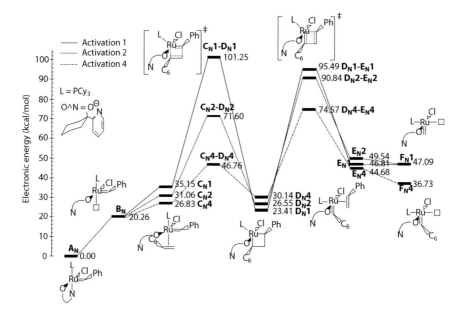

FIGURE 12.2 Electronic energy profiles of the activation steps in the productive 1-octene metathesis using **III** (only the C_N4 to F_N4 structures are shown) (C_p mode).

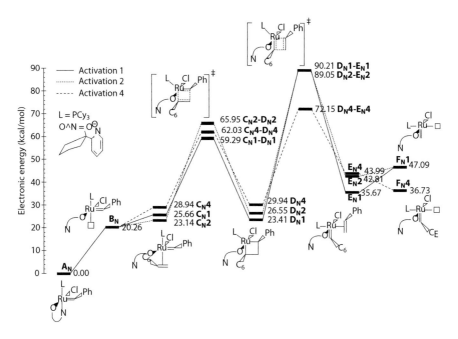

FIGURE 12.3 Electronic energy profiles of the activation steps in the productive 1-octene metathesis using **III** (only the C_N4 to F_N4 structures are shown) (C_{II} mode).

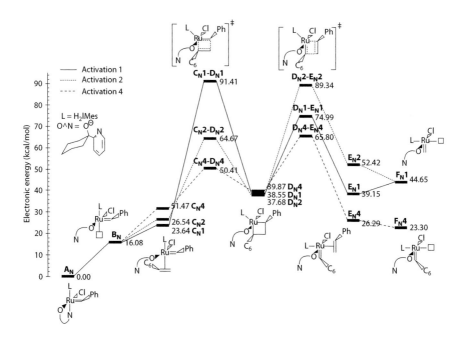

FIGURE 12.4 Electronic energy profiles of the activation steps in the productive 1-octene metathesis using **IV** (only the C_N4 to F_N4 structures are shown) (C_p mode).

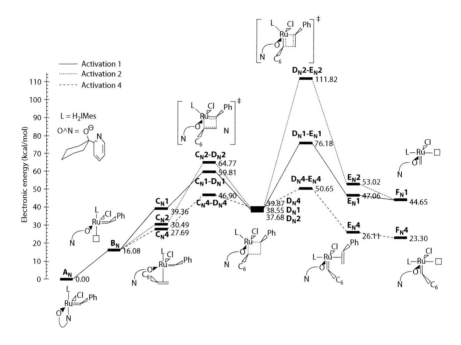

FIGURE 12.5 Electronic energy profiles of the activation steps in the productive 1-octene metathesis using **IV** (only the C_N4 to F_N4 structures are shown) (C_{II} mode).

Activation step 4 (B_N to C_N4 to F_N4) is thermodynamically and kinetically more favorable than activation steps 1 (B_N to C_N1 to F_N1) and 2 (B_N to C_N2 to F_N1; note $F_N2 = F_N1$) for both systems in the C_p and C_{II} modes (Figures 12.2 through 12.5). However, the modeling results for **IV** cannot explain the observed ^{1}H NMR results, in which the carbene signal of F_N1 from **IV** is observed to keep growing without being depleted.[25] This indicates that the complete mechanism, with the inclusion of the catalytic cycle, should be investigated before making any concluding remarks. It has, however, been demonstrated by Halpern[36] that the most abundant complex in a solution need not be a part of the most favorable catalytic cycle; it may represent, in fact, a dead end. Therefore, the absence of F_N1 in the ^{1}H NMR study of the 1-octene metathesis reaction with **III**[25] might suggest that a different mechanistic pathway is followed.

The calculated reaction energy for the formation of the ruthenacyclobutane intermediate (**III**–D_N4) from the π-complex **III**–C_N4 in the presence of **III** is endothermic (3.31 kcal/mol) for the C_p mode with activation energy of 31 kcal/mol (Figure 12.2). In contrast, the formation of "D_N" from the π-complex "**C**" is exothermic for both activation steps 1 (–11.95 kcal/mol) and 2 (–4.51 kcal/mol) with activation energies of 66 kcal/mol and 45 kcal/mol, respectively. The decomposition of the ruthenacyclobutane ("D_N" to "E_N") is endothermic for both the heptylidene (15 kcal/mol) and methylidene (*ca.* 23 kcal/mol) formation steps with activation energies of about 45 kcal/mol and 70 kcal/mol, respectively. The coordination of the 1-octene in the C_{II} mode has an about 2 kcal/mol–5 kcal/mol decrease in $\Delta E_o^{\ddagger}$ for "D_N" to "E_N"

TABLE 12.2

Calculated Electronic (ΔE) Energies and Selected Activation ($\Delta E_0^{\ddagger}$) Energies (in kcal/mol) for the Metathesis Mechanistic Sequence A_N–F_N as Catalyzed by III and IV

Mechanistic Sequence No.	III				IV			
	$\Delta E\,C_p$	$\Delta E_0^{\ddagger}\,C_p^{a}$	$\Delta E\,C_{II}$	$\Delta E_0^{\ddagger}\,C_{II}^{a}$	$\Delta E\,C_p$	$\Delta E_0^{\ddagger}\,C_p^{a}$	$\Delta E\,C_{II}$	$\Delta E_0^{\ddagger}\,C_{II}^{a}$
A_N	0.00		0.00		0.00		0.00	
B_N	20.26		20.26		16.08		16.08	
C_N1	35.16		26.66		23.64		39.36	
C_N2	31.06		23.14		26.54		30.49	
C_N4	26.83		28.94		31.47		27.69	
C_N1–D_N1	101.21	66.05	59.29	33.63	91.41	67.77	59.81	20.45
C_N2–D_N2	71.60	40.54	65.95	42.81	64.67	38.13	64.77	34.28
C_N4–D_N4	46.76	19.93	62.03	33.09	50.41	18.94	46.90	19.21
D_N1	23.41		23.41		38.55		38.55	
D_N2	26.55		26.55		37.68		37.68	
D_N4	30.14		29.94		39.87		39.87	
D_N1–E_N1	95.49	72.08	90.21	66.80	74.99	36.44	76.18	39.63
D_N2–E_N2	90.84	64.29	89.05	62.50	89.34	51.66	111.82	74.14
D_N4–E_N4	74.57	44.43	72.15	42.21	65.80	25.93	50.65	10.78
E_N1	46.81		35.67		39.15		47.06	
E_N2	49.54		42.81		52.42		53.02	
E_N4	44.68		43.99		26.29		26.11	
F_N1	47.09		47.09		44.65		44.65	
F_N4	36.73		36.73		23.30		23.30	

Note: All energies are reported relative to the respective precatalysts "A" and balanced with the energies of free ligand where necessary.

[a] $\Delta E_0^{\ddagger}$ calculated for the various transition states, for example, $\Delta E_0^{\ddagger}(C_N1 - D_N1) = \Delta E(D_N1) - \Delta E(C_N1)$.

for all three activation steps, as well as for "C_N" to "D_N" in step 2 (Figure 12.3). Additionally, a 30 kcal/mol decrease in $\Delta E_0^{\ddagger}$ is observed for the formation of the ruthenacyclobutane in step 1, whereas step 4 shows a 2 kcal/mol increase. This indicates that the alkene prefers to coordinate to **III** in the C_{II} mode, in which the steric bulk around the Ru-center is relieved in steps 1 and 2. This is because the alkyl chain of 1-octene in steps 1 and 2 is no longer parallel to the Cl–Ru–O line, in which the alkyl chain was directly below the O,N-ligand and thereby adding to the steric bulk around the Ru-center. The rate-limiting step for the formation of the heptylidene, as well as the methylidene species in the presence of **III**, is the decomposition of the ruthenacyclobutane ("D_N" to "E_N") for both coordination modes.

For **IV**, the decomposition of the ruthenacyclobutane ("D_N" to "E_N") in step 4 for both coordination modes (Figures 12.4 and 12.5), as well as the formation of

"D_N" from "C_N" in step 1 for the C_{II} mode (Figure 12.5), is exothermic, that is, approximately −13 kcal/mol and −1 kcal/mol, respectively. All the other formation or decomposition steps are endothermic for both the C_p and C_{II} modes. In contrast to **III**, different rate-limiting steps are involved for the formation of the heptylidene species from **IV** in the C_p and C_{II} modes, that is, "D_N" to "E_N" for C_p (Figure 12.4) and "C_N" to "D_N" for C_{II} (Figure 12.5) with activation energies of 25.93 and 19.21 kcal/mol, respectively. The decomposition of the ruthenacyclobutane ("D_N" to "E_N") is considered the rate-limiting step for the formation of the methylidene species in steps 1 ($\Delta E^{\ddagger}_{\text{o}D1-E1} = 37.63$ kcal/mol) and 2 ($\Delta E^{\ddagger}_{\text{o}D2-E2} = 74.14$ kcal/mol) in the C_{II} mode (Figure 12.5) as well as step 2 ($\Delta E^{\ddagger}_{\text{o}D2-E2} = 51.66$ kcal/mol) in the C_p mode (Figure 12.4). The rate-limiting step for the formation of the methylidene species in step 1 in the C_p mode is the formation of the ruthenacyclobutane with an activation energy of 67.77 kcal/mol. No conclusions can be made regarding the preferred coordination mode of the alkene to **IV** because mixed increasing and decreasing effects are observed with regard to the TSs. This might be due to the fact that the preliminary frequency analysis on various TSs indicated that only D_N1-E_N1 (263i cm^{-1}) and C_N4-D_N4 (120i cm^{-1}) in the C_p mode (Figure 12.4) are close to a transition state. The overall energy change from "B_N" to "F_N" for both the C_p and C_{II} modes for heptylidene formation in the presence of **III** and **IV** is 16.47 kcal/mol and 7.22 kcal/mol, respectively. For methylidene formation, it is between 27 kcal/mol and 29 kcal/mol, respectively. The activation phase during the 1-octene metathesis reaction in the presence of **III** and **IV** is, therefore, strongly endothermic compared with **I** and **II**,[8] which indicates that the hemilabile Ru–carbene complexes are relatively more stable compared with the Grubbs carbenes. Only after full frequency calculations have been performed to obtain the Gibbs free energies will kinetic or thermodynamic stability of these complexes be confirmed.

Comparison of the activation steps of **III** and **IV** for the formation of the heptylidene species in the C_{II} mode (which is more favorable than the C_p mode) (see Figure 12.6) suggests that **IV** is more active than **III**. Due to the endothermic nature of the sequence "A_N" → "B_N" → "C_N" → "C_N-D_N" (Figure 12.7) for both systems, the total barrier for ruthenacyclobutane ("D_N") formation may be correlated to the energy change of "A_N" → "C_N-D_N."[12] The largest barrier is calculated for **III** (62.03 kcal/mol), with **IV** (46.90 kcal/mol) exhibiting a much lower barrier. Therefore, relative reaction rates for these catalysts in decreasing order are suggested: **IV** > **III**, which is in qualitative agreement with the relative turnover numbers of these systems.[25]

12.3.2.2   Dissociation of Ligand L from III and IV

The possibility of phosphine or NHC ligand dissociation from the respective first- and second-generation hemilabile carbenes should not be excluded, because it is generally assumed that ruthenium-catalyzed metathesis reactions proceed through 14-electron intermediates.[6,15,16,35] Consequently, we have postulated a mechanism for the first- and second-generation hemilabile-catalyzed 1-octene metathesis reaction (Scheme 12.3) whereby the O,N-ligand remains attached to the Ru-center.

Experimental[15,35] and theoretical evidence[4–7,37] has shown that the phosphine, and not the NHC carbine, dissociates from **II**, indicating that the intermediates for catalysis by first- and second-generation catalysts are different. Due to the higher binding

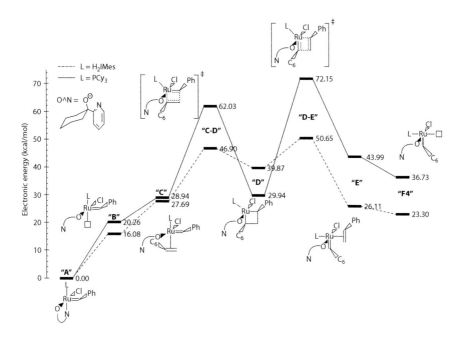

FIGURE 12.6 Comparison of the electronic energy profiles of the activation steps of **III** and **IV** in the productive 1-octene metathesis (C_{II} mode).

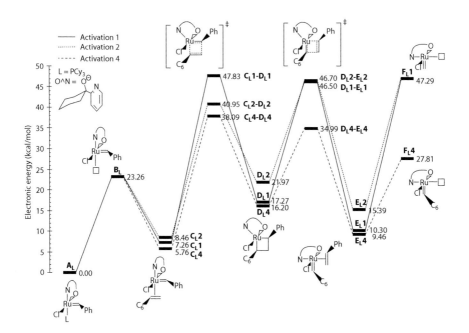

FIGURE 12.7 Electronic energy profiles of the activation steps in the productive 1-octene metathesis using **III** (only the C_L4 to F_L4 structures are shown) (C_p mode).

energy of NHC ligands in comparison to phosphine ligands, the dissociation of the NHC ligand during the initiation step of the mechanistic cycle will result in a higher dissociation energy.[15,37] This is most probably due to the increased σ-donor capability and the reduced π-acidity of the NHC ligand[38–40] in comparison to PR_3 ligands, which increase the stability of the second-generation catalysts. When applied to the hemilabile second-generation systems, it is clearly shown that the dissociation of the NHC ligand, which is 45 kcal/mol higher than the dissociation of the labile N-atom, will be improbable. Therefore, the activation phase resulting from initiation via dissociation of H_2IMes from **4** will not be considered further.

The activation steps for the C_p mode of **III** are graphically represented in Figure 12.7, whereas Figure 12.8 illustrates the steps for the C_{II} mode according to Scheme 12.3. The relative energies for the optimized structures are summarized in Table 12.3. Activation step 3 is not shown due to its similarity to activation step 4. No energy barriers were calculated for the dissociation of ligand L (L = PCy_3 or H_2IMes) or the association of the alkene because it proceeds without considerable rearrangement of the complex.

Although the dissociation of PCy_3 from **III** is 3 kcal/mol higher compared with the dissociation of the labile N-atom of the O,N-ligand, more stable intermediates form during the activation phase for both the C_p and C_{II} modes (Figures 12.7 and 12.8). A decrease of approximately 15 kcal/mol–30 kcal/mol is observed in the energies of the alkene-coordinated π-complexes, together with a 5 kcal/mol–15 kcal/mol

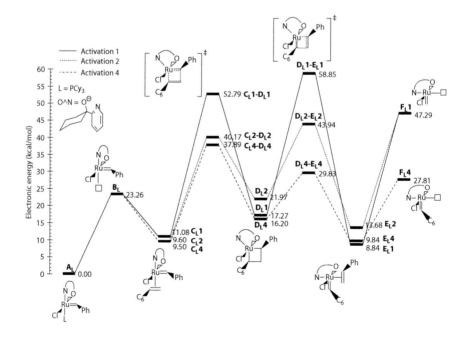

FIGURE 12.8 Electronic energy profiles of the activation steps in the productive 1-octene metathesis using **III** (only the C_L4 to F_L4 structures are shown) (C_{II} mode).

TABLE 12.3

Calculated Electronic (ΔE) Energies (in kcal/mol) for the Metathesis Mechanistic Sequence A_L–F_L as Catalyzed by III

	III			
	$\Delta E\ C_p$	$\Delta E_o^{\ddagger}\ C_p{}^a$	$\Delta E\ C_{II}$	$\Delta E_o^{\ddagger}\ C_{II}{}^a$
A_L	0.00		0.00	
B_L	23.26		23.26	
C_L1	7.26		11.08	
C_L2	8.46		9.60	
C_L4	5.76		9.50	
C_L1–D_L1	47.83	40.57	52.79	41.71
C_L2–D_L2	40.95	32.49	40.17	30.57
C_L4–D_L4	38.09	32.33	37.89	28.39
D_L1	17.27		17.27	
D_L2	21.97		21.97	
D_L4	16.20		16.20	
D_L1–E_L1	46.50	29.23	58.85	41.58
D_L2–E_L2	46.70	24.73	43.94	21.97
D_L4–E_L4	34.99	18.79	29.83	13.63
E_L1	10.30		8.84	
E_L2	15.39		13.68	
E_L4	9.48		9.48	
F_L1	47.29		47.29	
F_L4	27.81		27.81	

Note: All energies are reported relative to the respective precatalysts "A" and balanced with the energies of free ligand where necessary.

^{a} $\Delta E_o^{\ddagger}$ calculated for the various transition states, for example, $\Delta E_o^{\ddagger}(C_L1 - D_L1) = \Delta E(D_L1) - \Delta E(C_L1)$.

decrease in the metallacyclobutane "D" species. A decrease was also noted in the transition states, which varied between 5 kcal/mol and 45 kcal/mol. Additionally, the overall energy changes from "B_L" to "F_L" for both the C_p and C_{II} modes decrease with 12 kcal/mol. This indicates that **III** most likely initiates according to Scheme 12.3 rather than Scheme 12.1, which explains why only three carbene species were observed during the ^{1}H NMR investigation.[25]

Activation step 4 (B_L to C_L4 to F_L4) is thermodynamically and kinetically more favorable than activation steps 1 (B_L to C_L1 to F_L1) and 2 (B_L to C_L2 to F_L1; note $F_L2 = F_L1$) for **III** in the C_p and C_{II} modes. The calculated reaction energy for the formation of the ruthenacyclobutane intermediate ("D_L") from the π-complex "C_L" in the presence of **III** is endothermic for the C_p (about 11 kcal/mol) and C_{II} (about 9 kcal/mol) modes for all the activation steps. In contrast, the formation of "D_L" from the π-complex "C_L" is exothermic for all the activation steps in the C_p

(about -7 kcal/mol) and $\mathbf{C_{II}}$ (about -8 kcal/mol) modes. The coordination of the 1-octene in the $\mathbf{C_{II}}$ mode shows only a slight decrease in the activation energy of the ruthenacyclobutane formation and decomposition steps for activation steps 2 (1.92 kcal/mol decrease) and 4 (3.94 kcal/mol decrease), with a 1 kcal/mol increase for step 1. In contrast, a 12.59 kcal/mol increase was observed in $\Delta E_o^{\ddagger}$ for "$\mathbf{D_L}$" to "$\mathbf{E_L}$" for step 1. This indicates that the alkene can coordinate to $\mathbf{III}$ in either mode. The rate-limiting step for the formation of the heptylidene ($\Delta E_o^{\ddagger} \approx 30$ kcal/mol) as well as the methylidene ($\Delta E_o^{\ddagger} \approx 30$ kcal/mol to 42 kcal/mol) species in the presence of $\mathbf{III}$, from which PCy_3 has dissociated, is the formation of the ruthenacyclobutane ("$\mathbf{C_L}$" to "$\mathbf{D_L}$") for both coordination modes.

12.4   CONCLUSIONS

The initiation and activation steps of $\mathbf{III}$ and $\mathbf{IV}$, with 1-octene as the substrate, were investigated to determine the catalytically active species that preferentially forms from the benzylidene precatalyst. Although it is generally accepted that the Grubbs carbenes initiate with the dissociation of PCy_3, this is not necessarily the case for the hemilabile complexes because only one or no PCy_3 is present in the system. Therefore, these systems initiate either through the dissociation of the labile N-atom of the pyridine ring or through the dissociation of L. Because hemilabile ligands are known to release a free coordination site "on demand" of competing substrates such as an alkene and occupy it otherwise, the dissociation of the N-atom was considered as a viable route of initiation. Modeling results indicated that the dissociation of the labile N-atom of the pyridine ring for both hemilabile systems was more favorable than the dissociation of L. However, for $\mathbf{III}$, this dissociation is only 3 kcal/mol more favorable than the dissociation of L compared with 46 kcal/mol for $\mathbf{IV}$. This indicates that the possibility of PCy_3 dissociation from $\mathbf{III}$ should not be excluded.

Modeling results of the activation steps for $\mathbf{III}$ and $\mathbf{IV}$ indicated that the formation of the methylidene was thermodynamically favored, whereas the kinetically favored species was predicted to be the heptylidene. The computational results are in agreement with the experimental results obtained with NMR for the 1-octene metathesis reaction with $\mathbf{IV}$, in which the hemilability of the bidentate ligand is preferred above the dissociation of L.[26-28] However, further investigation into the metathesis mechanism with $\mathbf{III}$ is needed to determine whether a combination between phosphine ligand dissociation and hemilability of the bidentate ligand exists since experimental and computational results do not agree. Although good correlations with experimental data were obtained in this study, more realistic correlations should be obtained with the completion of all the frequency analyses and if solvation effects are taken into account.

ACKNOWLEDGMENTS

We would like to express our gratitude toward Sasol Technology R&D and the Technology and Human Resources for Industry Programme (Grant No. 2789) of the National Research Foundation of South Africa for their financial contribution toward the studies of M.J.

REFERENCES

1. H. Kawamura-Kuribayashi, N. Koga and K. Morokuma. An *ab initio* MO study on ethylene and propylene insertion into the titanium–methyl bond in CH_3TiCl^{2+} as a model of homogeneous olefin polymerization. *J. Am. Chem. Soc.* **114**, 1992, 2359–2366.

2. T. Ziegler, E. Folga and A. Berces. A density functional study on the activation of hydrogen–hydrogen and hydrogen–carbon bonds by $Cp_2Sc–H$ and $Cp_2Sc–CH_3$. *J. Am. Chem. Soc.* **115**, 1993, 636–646.

3. G. Sini, S. A. Macgregor, O. Eisenstein and J. H. Teuben. Why is beta-Me elimination only observed in d0 early-transition-metal complexes? An organometallic hyperconjugation effect with consequences for the termination step in Ziegler–Natta catalysis. *Organometallics* **13**, 1994, 1049–1051.

4. C. Adlhart and P. Chen. Comparing intrinsic reactivities of the first- and second-generation ruthenium metathesis catalysts in the gas phase. *Helv. Chim. Acta* **86**, 2003, 941–949.

5. C. Adlhart and P. Chen. Ligand rotation distinguishes first- and second-generation ruthenium metathesis catalysts. *Angew. Chem., Int. Ed.* **41**, 2002, 4484–4487.

6. C. Adlhart and P. Chen. Mechanism and activity of ruthenium olefin metathesis catalysts: The role of ligands and substrates from a theoretical perspective. *J. Am. Chem. Soc.* **126**, 2004, 3496–3510.

7. L. Cavallo. Mechanism of ruthenium-catalyzed olefin metathesis reactions from a theoretical perspective. *J. Am. Chem. Soc.* **124**, 2002, 8965–8973.

8. M. Jordaan, P. van Helden, C. G. C. E. van Sittert and H. C. M. Vosloo. Experimental and DFT investigation of the 1-octene metathesis reaction mechanism with the Grubbs 1 precatalyst. *J. Mol. Catal. A: Chem.* **254**, 2006, 145–154.

9. S. F. Vyboishchikov, M. Bühl and W. Thiel. Mechanism of olefin metathesis with catalysis by ruthenium carbene complexes: Density functional studies on model systems. *Chem. Eur. J.* **8**, 2002, 3962–3975.

10. W. Janse van Rensburg, P. J. Steynberg, M. M. Kirk, W. H. Meyer and G. S. Forman. Mechanistic comparison of ruthenium olefin metathesis catalysts: DFT insight into relative reactivity and decomposition behavior. *J. Organomet. Chem.* **691**, 2006, 5312–5325.

11. W. Janse van Rensburg, P. J. Steynberg, W. H. Meyer, M. M. Kirk and G. S. Forman. DFT prediction and experimental observation of substrate-induced catalyst decomposition in ruthenium-catalyzed olefin metathesis. *J. Am. Chem. Soc.* **126**, 2004, 14332–14333.

12. S. Fomine, S. M. Vargas and M. A. Tlenkopatchev. Molecular modeling of ruthenium alkylidene mediated olefin metathesis reactions. DFT study of reaction pathways. *Organometallics* **22**, 2003, 93–99.

13. G. S. Forman, A. E. McConnell, R. P. Tooze, W. Janse van Rensburg, W. H. Meyer, M. M. Kirk, C. L. Dwyer and D. W. Serfontein. A convenient system for improving the efficiency of first-generation ruthenium olefin metathesis catalysts. *Organometallics* **24**, 2005, 4528–4542.

14. C. Adlhart. Intrinsic reactivity of ruthenium carbenes: A combined gas phase and computational study. PhD Thesis, Eidgenossische Technische Hochschule, Zurich, 2003.

15. M. S. Sanford, J. A. Love and R. H. Grubbs. Mechanism and activity of ruthenium olefin metathesis catalysts. *J. Am. Chem. Soc.* **123**, 2001, 6543–6554.

16. C. Adlhart, C. Hinderling, H. Baumann and P. Chen. Mechanistic studies of olefin metathesis by ruthenium carbene complexes using electrospray ionization tandem mass spectrometry. *J. Am. Chem. Soc.* **122**, 2000, 8204–8214.

17. J. Louie and R. H. Grubbs. Metathesis of electron-rich olefins: Structure and reactivity of electron-rich carbene complexes. *Organometallics* **21**, 2002, 2153–2164.

18. J. Huang, E. D. Stevens, S. P. Nolan and J. L. Petersen. Olefin metathesis-active ruthenium complexes bearing a nucleophilic carbene ligand. *J. Am. Chem. Soc.* **121**, 1999, 2674–2678.
19. K. A. Burdett, L. D. Harris, P. Margl, B. R. Maughon, T. Mokhtar-Zadeh, P. C. Saucier and E. P. Wasserman. Renewable monomer feedstocks via olefin metathesis: Fundamental mechanistic studies of methyl oleate ethenolysis with the first-generation Grubbs catalyst. *Organometallics* **23**, 2004, 2027–2047.
20. F. Bernardi, A. Bottoni and G. P. Miscione. DFT study of the olefin metathesis catalyzed by ruthenium complexes. *Organometallics* **22**, 2003, 940–947.
21. K. Denk, J. Fridgen and W. A. Herrmann. *N*-heterocyclic carbenes. Part 33. Combining stable NHC and chelating pyridinyl–alcoholato ligands: A ruthenium catalyst for applications at elevated temperatures. *Adv. Synth. Catal.* **344**, 2002, 666–670.
22. C. W. Bielawski. Tailoring polymer synthesis with designer ruthenium catalysts. PhD Thesis, California Institute of Technology, California, 2003.
23. C. S. Slone, D. A. Weinberger and C. A. Mirkin. The transition metal coordination chemistry of hemilabile ligands. *Prog. Inorg. Chem.* **48**, 1999, 233–250.
24. W. H. Meyer, R. Brull, H. G. Raubenheimer, C. Thompson and G. J. Kruger. Thioethercarboxylates in palladium chemistry: First proof of hemilabile properties of S–O ligands. *J. Organomet. Chem.* **553**, 1998, 83–90.
25. M. Jordaan and H. C. M. Vosloo. Ruthenium catalyst with a chelating pyridinyl-alcoholato ligand for application in linear alkene metathesis. *Adv. Synth. Catal.* **349**, 2007, 184–192.
26. B. Delley. An all-electron numerical method for solving the local density functional for polyatomic molecules. *J. Chem. Phys.* **92**, 1990, 508–517.
27. B. Delley. From molecules to solids with the DMol3 approach. *J. Chem. Phys.* **113**, 2000, 7756–7764.
28. B. Delley. Fast calculation of electrostatics in crystals and large molecules. *J. Phys. Chem.* **100**, 1996, 6107–6110.
29. http://www.accelrys.com/.
30. J. P. Perdew and Y. Wang. Accurate and simple analytic representation of the electron–gas correlation energy. *Phys. Rev. B* **45**, 1992, 13244–13249.
31. B. Delley. *DMol*, a standard tool for density functional calculations: Review and advances, In *Modern Density Functional Theory: A Tool for Chemistry, Theoretical and Computational Chemistry*, Volume 2, edited by J. M. Seminario and P. Politzer. Amsterdam: Elsevier, 1995, 221–254.
32. J. Andzelm, R. D. King-Smith and G. Fitzgerald. Geometry optimization of solids using delocalized internal coordinates. *Chem. Phys. Lett.* **335**, 2001, 321–326.
33. W. Janse van Rensburg, C. Grove, J. P. Steynberg, K. B. Stark, J. J. Huyser and P. J. Steynberg. A DFT study toward the mechanism of chromium-catalyzed ethylene trimerization. *Organometallics* **23**, 2004, 1207–1222.
34. G. Henkelman and H. Jonsson. Improved tangent estimate in the nudged elastic band method for finding minimum energy paths and saddle points. *J. Chem. Phys.* **113**, 2000, 9978–9985.
35. M. S. Sanford, M. Ulman and R. H. Grubbs. New insights into the mechanism of ruthenium-catalyzed olefin metathesis reactions. *J. Am. Chem. Soc.* **123**, 2001, 749–750.
36. J. Halpern. Mechanism and stereoselectivity of asymmetric hydrogenation. *Science* **217**, 1982, 401–407.
37. T. Weskamp, F. J. Kohl, W. Hieringer, D. Gleich and W. A. Herrmann. Highly active ruthenium catalysts for olefin metathesis: The synergy of *N*-heterocyclic carbenes and coordinatively labile ligands. *Angew. Chem., Int. Ed.* **38**, 1999, 2416–2419.
38. J. P. Collman, L. S. Hegedus, J. R. Norton and R. G. Finke. *Principles and Applications of Organotransition Metal Chemistry*. Mill Valley, CA: University Science Books, 1987.

39. A. J. Arduengo III. Looking for stable carbenes: The difficulty in starting anew. *Acc. Chem., Res.* **32**, 1999, 913–921.

40. L. Perrin, E. Clot, O. Eisenstein, J. Loch and R. H. Crabtree. Computed ligand electronic parameters from quantum chemistry and their relation to Tolman parameters, Lever parameters, and Hammett constants. *Inorg. Chem.* **40**, 2001, 5806–5811.

13 Empirical Molecular Modeling of Suspension Stabilization with Polysorbate 80

Jamie T. Konkel and Allan S. Myerson

CONTENTS

13.1 INTRODUCTION

The formulation of poorly water-soluble drugs into crystalline nanosuspensions is an area of active pharmaceutical research [1,2]. As the particle size is reduced, the surface area of the drug is increased, leading to greater bioavailability [3]. However, the stabilization of crystalline nanosuspensions is nontrivial, especially given the limited number of surfactants that are also pharmaceutically acceptable excipients. Regardless, we have identified several surfactant combinations that can stabilize a variety of crystalline nanosuspensions. However, each new drug compound needs to be screened empirically with the various surfactant combinations to find the most stable formulation. This is time-consuming and can require a prohibitive amount of drug in early development phases [4]. This work was undertaken to develop a molecular modeling approach to rapid surfactant screening, which could identify stabilizing surfactant systems for a particular drug, based on its crystal structure.

In this paper, we present the results of screening five model pharmaceutical drug crystals with the surfactant, Polysorbate 80. The model drugs were selected to offer a range of degree of stabilization with Polysorbate 80, and also had crystal structures available in the Cambridge Structural Database (CSD).

13.2 METHODS

13.2.1 EXPERIMENTAL

Crystalline nanosuspensions of the model drugs, nabumetone, carbamazepine (Polymorph III), celecoxib, fluorometholone, and Compound A, an internal drug candidate, were prepared by high-pressure piston gap homogenization (Avestin, Emulsiflex C5). With the exception of fluorometholone, each suspension was prepared by dispersing 1% (w/v) of crystalline drug in an aqueous surfactant solution containing 0.25% (w/v) of the surfactant Polysorbate 80, 0.5% (w/v) of the polymer Poloxamer 188, and 2.25% (w/v) glycerin for tonicity adjustment. The fluorometholone suspension was prepared with 0.25% (w/v) of the drug, 0.025% (w/v) Polysorbate 80, 0.05% Poloxamer 188, and 0.2% (w/v) glycerin. The coarse suspension was homogenized at a pressure of 20 kpsi until the target particle size of approximately 1 μm was reached. The particle size distribution was measured by static light scattering with a Horiba LA-920 particle size analyzer. Suspensions were also examined by light microscopy to ensure the suspensions were well dispersed. Aliquots of each suspension were stress-tested to rapidly assess the physical stability of the formulation. The stress tests included centrifugation, a freeze–thaw cycle, three days of temperature cycling between 5°C and 40°C, and three days of shaking. The stability of some of the more promising suspensions was also determined in real time at several temperatures.

13.2.2 MOLECULAR MODELING

All molecular modeling was performed in Materials Studio®, version 4.2 (Accelrys).

13.2.2.1 Crystal Model

The growth morphology for each drug was calculated using the Morphology module of Materials Studio, using the Forcite energy method [5]. The COMPASS [6] force field was used for nabumetone, carbamazepine, and fluorometholone. The consistent valence force field (CVFF) [7] was used for celecoxib and Compound A because they contained atoms not defined in COMPASS. COMPASS and CVFF were used because they can be used in Morphology, Forcite, and Discover modules in Materials Studio. The crystal structures were obtained from the CSD [8] and were minimized with cell optimization prior to morphology calculation. The optimized cells were compared with their experimental structures to ensure that none had changed significantly, and thus that the force field used was appropriate.

For each drug, at least one crystal face was selected for modeling. To select a face for modeling, the morphology of the crystal was calculated using the attachment energy method [9]. The faces with significant facet area and the highest attachment energy were selected first. These are the faces most likely to grow if the crystals

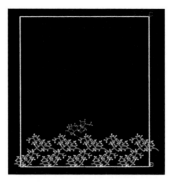

FIGURE 13.1   (**See color insert.**) Fluorometholone molecule with the (020) face of fluorometholone.

ripen. A relatively flat face was also desirable for modeling. To prepare a face for modeling, the face was recleaved to give at least two layers of the crystal surface. Next, the surface was expanded by symmetry to give enough surface area to model the interaction with polysorbate 80, at least 60 Å × 60 Å. Finally, a vacuum slab was prepared with at least 50 Å of space over the crystal face. See Figures 13.1 and 13.2 for an example of docking of both polysorbate 80 and fluorometholone on the (020) face of fluorometholone.

13.2.2.2   Surfactant Model

Polysorbate 80 is a mixture of compounds with the general structure shown in Figure 13.3. The structure has three arms consisting of chains of hydrophilic oxyethylene and a fourth arm that terminates with a hydrophobic oleate tail (Figure 13.3). For modeling purposes, a single molecule of polysorbate 80 with equal-length arms was used ($w, x, y, z = 5$). The molecule of polysorbate 80 was minimized using the same energy parameters as the drug crystal. A molecule of each drug was also minimized using the same energy parameters.

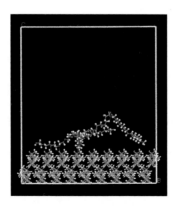

FIGURE 13.2   (**See color insert.**) Polysorbate 80 with the (020) face of fluorometholone.

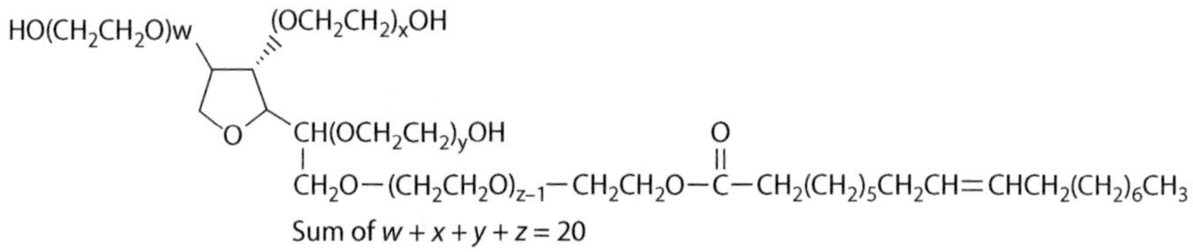

FIGURE 13.3 Structure of polysorbate 80.

13.2.2.3 Adsorption Model

The adsorption of both polysorbate 80 and a molecule of the drug was modeled on each face. The Cartesian position of each atom of the crystal was fixed for the simulation. A molecule of either polysorbate 80 or the drug was positioned over the crystal face in a position where the additive molecule could interact with the surface. Hydrogen bonding was monitored to aid in the initial positioning of the additive. Next, a short molecular dynamics simulation was performed using the Discover module, usually a minimum of 500 steps, with at least 10 frames of output for each run. For each simulation, various starting conditions were used to find the lowest energy configuration. Forcite was used to calculate the energies of the total system, the additive alone, and the crystal alone from the frame with the lowest energy.

This model of adsorption is greatly simplified and provides only a rapid screening tool to aid in the design of empirical formulation studies. While the effects of the aqueous environment are critical to the adsorption of nonpolar surfactants, they are ignored in this model. The simulations used only one molecule of one isomer of polysorbate 80 and did not take into account various possible conformations of the oxyethylene side chains on the surface of the drug crystals. While these simulations do not give a complete picture of the adsorption of polysorbate 80 on the crystal surfaces, they do give a prediction of the relative stability of the suspension in a matter of hours using a Windows-based desktop computer.

13.3 RESULTS

13.3.1 EXPERIMENTAL RESULTS

The model drugs represented a range of stabilization with the polysorbate 80/poloxamer 188 surfactant combination. Crystal nanosuspensions of carbamazepine and compound A were least stable with polysorbate 80. Because of the aggregation of the initial suspension, and a limited amount of drug material available, the suspension of compound A was not subjected to further stability testing. The particle size results after stress testing for the other drugs are shown in Table 13.1. Celecoxib nanosuspensions stabilized with polysorbate 80 were moderately stable. The polysorbate 80-stabilized nanosuspension of fluorometholone was stable under most stress conditions. The stability of nabumetone nanosuspensions was the most enhanced by the polysorbate 80 surfactant. The particle size of this formulation of nabumetone nanosuspension was stable for three months at $-20°C$ and $5°C$. After six months of storage, ripening was observed microscopically by the presence of large needle-shaped

TABLE 13.1
Particle Size Stability of Model Drugs

Drug Particle Size (μm)	Initial		Temperature Cycling		Shaking		Freeze–Thaw		Centrifugation	
	Mean	99th %ile	Mean	99th %ile	Mean	99th %ile	Mean	99th %ile	Mean	99th %ile
Nabumetone	0.98	2.15	1.23	2.94	1.12	2.67	1.06	2.50	1.08	2.59
Fluorometholone	0.58	1.10	0.53	0.98	0.52	0.96	4.51	13.04	0.56	1.34
Celecoxib	1.24	2.97	2.01	6.28	1.74	5.27	1.41	3.84	1.73	5.97
Carbamazepine	1.32	4.65	1.56	6.21	2.41	12.32	1.35	4.45	1.42	4.50

crystals. It is important to note that we have found nabumetone to be a worst-case model drug for stabilization of nanosuspensions. The reported surfactant system is the only one that we have found that can stabilize nabumetone nanosuspensions for any significant length of time.

While this paper focused solely on the interactions of the polysorbate 80 with the model drug compounds, it is important to note that poloxamer 188 is also a surface active polymer present in each of the model suspensions. As a control, the model suspensions were also prepared with the poloxamer 188 alone. The control suspensions prepared without polysorbate 80 had significant aggregation in the initial suspensions, which worsened upon stress testing (data not shown).

13.3.2 MOLECULAR MODELING RESULTS

The binding energy of each drug with its additives was calculated as follows:

$$E_{binding} = E_{total} - (E_{crystal} + E_{additive}),$$

where E_{total} is the energy of the system with the additive adsorbed, $E_{crystal}$ is the energy of the crystal surface alone, and $E_{additive}$ is the energy of the molecule of polysorbate 80 or drug alone. The binding energies of each drug crystal and its additives are shown in Table 13.2. A more negative binding energy corresponds to a more favorable interaction. More importantly, the ratio of the polysorbate 80 binding energy to that of the drug correlated well with the rank order stabilization of the crystalline suspension. These stabilization ratios are also presented in Table 13.2. A ratio of approximately 3 correlated to moderate stability; a ratio larger than 3, such as 9.40 for the nabumetone, correlated to excellent stabilization. For carbamazepine and compound A (the drugs with poor stabilization by polysorbate 80), the ratio was near 1. In fact, for these systems, it took several simulations to obtain a negative binding energy result for both the polysorbate 80 and the drug.

The face of the crystal used for simulations had a significant impact on the results. For example, nabumetone's fastest growing face (011) had the highest $E_{binding}$ with the surfactant. The slowest growing face (110) had much lower $E_{binding}$ with polysorbate 80, but comparable $E_{binding}$ with the drug. However, the fastest growing face

TABLE 13.2

Molecular Modeling $E_{binding}$ Results

Drug	Crystal Face	$E_{binding}$ with Polysorbate 80 (kJ/mol)	$E_{binding}$ with Drug Molecule (kJ/mol)	Ratio of $E_{binding}$ Polysorbate 80/drug
Nabumetone	0 1 1	−414.47	−44.10	9.40
Nabumetone	1 1 0	−109.19	−52.55	2.08
Fluorometholone	0 2 0	−185.14	−40.88	4.53
Celecoxib	0 0 1	−122.51	−32.43	3.78
Celecoxib	1 0 −1	−235.68	−183.97	1.28
Compound A	1 1 −1	−141.84	−73.60	1.93
Compound A	1 0 −2	−85.44	−99.50	0.86
Carbamazepine	1 1 0	−33.89	−39.71	0.85

is the one that needs the most stabilization to give a stable suspension. For cele-coxib, the (001) face was the fastest growing and showed a stabilization ratio of 3.78 with the polysorbate 80. However, this face only represented 1% of the surface area of the calculated crystal morphology. The next fastest growing face (10−1) was 7% of the surface area and only had a stabilization ratio of 1.28. In the empirical study, the addition of polysorbate 80 did greatly reduce the aggregation of the celecoxib suspension compared with the control suspension with poloxamer 188 alone, but it did not completely inhibit the particle growth.

The faces with the greatest interaction with the surfactant were the ones with the most polar groups on the surface that could hydrogen-bond with the oxyethylene arms of the polysorbate 80 molecule. Figure 13.2 shows polysorbate 80 with the (020) face of fluorometholone. This surface has both alcohol and amide groups eas-ily accessible for hydrogen bonding with the oxyethylene groups. For comparison, Figure 13.4 shows polysorbate 80 with the (110) face of carbamazepine. The surface of this face consists of hydrocarbons and cannot hydrogen-bond with the surfactant.

FIGURE 13.4   (See color insert.) Polysorbate 80 with the nonpolar (110) face of carbamazepine.

13.4   SUMMARY AND CONCLUSION

In summary, we have used molecular modeling to predict the stabilization of crystalline nanosuspensions by the surfactant, polysorbate 80. The binding energies of surfactant and drug molecules with crystal surfaces were calculated using Materials Studio. The ratio of the binding energy of surfactant to that of the drug gives a stabilization ratio that correlates well with the stabilization of the drug suspension.

The use of this technique for preliminary formulation screening could be beneficial because the molecular modeling experiments are less time-consuming and do not use valuable and limited drug material. The technique could be expanded for use with different surfactant compounds. This could help to prioritize the formulations screened by actual laboratory experiments, so that only the most promising surfactant systems would need to be screened to find a stable formulation.

REFERENCES

1. B. Rabinow. Nanosuspensions in drug delivery. *Nat. Rev. Drug Discovery* **3**, 2004, 785–796.
2. T. Shah, D. Patel, J. Hirani and A. F. Amin. Nanosuspensions as a drug delivery system: A comprehensive review. *Drug Delivery Technol.* **7**, 2007, 42, 44, 46, 48–50, 52–53.
3. F. Keisisoglou, S. Panmai and Y. Wu. Application of nanoparticles in oral delivery of immediate release formulations. *Curr. Nanosci.* **3**, 2007, 183–190.
4. S. Branchu, P. G. Rogueda, A. P. Plumb and W. G. Cook. A decision-support tool for the formulation of orally active, poorly soluble compounds. *Eur. J. Pharm. Sci.* **32**, 2007, 128–139.
5. A. K. Rappe, C. J. Casewit, K. S. Colwell, W. A. Goddard and W. M. Skiff. UFF, a full periodic table force field for molecular mechanics and molecular dynamics simulations. *J. Am. Chem. Soc.* **114**, 1992, 10024–10035.
6. H. Sun. COMPASS: An ab initio force-field optimized for condensed-phase applications—Overview with details on alkane and benzene compounds. *J. Phys. Chem.* B **102**, 1998, 7338–7364.
7. P. Dauber-Osguthorpe, V. A. Roberts, D. J. Osguthorpe, J. Wolff, M. Genest and A. T. Hagler. Structure and energetics of ligand binding to proteins: *Escherichia coli* dihydrofolate reductase–trimethoprim, a drug-receptor system. *Proteins* **4**, 1988, 31–47.
8. F. H. Allen. The Cambridge Structural Database: A quarter of a million crystal structures and rising. *Acta Cryst.* **B58**, 2002, 380–388.
9. A. S. Myerson. *Molecular Modeling Applications in Crystallization.* New York: Cambridge University Press, 1999.

14 Multiscale Modeling of the Adsorption Interaction between Model Bitumen Compounds and Zeolite Nanoparticles in Gas and Liquid Phases

Stanislav R. Stoyanov, Sergey Gusarov, and Andriy Kovalenko

CONTENTS

14.1 INTRODUCTION

Canada has extensive bitumen resources that are at par with Saudi Arabia's conventional oil reserves.[1] Oil sands are unconsolidated deposits of very heavy hydrocarbon bitumen and require multiple stages of processing before refining.[2] Bitumen molecules contain highly unsaturated hydrocarbons and large amounts of chemical impurities, such as sulfur, nitrogen, nickel, and vanadium. Bitumen upgrading requires complex catalysts and special reaction conditions. Zeolites are well-known adsorbents and are used for size-selective cracking of hydrocarbons in zeolite pores.[3] The large bitumen molecules cannot enter the zeolite pores but mainly adsorb on zeolite surfaces, which has tremendous potential for bitumen upgrading. Ng et al. have demonstrated that a zeolite Y-based catalyst containing active matrix decreases the N and S contents while increasing the vacuum gas oil conversion rate.[4] Zeolite-based catalytic *nanoparticles* can be derived from inexpensive natural minerals by engineering their surface nanostructure, so as to coordinate, purify, and crack the large and complex bitumen molecules.[5] Their reactive surface nanostructure is modified to have large accessible area and to provide highly specific attachment of bitumen fragments. Zeolite-based catalysts derived from inexpensive natural minerals optimized by modifying the zeolite reactive surface can be used for efficient bitumen cracking, removing undesired additives, and extraction. Recently, Kuznicki et al. have reported that the catalytic and adsorptive properties of the modified natural zeolite chabazite reduce bitumen viscosity dramatically and remove much of the metals, sulfur, and nitrogen at 400°C.[5]

A crucial factor in bitumen upgrading is the precise control of reaction conditions. Bitumen hydrotreatment tests performed in *critical conditions* show superior performance with enhanced hydrogen transport to the catalyst surface and reduced zeolite coking, thus promoting the desired reaction pathways.[6] Modeling of processes in *critical conditions* can be done using our development of the statistical–mechanical, three-dimensional reference interaction site model (3D-RISM) molecular theory of solvation with the Kovalenko–Hirata (KH) closure.[7,8] We have coupled 3D-RISM–KH with Kohn–Sham density functional theory (KS-DFT) in a self-consistent field (SCF) multiscale formalism for electronic structure in solution[7,8] and implemented this method (including analytical gradients of solvation forces) in the Amsterdam Density Functional (ADF) quantum chemistry software package.[9,10] This development allows one to model chemical reactions in solution described by NVT ensemble (N = number of particles, V = volume, T = temperature). We have applied the 3D-RISM–KH molecular theory of solvation to a wide range of problems, including solvation thermodynamics effects on synthetic organic supramolecular rosette nanotube architectures[11,12] and biomolecular architectures,[13–15] as well as for prediction of asphaltene aggregation.[16] We have also implemented in ADF the self-consistent combination[17] of 3D-RISM–KH with the orbital-free embedding (OFE) method for the electron density of a macromolecule in that of its environment (OFE-DFT/3D-RISM–KH)[18] that is used to generate the average embedding potential and to subsequently calculate electronic excited states for prediction of solvatochromic shifts.

The preferred sites for metal ions in zeolite frameworks have been investigated by using different theoretical methods. Monte Carlo-simulated annealing modeling results of Na^+ sitting in mordenite are in a very good agreement with experimental results.[19] First-principles molecular dynamics simulation of metal cations in the zeolite offretite has shown that Cu^+ is more tightly bound than alkali metals.[20] Zakharov et al. have modeled the decomposition of NO in a small zeolite cluster that contains Cu^+.[21] High-level cluster MP2 calculations indicate that both NO and N_2O bind stronger to Cu^+-exchanged zeolite than to Ag^+.[22] A number of recent reports make use of periodic boundary condition density functional theory (PBC-DFT) codes. Civalleri et al. have applied PBC-DFT as implemented in the CRYSTAL code[23] to the investigation of Li^+-, Na^+-, and K^+-preferred sites in bulk chabazite.[24] These alkali ions form very weak covalent bonds with O atoms. On the other hand, transition metal ions, such as Cu^+, form strong covalent bonds to one or several zeolite framework O atoms.[25] Benco et al. apply the VASP code[26–29] and report that Cu^+ and Ag^+ represent particularly strong centers for H_2 adsorption in mordenite channels.[30]

Global and local reactivity indices based on DFT provide a highly accurate insight in the reaction site preference prediction that comes at low computational cost.[31,32] Fukui functions have been applied for prediction of inorganic[33,34] and organic reaction sites.[35] In a PBC-DFT study using DMol[3], the effect of organic moieties incorporation in chabazite framework on Brønsted acidity has been investigated using several reactivity descriptors.[36] Chatterjee has proposed a semiquantitative scale for correlation of Brønsted and Lewis acidity for a range of bi- and trivalent metal dopants in aluminophosphates.[37] Bifunctional acid–base catalyzed reactions in zeolites have been modeled using the principle of maximum hardness.[38]

The theoretical reports on thiophene adsorption on zeolites are limited to cluster models. DFT studies suggest that the thiophene cracking reaction could be catalyzed by Lewis instead of Brønsted sites.[39] The weak interaction between thiophene and zeolite cluster Brønsted sites has been shown to lead to the formation of van der Waals complexes.[40] Yang and coworkers have shown by elution experiment and DFT calculation of a small zeolite cluster that Cu^+- and Ag^+-exchanged zeolite Y selectively adsorbs both sulfur- and nitrogen-containing heterocycles from diesel fuels under ambient conditions.[41,42] Nitrogen-containing heterocycles reduce the fuel desulfurization capacity of Cu(I)Y zeolite because of competitive adsorption.[43,44]

Theoretical studies of catalytic activity involve several important steps: substrate modeling, prediction of preferred binding sites and configurations, calculation of adsorption or binding energies, reaction paths, reaction free energies and barriers, kinetic and thermodynamic effects, effects of spectator species, and catalyst contamination. Recently, we have presented a PBC-DFT (DMol[3]) and quantum mechanics/molecular mechanics (QM/MM) investigation of zeolite nanoparticle surface acidity. The PBC-DFT results suggest that aluminum substitution near chabazite "open" surface creates stable acid sites that are accessible to bitumen fragments. The most stable Brønsted sites contain protons at the O1 and O3 positions for both bulk and slab chabazite models, in agreement with the experiment. Fukui functions are applied for nanoparticle reactivity prediction and are considered useful for the development of surface reactivity maps that could be applied for the design of nanoparticles with optimal functionality.[45] We have also used Fukui functions to explore the binding interaction preferences of heavy oil fragments containing Ni(II) and vanadyl porphyrins.[46] The binding interaction of fused thiophene bitumen fragments to ion-exchanged chabazite surfaces has been studied using PBC-DFT in the gas phase to predict the adsorption configurations and energies.[47] Here, we extend this study to include the effect of solvation on the adsorption interaction of sulfur-containing bitumen fragments on metal ion-exchanged zeolite nanostructures in liquid phase using the statistical–mechanical 3D-RISM–KH molecular theory of solvation.

14.2 COMPUTATIONAL TECHNIQUE

For the PBC-DFT modeling using the DMol[3] software from Accelrys, Materials Studio® 4.0,[48] we use the PW91 density functional.[49] Geometries optimized using the PW91 functional are in a better agreement with the experiment than those from other functionals, such as the widely used BLYP functional.[50,51] It has also been shown that PW91 is capable of modeling van der Waals interactions, although dispersive interactions are not physically included in it.[52] We use the all-electron double numerical basis set with polarization function (DNP).[53] The DNP basis set is comparable to the 6–31G** basis set. However, the numerical basis set performs better than a Gaussian basis set of the same size for molecules, semiconductors, and insulators.[54] Brillouin-zone sampling is restricted to the Γ point because the zeolite slab is an insulator and sampling of the five-slab irreducible k-points changes the E_{ADS} values by less than 1 kJ/mol. Moreover, Fukui function calculations are not enabled for multiple k-points in DMol[3]. We have used the extra-fine numerical integration grid that contains 474 angular points. The number of radial points is a function of the nuclear charge and varies with the periodic system size.[53,55] The electronic singlet states are treated using single determinants. The initial periodic unit cell zeolite geometries are obtained from the Accelrys Materials Studio[48] database and are fully optimized. The chabazite lattice parameters are taken from the x-ray report[56] and are not optimized. The real space cutoff for the atomic numerical basis set calculation has been set to 4.0 Å. The geometry optimization convergence is achieved when the energy, gradient, and displacement are lower than 1×10^{-5} Ha, 1×10^{-3} Ha/Å, and 1×10^{-3} Å, respectively. For solvation modeling in DMol[3], we use the Conductor-like Screening Model (COSMO)[57,58] and calculate the energies at each disaggregation step in quinoline ($\varepsilon = 9.16$) and benzene ($\varepsilon = 2.284$) solvents with the isosurface–ball–stick (IBS) method.

For the 3D-RISM calculations, the starting geometries are the geometries optimized using DMol[3]/GGA-PW91/DNP and contain bitumen fragments adsorbed on the ion-exchanged chabazite surfaces. The periodic unit cells are extended along the z-axis to $9.42 \times 9.42 \times 64.00$ Å to allow modeling of disaggregation within the unit cell. The disaggregation configurations are obtained by displacement of the bitumen fragments along the z-axis from -1.2 Å to 20.0 Å relative to the optimized geometry. For each displacement step, we calculate the total electronic energy using the DMol[3]/GGA-PW91/DNP method, the solvation free energy using the 3D-RISM–KH theory of solvation, and the COSMO solvation energy using DMol[3]/GGA-PW91/DNP. Alternative disaggregation paths can be followed using computational techniques, such as linear and quadratic transit methods.

The solvent site–site radial correlation functions of bulk solvent are obtained from the dielectrically consistent RISM theory (DRISM).[59–61] The number of grid points is 8192 and the spacing between them is 0.05 Å. The solvent systems pure benzene, 0.0 M thiophene in benzene, 0.0 M benzothiophene in benzene, 0.0 M dibenzothiophene in benzene, and pure quinoline are modeled at 298 K. The solutions of thiophene, benzothiophene, and dibenzothiophene solutions in benzene are for infinite dilution. The dielectric constants of benzene and quinoline at 298 K are 2.2736 and 9.0, respectively.[62,63] The densities of benzene and quinoline at 298 K are 0.87367 g/cm[3] and 1.093 g/cm[3], respectively. The potential parameters σ and ε for every atom of benzene and quinoline are taken from the Universal Force Field (UFF) file[64] in Materials Studio 4.4. For the thiophene and dibenzothiophene dissolved in benzene, the united atom (UA) model is applied as follows: for thiophene, each pair of symmetry-equivalent CH groups is treated as one site and the S atom is treated explicitly; in dibenzothiophene, each pair of symmetry-equivalent CH groups is treated as one site, each pair of symmetry-equivalent C atoms of the thiophene moiety is treated as one site, and the S atom is treated explicitly. Thus, thiophene has three sites and dibenzothiophene has seven sites. The potential parameters σ and ε for C and S are taken from UFF. The potential parameters σ and ε for the CH site are σ_C(UFF) and ε_C(UFF) scaled by the ratios σ_{CH}/σ_C and $\varepsilon_{CH}/\varepsilon_C$.[65,66] All atomic charges are obtained using the QEq charge equilibration method version 1.1[67] implemented in Materials Studio Visualizer v. 5.0. Site charges are calculated as a sum of all atomic charges. The solvent–solute interaction, including solvation thermodynamics and structure, is calculated using the 3D-RISM–KH theory. In the 3D-RISM–KH calculation, all solute atoms are treated explicitly. The solute potential parameters σ and ε are taken from UFF.[64] The solute atomic charges are calculated using the QEq equilibration method.

14.3   KOHN–SHAM DFT AND 3D-RISM–KH THEORY OVERVIEW

14.3.1   KOHN–SHAM DFT IN THE PRESENCE OF SOLVENT

The electronic structure of the solute is calculated from the self-consistent KS-DFT equations modified to include the presence of solvent. The total system of the solute and solvent has the Helmholtz free energy defined as

$$A[n_e(\mathbf{r}),\{\rho_\gamma(\mathbf{r})\}] = E_{\text{solute}}[n_e(\mathbf{r})] + \Delta\mu_{\text{solv}}[n_e(\mathbf{r}),\{\rho_\gamma(\mathbf{r})\}], \tag{14.1}$$

where E_{solute} is the electronic energy of the solute consisting of the standard components,[68] $\Delta\mu_{\text{solv}}$ is the excess chemical potential of solvation coming from the solute–solvent interaction and solvent reorganization due to the presence of the solute, $n_e(\mathbf{r})$ is the electron density distribution, and $\rho_\gamma(\mathbf{r})$ is the classical density distribution of interaction sites $\gamma = 1,\ldots,s$ of the solvent molecule. The solute energy is determined by the standard KS-DFT expression written in atomic units as

$$E_{\text{solute}}[n_e(\mathbf{r})] = T_s[n_e(\mathbf{r})] + \int d\mathbf{r}\, n_e(\mathbf{r})\upsilon_i(\mathbf{r}) + \frac{1}{2}\int \frac{n_e(\mathbf{r})n_e(\mathbf{r}')}{d\mathbf{r}\,d\mathbf{r}'\,|\mathbf{r}-\mathbf{r}'|} + E_{\text{XC}}[n_e(\mathbf{r})], \tag{14.2}$$

where $T_s[n_e(\mathbf{r})]$ is the kinetic energy of a noninteracting electron gas in its ground state with density distribution $n_e(\mathbf{r})$, $E_{\text{XC}}[n_e(\mathbf{r})]$ is the exchange-correlation energy, and $\upsilon_i(\mathbf{r})$ comprises the external potential and the nuclear attractive potential.

From the minimization of the free-energy functional (Equation 14.1)

$$\frac{\delta A[n_e(\mathbf{r}),\{\rho_\gamma(\mathbf{r})\}]}{\delta n_e(\mathbf{r})} = 0, \tag{14.3}$$

subject to the normalization condition for N_e valence electrons of solute

$$\int d\mathbf{r} n_e(\mathbf{r}) = N_e, \tag{14.4}$$

we obtain the self-consistent KS equation modified due to the presence of solvent[7,8]

$$\left[-\frac{1}{2}\Delta + \upsilon_i(\mathbf{r}) + \upsilon_h(\mathbf{r}) + \upsilon_{XC}(\mathbf{r}) + + \upsilon_{solv}(\mathbf{r}) \right]\psi_j(\mathbf{r}) = \varepsilon_j\psi_j(\mathbf{r}), \tag{14.5}$$

where the Hartree potential is

$$\upsilon_h(\mathbf{r}) = \int d\mathbf{r}' \frac{n_e(\mathbf{r})}{|\mathbf{r} - \mathbf{r}'|}. \tag{14.6}$$

The electron density distribution is determined by summation over the N_e lowest occupied eigenstates with allowance for their double occupancy by electrons with opposed spins

$$n_e(\mathbf{r}) = \sum_{j=1}^{N_e} |\psi_j(\mathbf{r})|^2, \tag{14.7}$$

the exchange-correlation potential is the functional derivative

$$\upsilon_{XC}(\mathbf{r}) = \frac{\delta E_{XC}[n_e(\mathbf{r})]}{\delta n_e(\mathbf{r})}, \tag{14.8}$$

and the solvent potential is defined as

$$\upsilon_{solv}(\mathbf{r}) = \frac{\delta \Delta \mu_{solv}[n_e(\mathbf{r}), \{\rho(\mathbf{r})\}]}{\delta n_e(\mathbf{r})}. \tag{14.9}$$

To simplify the calculation of $\upsilon_h(\mathbf{r})$ in ADF, we use the fitted density

$$n_e(\mathbf{r}) \approx \sum_{j=1}^{N_e} c_a f_a(\mathbf{r}), \tag{14.10}$$

which after substitution in Equation 14.6 yields the fitted potential. Here f_a is the set of the single-center Slater functions and the coefficients c_a are determined by least-squares fitting.[69,70] Together with using the locality properties,[71] this allows one to dramatically reduce the amount of calculations necessary for evaluation of the potentials and matrix elements.[9]

The total free energy is calculated as

$$A_{tot} = \sum_{j=1}^{N_e} \varepsilon_j - \frac{1}{2} \int d\mathbf{r}\, d\mathbf{r}' \, \frac{n_e(\mathbf{r})n_e(\mathbf{r}')}{|\mathbf{r}-\mathbf{r}'|} + E_{XC}[n_e(\mathbf{r})] - \int d\mathbf{r}\upsilon_{XC}(\mathbf{r})n_e(\mathbf{r})$$

$$+ \Delta\mu_{solv}[n_e(\mathbf{r}),\{\rho_\gamma(\mathbf{r})\}] - \int d\mathbf{r}\upsilon_{solv}(\mathbf{r})n_e(\mathbf{r}). \tag{14.11}$$

14.3.2   Three-Dimensional RISM Theory with Kovalenko–Hirata Closure

The 3D-RISM integral Equation 14.12 coupled with the Kovalenko–Hirata (KH) closure Equation 14.13 read

$$h_\gamma(\mathbf{r}) = \sum_{\gamma'} \int d\mathbf{r}' c_{\gamma'}(\mathbf{r}') \chi_{\gamma'\gamma}(|\mathbf{r}-\mathbf{r}'|) \tag{14.12}$$

$$h_\gamma(\mathbf{r}) = \begin{cases} \exp(d_\gamma(\mathbf{r})) - 1 & \text{for} \quad d_\gamma(\mathbf{r}) \leq 0 \\ d_\gamma(\mathbf{r}) & \text{for} \quad d_\gamma(\mathbf{r}) > 0 \end{cases} \tag{14.13}$$

$$d_\gamma(\mathbf{r}) = -u_\gamma(\mathbf{r})/(k_B T) + h_\gamma(\mathbf{r}) - c_\gamma(\mathbf{r}),$$

where $h_\gamma(\mathbf{r})$ is the total correlation function related to the density distribution function $g_\gamma(\mathbf{r}) = h_\gamma(\mathbf{r}) + 1$, and $c_\gamma(\mathbf{r})$ is the 3D direct correlation function reflecting "direct interactions" in solution and having the asymptotics of the interaction potential $u_\gamma(\mathbf{r}), c_\gamma(\mathbf{r}) \sim -u_\gamma(\mathbf{r})/(k_B T)$, where $u_\gamma(\mathbf{r})$ is the 3D interaction potential between the whole solute molecule and solvent site γ specified by a molecular force field, and $k_B T$ is the Boltzmann constant times the solution temperature. The site–site susceptibility of pure solvent $\chi_{\gamma'\gamma}(\mathbf{r})$ is an input to the 3D-RISM theory, and site indices γ' and γ enumerate all sites on all sorts of solvent species[72] (in the case of a multicomponent solvent such as thiophene in benzene in this work). The 3D-RISM theory with the KH closure approximation reproduces various phase and structural transitions in complex liquids and mixtures and properly accounts for both hydrogen bonding and hydrophobic interactions.[13–15,73,74] The solvation free energy (subscript "u" for solute and "V" for volume) is defined as

$$\mu_u = \varepsilon_u - T\Delta S_{u,v}, \tag{14.14}$$

and is obtained from the 3D-RISM–KH theory in a closed analytical form in terms of the correlation functions

$$\Delta\mu_{solv}^{KH} = k_B T \sum_\gamma \rho_\gamma \int d\mathbf{r} \left(\frac{1}{2}(h_\gamma(\mathbf{r}))^2 \Theta(-h_\gamma(\mathbf{r})) - c_\gamma(\mathbf{r}) - \frac{1}{2} h_\gamma(\mathbf{r}) c_\gamma(\mathbf{r}) \right), \tag{14.15}$$

where ρ_γ is the number density of bulk solvent at temperature T. Under isochoric condition, the solvation free energy is decomposed into the excess partial molecular entropy and the excess partial solvation energy, as follows:

$$\Delta S_{u,v} = -\left(\frac{\partial \mu_u}{\partial T}\right)_{\rho_V} \text{ and } \varepsilon_u = \mu_u + T\Delta S_{u,v}. \tag{14.16}$$

The potential of mean force (PMF) between the molecule and the zeolite at separation d Å (Figure 14.2) is defined as the difference between the gas phase (E_{PW91}) and solvation free energy terms and those at $d_0 = 20$ Å

$$\text{PMF}(d) = \left(E_{PW91}(d) + \Delta\mu_{solv}^{KH}(d)\right) - \left(E_{PW91}(d_0) + \Delta\mu_{solv}^{KH}(d_0)\right). \tag{14.17}$$

The partial molar volume (PMV) is expressed in terms of the solvent–solute 3D correlation function as follows:

$$\text{PMV} = \kappa_B T \chi_T - \int d\mathbf{r} h_\gamma(\mathbf{r}), \tag{14.18}$$

where χ_T is the isothermal compressibility of pure solvent. In Equation 14.18, the first term is the ideal volume term of the PMV, whereas the second term is the excess term for the solvent–solute interactions.[75] In 3D-RISM, the solvation free energy is obtained in a straightforward manner from the first principles of statistical mechanics, starting from the force field (molecular geometry, Lennard–Jones parameters, and electrostatic charges). Full details of the 3D-RISM–KH theory are available elsewhere.[7–10]

14.3.3 Effective Potentials and Analytical Gradients

The classical effective potential energy of the solute acting on solvent site γ_α is broken up into the short-range interaction $u_\gamma^{(sr)}(\mathbf{r})$ between the solvent site and the whole solute and the electrostatic energy of the solvent site effective charge q in the fields of the solute nuclei $\varphi^{(n)}(\mathbf{r}) = \sum_i Z_i/|\mathbf{r} - \mathbf{R}_i|$ and electrons $\phi^{(e)}(\mathbf{r})$

$$u_\gamma(\mathbf{r}) = u_\gamma^{(sr)}(\mathbf{r}) + q_\gamma\left(\varphi^{(n)}(\mathbf{r}) + \varphi^{(e)}(\mathbf{r})\right). \tag{14.19}$$

This short-range part is represented by the sum of the 12–6 Lennard–Jones potentials over the solute sites

$$u_\gamma^{(sr)}(\mathbf{r}) = \sum_i 4\varepsilon_{i\gamma}\left[\left(\frac{\sigma_{i\gamma}}{r_i}\right)^{12} - \left(\frac{\sigma_{i\gamma}}{r_i}\right)^6\right], \tag{14.20}$$

where $r_i = |\mathbf{r} - \mathbf{R}_i|$ is the separation between the solute nucleus i and solvent site γ and $\sigma_{i\gamma}$ and $\varepsilon_{i\gamma}$ are the LJ diameter and energy parameters, respectively. The potential of valence electrons acting on a single solvent site, $\varphi^{(e)}(\mathbf{r})$, is calculated in the density fitting procedure.[70]

The effective potential of solvent acting on the solute electrons, $\upsilon_{solv}(\mathbf{r})$, is the functional derivative of the excess chemical potential of solvation with respect to the electron density distribution of the solute. In the 3D-KH form of the excess chemical potential, Equation 14.13, this leads to the expression[7,8]

$$\upsilon_{solv}(\mathbf{r}) = \frac{\delta\Delta\mu_{solv}}{\delta n_e(\mathbf{r})} = \rho \sum_\gamma \int d\mathbf{r}' h_\gamma(\mathbf{r}) \upsilon_\gamma^{ps}(|\mathbf{r} - \mathbf{r}'|), \tag{14.21}$$

where $\upsilon_\gamma^{ps}(|\mathbf{r} - \mathbf{r}'|)$ is the contribution of site γ into the pseudopotential of a solvent molecule acting on an external electron that is given by the variational derivative of the classical site potential with respect to the valence electron density[8]

$$\upsilon_\gamma^{ps}(|\mathbf{r} - \mathbf{r}'|) = \frac{\delta u_\gamma(\mathbf{r})}{\delta n^{(e)}(\mathbf{r}')}. \tag{14.22}$$

Potential 14.21 signifies the mean field approximation, which follows essentially from the use of the solvation free energy in the form of Equation 14.15. The analytical first derivative with respect to the nuclear coordinates $\mathbf{R}_i$ is obtained by differentiation of the free energy

$$\frac{dA[n_e(\mathbf{r}), \{\rho_\gamma(\mathbf{r})\}]}{d\mathbf{R}_i} = \frac{dE_{solute}[n_e(\mathbf{r})]}{d\mathbf{R}_i} + \frac{d(\Delta\mu_{solv}[n_e(\mathbf{r}), \{\rho_\gamma(\mathbf{r})\}]}{d\mathbf{R}_i}, \tag{14.23}$$

where the former term has the same structure as in the gas phase case. The latter term is derived from the excess chemical potential in the form of Equation 14.15. Its variation can be written as[7,8]

$$\Delta\mu_{solv} = kT\rho^v \sum_\gamma \int d\mathbf{r} \left[h_\gamma(\mathbf{r})\delta h_\gamma(\mathbf{r})\Theta(-h_\gamma(\mathbf{r})) - \delta c_\gamma(\mathbf{r}) - \frac{1}{2}\delta(h_\gamma(\mathbf{r})c_\gamma(\mathbf{r})) \right]$$

$$= \rho^v \sum_\gamma \int d\mathbf{r} g_\gamma(\mathbf{r})\delta u_\gamma(\mathbf{r}). \tag{14.24}$$

Similar to Kovalenko and Hirata,[7,8] we have the following variation of the classical potential between the solute and solvent site γ:

$$\delta u_\gamma = \int d\mathbf{r} \left[\left(\frac{\delta u_\gamma(\mathbf{r})}{\delta n^{(e)}(\mathbf{r}')} \right) \delta n^{(e)}(\mathbf{r}') + \left(\frac{\delta u_\gamma(\mathbf{r})}{\delta n_N(\mathbf{r}')} \right) \delta n_N(\mathbf{r}') \right]$$

$$= \int d\mathbf{r}' \upsilon_\gamma^{ps}(|\mathbf{r} - \mathbf{r}'|)[\delta n_N(\mathbf{r}') + \delta n^{(e)}(\mathbf{r}'). \tag{14.25}$$

Substituting Equation 14.24 into Equation 14.25 and using $n_N(\mathbf{r}) = \sum_j q_j \delta(\mathbf{r} - \mathbf{R}_j)$ give the solvation contribution to the free energy gradients

$$\frac{d(\Delta\mu_{solv})}{d\mathbf{R}_i} = \int d\mathbf{r} \left[\rho^v g_\gamma(\mathbf{r}) \frac{\partial u_\gamma^{(sr)}(\mathbf{r})}{\partial \mathbf{R}_i} + \upsilon_{solv}(\mathbf{r}) \frac{\partial n^{(e)}}{\partial \mathbf{R}_i} \right] + q_i \left. \frac{\partial \upsilon_{solv}(\mathbf{r})}{\partial \mathbf{r}} \right|_{\mathbf{r}=\mathbf{R}_i}. \qquad (14.26)$$

Calculation of Expression 14.26 does not require much computational effort. The second term in the square brackets is calculated together with the gradients of the exchange-correlation potential,[76] and the rest is calculated using the 3D-RISM procedure. Notice that the first term in the square brackets contributes little to the gradients because large values of the derivatives of the short-range potential $u_\gamma^{(sr)}(\mathbf{r})$ around the core are suppressed by the distribution function $g\gamma(r)$ exponentially decaying in that region. We emphasize that Formula 26 differs from the analytical free energy derivative following from the 1D-RISM scheme[77] by the absence of the term representing the change in the solute–solvent correlations with the intramolecular distribution functions (intermolecular matrix).[9]

The 3D-RISM–KH solvation method can be combined in a self-consistent field approach with any multireference electronic structure theory. Sato et al. pioneered such a combination of 3D-RISM–KH and *ab initio* complete active space SCF method.[77]

14.4   ION EXCHANGE

The metal ion exchange of zeolite Brønsted protons leads to the formation of Lewis acid sites that are the strongest adsorption centers in zeolites. Lewis sites exhibit the highest adsorption energies,[78] and molecules adsorbed on these sites yield the largest molecular stretching mode frequency shifts.[79] This is because the coordination by zeolite framework O atoms cannot provide complete coordination to the metal ions. Thus, cations move closer to the framework and exhibit extraordinary adsorption power.[23] Metal cations can be reduced and form metal clusters embedded in the zeolite framework.[80,81] The interaction between metal cations and zeolites is a combination of three factors: covalent bonding between the cation and the framework O atoms, electrostatic bonding between the cation and the partially ionic O atoms adjacent to the Al substitution site, and electrostatic interaction between the cation and the positively charged Al center.[24]

In Figure 14.1 (left), we show the preferred sites for Ag^+ and Cu^+ in bulk chabazite. For Ag^+, we found three preferred sites—near the center of the six-membered ring, near the center of the hexagonal prism, and in the eight-membered ring, with relative energies of 0, –2, and –4 kJ/mol, respectively. These small energy differences show that the three sites have similar stability. For comparison, x-ray crystallographic determination of fully Ag^+-exchanged chabazite with Si/Al ratio of 2.18 also found Ag^+ in these three sites.[82] The bond alternation determined from the x-ray[82] is in agreement with our modeling results. For Cu^+, we found two preferred sites—near the center of the six-membered ring and in the eight-membered

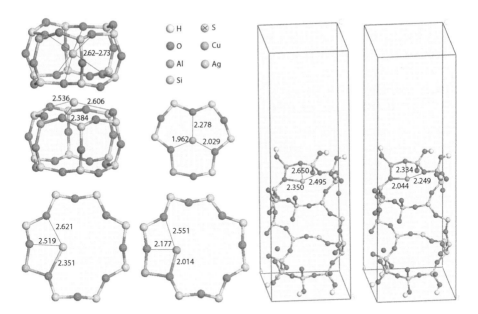

FIGURE 14.1   Ag⁺- and Cu⁺-exchanged bulk (left) and slab chabazite (right).[47]

ring, with relative energies of −43 kJ/mol and 0 kJ/mol, respectively.[47] Analysis of Cu K-edge x-ray absorption spectroscopy (extended x-ray absorption fine structure) data for Cu⁺-exchanged faujasite has given average Cu–O distance of 1.99 Å.[83] This value is in relatively good agreement with our modeling results.[47] For comparison, Solans-Monfort et al. report two Cu⁺-preferred sites from periodic DFT calculations. These authors obtain Cu–O bond lengths of 2.02, 2.10, and 2.25 Å near the center of the six-membered ring, whereas in the eight-membered ring, the Cu–O bond lengths are 2.06 Å and 2.22 Å. These authors also found that Cu⁺ in the eight-membered ring acts as a stronger H_2 adsorbent than near the center of the six-membered ring.[25] The B3LYP functional used in the latter study is known to yield longer metal–ligand bonds relative to the experiment.[84,85]

We have modeled Brønsted acid sites on chabazite nanoparticle surface by using periodic cells designed by termination along the (00–3) and (003) planes and Al substitution. Our results have shown that the most stable structure contains Al on the surface and has a periodic cell designed from 1 × 1 × 1.6 chabazite unit cells (9.42 × 9.42 × 15 Å) cut along the (003) plane and a 13.3-Å-thick vacuum slab.[45] We also found that the most stable Brønsted site contains a proton in the eight-membered ring near the surface termination.[45] This is important because we expect that bitumen fragment adsorption would occur on the chabazite surface due to the large fragment size. Hence, we selected this site for the bitumen fragment adsorption modeling.[47]

In Figure 14.1 (right), we show the optimized geometry of Ag⁺ and Cu⁺ exchanged at the most stable Brønsted site. Metal ions ion-exchanged on this site are undercoordinated and would be highly adsorptive.[23] It is interesting to note that within the eight-membered surface pore mouth, Cu⁺ forms three Cu–O bonds compared to two for the bulk. Calligaris et al. describe several coordination configurations of Ag⁺ in the eight-membered ring that resemble our slab model.[82] Overall, the slab Cu–O and Ag–O bond lengths are slightly longer than in bulk because of the reduced structure rigidity near the surface termination.[47]

14.5 PREDICTION OF BITUMEN FRAGMENT ADSORPTION CONFIGURATION

We predict bitumen fragment adsorption configurations by using global softness and Fukui functions. Using the finite difference approximation, the global softness is evaluated as $S = 1/(\text{IP} - \text{EA})$, where IP is the ionization potential and EA is the electron affinity.[86,87] In this work, for a neutral system that contains N electrons, we obtain

$$S = 0.1/[(E_{(N-0.1)} - E_{(N)}) - (E_{(N)} - E_{(N+0.1)})], \tag{14.27}$$

where $E_{(N-0.1)}$, $E_{(N)}$, and $E_{(N+0.1)}$ are the ground state energies of the system with $N - 0.1$, N, and $N + 0.1$ electrons, respectively.

The Fukui function is motivated by the following consideration: if a fraction of an electron δ is transferred to an N electron molecule, it will tend to distribute so as to minimize the energy of the resulting $N + \delta$ electron system.[88] The resulting change in electron density $(\delta\rho(r)/\delta N)^+_{v(r)}$ defined in Equation 14.28 is the nucleophilic Fukui function. In analogy, the $(\delta\rho(r)/\delta N)^-_{v(r)}$ defined in Equation 14.29 is the electrophilic Fukui function. In this work, we use δ values of 0.1 and –0.1 electrons. It is notable that there are similarities between the interpretation of the Fukui functions and the frontier molecular orbital theory proposed by Fukui.[89,90]

$$f_{N^+}(r) = (\delta\rho(r)/\delta N)^+_{v(r)} = \rho_{N+0.1}(r) + \rho_N(r) \tag{14.28}$$

$$f_{N^+}(r) = (\delta\rho(r)/\delta N)^-_{v(r)} = \rho_N(r) + \rho_{N-0.1}(r), \tag{14.29}$$

where $\rho_{N-0.1}(r)$, $\rho_N(r)$, and $\rho_{N+0.1}(r)$ are the electron densities of the system with $N - 0.1$, N, and $N + 0.1$ electrons, respectively, all evaluated at the geometry of the N electron system.

In the finite difference approximation, the condensed Fukui functions[87] of atom x in a molecule that contains N electrons are defined in Equations 14.30 and 14.31.

$$f_x^+ = 10\left[q_x(N+0.1) - q_x(N)\right] \quad \text{(for nucleophilic attack)} \tag{14.30}$$

$$f_x^+ = 10\left[q_x(N) - q_x(N-0.1)\right] \quad \text{(for electrophilic attack)}, \tag{14.31}$$

where q_x is the charge of atom x. For q_x calculation, we use the Hirshfeld,[91] Delley,[92] and Mulliken population analysis[93,94] methods. A more detailed discussion on reactivity indices is presented elsewhere.[45]

In Figure 14.2, we show the Fukui functions for slab and bitumen fragments mapped on the electron density isosurface. The Fukui functions are defined and discussed elsewhere.[45] The Fukui function maxima and minima are mapped in red and blue, respectively. The nucleophilic Fukui function map of our Cu^+-exchanged chabazite model slab suggests the reaction configuration preference toward an attack by a nucleophile, such as a bitumen fragment. The red region at the Cu atom indicates that this is the preferred nucleophilic attack site. The yellow–green region at the hydroxyl nest (shown toward the back) indicates that this is a less preferred site, whereas the remainder of the slab that is in blue is not a preferred site. Next, we consider the bitumen fragment reactivity preferences suggested by the electrophilic Fukui function maps. For thiophene, the preferred sites are the C atoms adjacent to the S atom. For benzo- and dibenzothiophene, the most preferred sites are the S atoms and the less preferred sites are the C atoms located at the yellow map regions. Thus, the preferred dibenzothiophene configuration would be at a small angle relative to the slab xy plane. This way, the S atom would interact with the metal atom, and the C atoms would interact with the hydroxyl nest for maximum overlap. The Fukui functions provide suggestions for the preferred adsorption configuration at the low computational cost of a single-point calculation at the optimized geometry of the reactants. For comparison, this configuration sampling could be obtained from quantum molecular dynamics at much higher computational cost.

In Table 14.1, we list selected reactivity indicators for Cu^+- and Ag^+-exchanged slabs as well as bitumen fragments. The global softness (S) values of the Cu^+-exchanged slab are higher than those of the Ag^+ slab, indicating that Cu^+ would be more reactive toward soft nucleophiles. The bitumen fragment global softness

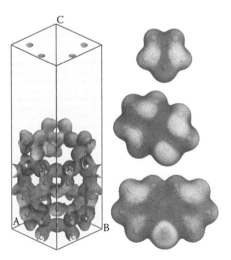

FIGURE 14.2 Nucleophilic Fukui function of Cu^+-exchanged chabazite slab (left) and electrophilic Fukui functions of thiophene, benzothiophene, and dibenzothiophene (right) mapped on the 0.02 $|e|/Å^3$ electron density surface.[47]

TABLE 14.1

Global Softness (S) and Condensed Nucleophilic (f^+) and Electrophilic (f^-) Fukui Functions for Metal Ion-Exchanged Chabazite Slab and Bitumen Fragments

	S, Ha^{-1}	f^+	f^-
Cu (Cu-slab)	8.94	0.515 (0.648)	0.562 (0.567)
Ag (Ag-slab)	7.22	0.612 (0.722)	0.076 (0.077)
Cu (CuCl)	9.99	0.722 (0.729)	0.570 (0.573)
Ag (AgCl)	9.25	0.739 (0.753)	0.418 (0.428)
S (T)	5.49	0.257 (0.253)	0.187 (0.197)
Cα (T)	–	0.139 (0.116)	0.170 (0.122)
S (BT)	6.73	0.146 (0.162)	0.223 (0.238)
S (DBT)	7.50	0.088 (0.114)	0.242 (0.259)
C (B)	4.90	0.106 (0.064)	0.110 (0.061)

Note: The condensed Fukui function values are calculated by using Hirshfeld and Delley (Mulliken) population analysis.[47] T = thiophene, BT = benzothiophene, DBT = dibenzothiophene, B = benzene.

increases in the order benzene < thiophene < benzothiophene < dibenzothiophene. We expect that the strength of bitumen fragment adsorption would also increase in this order.[47]

Condensed Fukui functions provide quantitative predictions on reaction regioselectivity. In the metal cation-exchanged slabs, both f^- and f^+ of Cu$^+$ are higher than for the other atoms. Moreover, the values of f^- and f^+ are comparable, with the Hirshfeld and Delley method yielding higher f^- and Mulliken yielding higher f^+. This clearly suggests that Cu$^+$ can act as both an electron donor and an electron acceptor, as it is well known from the disproportionation reaction $2Cu^+ \rightarrow Cu + Cu^{2+}$. For Ag$^+$ in slab, we obtain f^+ value that is much higher than f^-, indicating that Ag$^+$ is a hard acid.[47]

The bitumen fragment condensed Fukui functions could help us to predict their reactivity. For thiophene, the S atom f^+ and f^- values are higher than the C atom adjacent to S (Cα). These values suggest that thiophene could react with S or Cα atom, depending on the reactivity of the other reactant. The partitioning of the molecular Fukui function to condensed (atomic) Fukui functions is strongly dependent on the population analysis method used. The performance of different condensed Fukui function calculation methods is discussed elsewhere.[45,95] For benzo- and dibenzothiophene, the S atom f^- values are larger than the respective f^+ values. The highest bitumen fragment f^- values increase in the order benzene < thiophene < benzothiophene < dibenzothiophene, giving a predicted reactivity order. The reaction site locations predicted from these values are in good agreement with the optimized structures of adsorbed thiophenes, as discussed below.[47]

14.6   BITUMEN FRAGMENT ADSORPTION IN GAS PHASE

In bitumen, sulfur is present primarily as thioethers, sulfides, and thiophene derivatives, with thiophene sulfur being particularly difficult to remove by hydrodesulfurization.[41] Transition metal ion-exchanged natural zeolites could be used as economical and disposable materials for bitumen desulfurization.[5]

In Figure 14.3, we show the optimized geometries and selected optimized bond length of bitumen fragments on Cu^+- and Ag^+-exchanged chabazite slabs that we show in Figure 14.1 (right). The bitumen fragment adsorption to the two metal atoms occurs in similar configurations. Upon bitumen fragment adsorption, the Cu–O and Ag–O bond lengths become shorter relative to the ion-exchanged slabs shown in Figure 14.1 (right). The ion-exchanged zeolite adsorption strength is evaluated from the adsorption energy $E_{ADS} = E_{Z\text{-}Ion\text{-}BF} - E_{Z\text{-}Ion} - E_{BF}$. The $E_{Z\text{-}Ion\text{-}M}$ is the total energy of the ion-exchanged zeolite unit cell with a bitumen fragment adsorbed on it. $E_{Z\text{-}Ion}$ and E_{BF} are the total energies of the ion-exchanged zeolite unit cell and the bitumen fragment, respectively.[47]

The benzene adsorption is relatively weak and does not cause changes in the strong Cu–O and Ag–O bonds (shown as colored sticks), whereas the weak ones (shown as black lines) become slightly shortened. The thiophene bitumen fragment adsorption is stronger than benzene, as demonstrated by the lower E_{ADS} values. For thiophene adsorption, the strong Cu–O and Ag–O bonds become shortened, whereas the weak bonds become elongated. We obtain E_{ADS} values of –150 kJ/mol and –159 kJ/mol for benzene and thiophene on Cu^+ ion-exchanged chabazite, respectively. This result indicates that thiophene adsorption is stronger than benzene. This result is in qualitative agreement with previous small-cluster DFT modeling reports.[42,43] The thiophene adsorption configurations to Ag^+ and Cu^+ via the S atom represent local minima on the potential energy surfaces. The optimized geometries of thiophene adsorbed to Cu^+- and Ag^+-exchanged slab via the S atom are 19 kJ/mol and 4 kJ/mol less stable than the respective structures shown in Figure 14.3. The thiophene adsorption via the C atom adjacent to S is in agreement with the thiophene electrophilic Fukui function prediction.

The benzo- and dibenzothiophene adsorption occurs via the S atoms. Both the metal–O and metal–S bonds are longer for dibenzothiophene relative to benzothiophene. The adsorption energy values for the bitumen fragments in this study suggest stronger adsorption to Cu^+ relative to Ag^+ slabs. This is in agreement with experimental adsorption studies.[42] For the Cu^+-exchanged slab, the E_{ADS} values increase in the order thiophene < benzothiophene < dibenzothiophene < benzene. For the Ag^+-exchanged slab, the E_{ADS} values increase in the order dibenzothiophene ~ benzothiophene < thiophene < benzene. Our benzene and thiophene adsorption strength orders are in agreement with the experimental results of Yang et al.[44] The different trends are due to the spatial constraints of the nanoparticle surface binding site. The larger ionic radius of Ag^+ allows it to sit above the pore mouth and provides for more effective adsorption of larger bitumen fragments.[47]

It would be instructive to compare our geometry optimization results to the report of Yang et al. for thiophene derivative adsorption to CuCl and the zeolite cluster $(OH)_3Si\text{-}OCu\text{-}Al(OH)_3$.[44] These authors obtain a cluster Cu–O distance of 2.14 Å

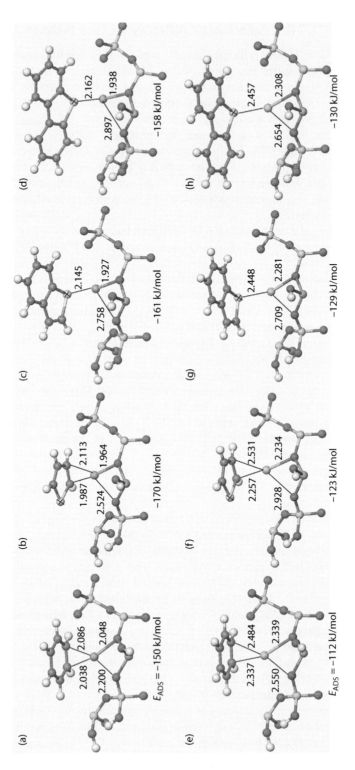

FIGURE 14.3 Enlarged view of the adsorption site of benzene and thiophenes adsorbed and fully optimized on the Cu⁺ (a–d) and Ag⁺-exchanged chabazite slabs (e–h) shown in Figure 14.1 (right).[47]

that is longer than our value of 2.044 Å (Figure 14.1, right). The thiophene adsorption is modeled via the S atom and the Cu–S distance of 2.50 Å is longer than our value of 2.151 Å. The Cu–S bond lengths for adsorption of benzo- and dibenzothiophene of 2.49 Å are also longer than ours. This could be due to the zeolite framework electron conjugation and long-range electrostatic effects that our periodic model accounts for. It is important to note that the cluster results are strongly dependent on the selection of cluster size and shape.[96] Moreover, small clusters are not reliable zeolite crystal models, especially when the dangling bonds are close to adsorbates or any reacting molecule.[97] On the other hand, periodic DFT does not require structure-related selections.

The preferred thiophene adsorption configuration on nanoparticles is not well understood. The thiophene highest occupied molecular orbital (HOMO) predicted from B3LYP calculations contains the π-component of the C=C bond but not the S atom lone pair, suggesting π-coordination to one or both of the C=C bonds. The LUMO is π-antibonding with respect to all adjacent atoms except for the distal C–C bond. The metal–S bond formation has been explained by considering a linear combination between the HOMO-1 and HOMO-2.[98] This frontier orbital description obtained is in agreement with our results. The literature on thiophene adsorption on zeolite surfaces is limited to small zeolite clusters. The long Cu–S distances produced by the cluster approach[44] suggest that a more accurate method is needed. Although the thiophene adsorption on Mo sulfide clusters is more extensively studied, the thiophene adsorption mode remains controversial. Frequency response experiments suggest that thiophene is first π-adsorbed on the transition metal ion in the zeolite Y cage and then large amounts of thiophene are adsorbed by directly forming metal–S bonds.[99] Raman and infrared spectroscopic studies suggest that thiophene is adsorbed via the S atom.[100]

In Figure 14.4, we show the HOMOs of Cu+-exchanged chabazite that contain adsorbed thiophene and benzene. The Cu+-exchanged chabazite HOMO (left) has Cu d_{xy} and O p character. The three Cu–O bonds have σ-antibonding character, as evidenced by the opposite eigenvalue signs, and the O atom with the largest node forms the longest Cu–O bond. Upon adsorption of thiophene (middle), one of the Cu–O bonds breaks. From the Hirshfeld, Delley, and Mulliken population analysis methods, we obtain that upon adsorption, thiophene donates 0.091 and 0.356 e,

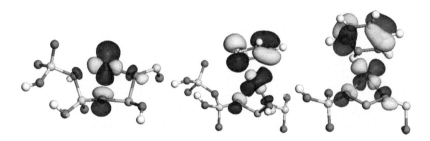

FIGURE 14.4 Schematic diagrams of the HOMOs of Cu+-exchanged chabazite slab (left) that contains adsorbed thiophene (middle) and benzene (right). The entire slab is as in Figure 14.1 (right). Isosurfaces of 0.03 |e|/Å³ and –0.03 |e|/Å³ are shown in dark and light gray, respectively.[47]

respectively. The electron donation from thiophene[98] changes the Cu frontier orbital occupancy. The Cu d_z^2 orbital forms two σ bonds to O atoms and a σ bond to the thiophene C atom adjacent to the S atom. The HOMO of the benzene adsorption (right) has Cu d_{xy} character similar to that of the Cu^+-exchanged chabazite (left) because benzene is a weaker electron donor. From the Hirshfeld, Delley, and Mulliken population analysis methods, we obtain that upon adsorption, benzene donates 0.020 and 0.302 e, respectively. The Cu d_{xy} orbital forms two σ bonds to O atoms and a π bond to a benzene C=C bond. The thiophene adsorption is stronger because it involves electron donation from thiophene to metal ion, as shown by Sargent and Titus for metal–S coordination.[98] The HOMOs for Ag^+ exchange and adsorption are qualitatively similar.[47]

14.7 BITUMEN FRAGMENT ADSORPTION IN BENZENE AND QUINOLINE SOLVENTS

In industrial conditions, the interaction of bitumen fragments with zeolite nanoparticles occurs in the presence of liquid phase, typically aromatic hydrocarbons. In this study, we model the liquid phase using the solvents benzene (nonpolar) and quinolone (polar). We calculate the solvation free energy for the adsorption of bitumen fragments on the ion-exchanged chabazite surfaces following a disaggregation path of displacement along the z-axis (Figure 14.5, right) in the presence of these solvents using the 3D-RISM–KH theory of solvation. The vacuum slab of the periodic unit cell is extended along the z-axis to 9.42 × 9.42 × 64.00 Å. This aggregation model in PBC represents a monolayer of bitumen fragments being adsorbed at each zeolite binding site. The 3D-RISM–KH theory allows prediction of processes occurring at large space and time scales in viscous solvents, and provides a more realistic description of the bitumen–zeolite interaction. For comparison, we include the results from a series of COSMO solvation energy calculations in benzene and quinoline solvents.

In Figure 14.5 (top), we show the dependence of the solvation free energy μ_{sol} and the COSMO solvation energy E_{COSMO} on the displacement distance for the Ag–T system in benzene and quinoline solvents. The shape of the μ_{sol} curve is very different from the E_{COSMO} one, with the first and second solvation shell minima and the solvent expulsion barrier in fact missing in the COSMO result. These essential molecular features of the solvation structure contribute significantly to the formation of the adsorption arrangements as well as to the adsorption thermodynamics and kinetics. The solvation free energies in benzene and quinoline solvents feature absolute maxima at 3.2 Å and 2.8 Å that correspond to the first solvation shell solvent expulsion barriers, respectively. The first solvation free energy maximum in the polar solvent quinoline has higher energy than in the nonpolar benzene. Second maxima are notable at 9.5 Å and 7.5 Å in benzene and quinoline solvents, respectively. The COSMO solvation energy dependence is very different from the μ_{sol} one. In benzene and quinoline solvents, the E_{COSMO} value decreases almost plainly with the disaggregation distance. Similar to μ_{sol}, the COSMO solvation energy E_{COSMO} in quinoline is higher than in benzene. The μ_{sol} curves for different bitumen fragments and metal ions have very similar shapes, and the same is observed for the E_{COSMO} curves.

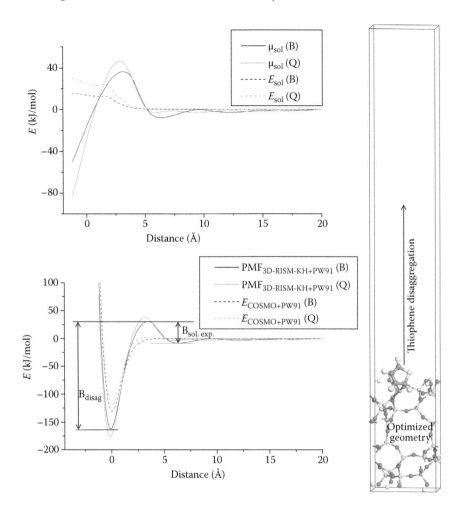

FIGURE 14.5   3D-RISM–KH solvation free energy μ_{sol} and the COSMO solvation energy E_{COSMO} in benzene (B) and quinoline solvents (Q) for disaggregation of a monolayer of thiophene from the surface of Ag+-exchanged chabazite surface (top). 3D-RISM–KH potential of mean force (PMF$_{3D\text{-}RISM\text{-}KH+PW91}$, Equation 14.17) and COSMO + PW91 energy ($E_{COSMO+PW91}$) in benzene (B) and quinoline (Q) for disaggregation of thiophene from Ag+-exchanged chabazite surface (bottom). The disaggregation distance along the z-axis (right panel) is defined relative to the optimized adsorption structure (Figure 14.3).

In Figure 14.5 (bottom), we compare the potential of mean force PMF$_{3D\text{-}RISM\text{-}KH+PW91}$ obtained using the 3D-RISM–KH approach and the COSMO energy $E_{COSMO+PW91}$ as a function of the displacement distance for the Ag–T system in benzene and quinoline solvents. The PMF$_{3D\text{-}RISM\text{-}KH+PW91}$ curve shape is very different from that of the COSMO + PW91 energy. The PMF features an absolute minimum near 0 Å, an absolute maximum at 3.6 Å, and a local minimum at 6.5 Å. All distances are along the z-axis relative to the optimized geometry. At distances larger than 3.6 Å, solvent

TABLE 14.2

Solvation Free Energy Maximum (μ_{sol}^{max}) and Potential of Mean Force (PMF) Minima and Maxima, Barriers for Disaggregation and Solvent Expulsion, Adsorption Energies in Benzene (Quinoline) Solvents That Include COSMO and PW91 Contributions ($E_{COSMO+PW91}$), and Adsorption Energy in the Gas Phase (E_{PW91}) (in kJ/mol)

	E_{PW91}	μ_{sol}^{max}	PMF_{max}^{1st}	PMF_{max}^{2nd}	PMF_{min}^{2nd}	Barrier for Disaggregation	Barrier for Solvent Expulsion	E_{COSMO} at 0.0 Å	$E_{COSMO+PW91}$
Ag-B	-112	32 (40)	-144 (-156)	27 (32)	-9 (-4)	171 (188)	36 (36)	14 (27)	-108 (-95)
Ag-T	-123	36 (46)	-162 (-177)	31 (32)	-8 (-4)	193 (205)	38 (36)	13 (24)	-131 (-95)
Ag-BT	-129	39 (49)	-151 (-149)	23 (29)	-15 (-11)	173 (178)	38 (40)	16 (29)	-125 (-112)
Ag-DBT	-130	40 (50)	-144 (-145)	30 (36)	-7 (-1)	174 (181)	37 (37)	18 (33)	-118 (-104)
Cu-B	-150	70 (74)	-147 (-171)	73 (77)	-7 (-3)	220 (248)	80 (80)	13 (25)	-135 (-123)
Cu-T	-170	73 (77)	-172 (-196)	76 (80)	-7 (-2)	248 (276)	83 (82)	13 (25)	-164 (-152)
Cu-BT	-161	75 (80)	-140 (-148)	80 (96)	-7 (-5)	220 (244)	87 (91)	16 (28)	-139 (-126)
Cu-DBT	-158	84 (93)	-117 (-122)	89 (97)	-8 (-5)	206 (219)	97 (102)	19 (34)	-128 (-113)

Note: The values of E_{PW91}, E_{COSMO}, and $E_{COSMO+PW91}$ are listed for the absolute minimum at 0.0 Å. The μ_{sol} and PMF are calculated using the 3D-RISM-KH method. T = thiophene, BT = benzothiophene, DBT = dibenzothiophene, B = benzene.

molecules enter between the disaggregating solute fragments. These extrema define the energy barriers for the main transitions that occur upon disaggregation. The barrier for disaggregation (B_{disag}) is the difference between the absolute maximum and absolute minimum. This is the energy required for disaggregation to occur in solution. The barrier of disaggregation can be correlated to adsorption energy as a higher barrier of disaggregation corresponds to stronger adsorption. The barrier of solvent expulsion ($B_{\text{sol. exp.}}$) is the difference between the absolute maximum and the second minimum. This is the energy required for the first solvation shell to be removed. These barriers are related to the liquid organization near the solid surface, as governed by the solid–liquid interaction,[65] depend on the aggregation path, and have to be overcome by bitumen molecules approaching the solid surface. In contrast to the PMF, the COSMO + PW91 energy features a Morse potential shape with a minimum at 0 Å.

In Table 14.2, we list the essential solvation and effective interaction energy values for all the eight systems studied. The values of the solvation free energy maxima increase as the solvent polarity and bitumen fragment size are increased as well as from Ag^+ to Cu^+. The $PMF_{\text{max}}^{\text{1st}}$ values decrease as the bitumen fragment size increases from thiophene to benzothiophene and dibenzothiophene. The $PMF_{\text{max}}^{\text{2nd}}$ values increase as solvent polarity is increased and are higher for the Cu^+-exchanged chabazite than for the Ag^+-exchanged chabazite. The $PMF_{\text{min}}^{\text{2nd}}$ values increase as solvent polarity is increased. The barrier for disaggregation of thiophene is higher than those for benzene, benzothiophene, and dibenzothiophene in both benzene and quinoline solvents. The COSMO solvation energies E_{COSMO} increase with solvent polarity. The COSMO energies $E_{\text{COSMO+PW91}}$ correlate very well with the barrier of disaggregation obtained using 3D-RISM–KH in showing that thiophene is the strongest adsorbing fragment and that all fragments adsorb stronger to Cu^+ than to Ag^+-exchanged chabazite.

In Figure 14.6, we present the isosurfaces of the statistical density distributions of a solvent containing thiophene (left) and dibenzothiophene (right) dissolved in benzene at infinite dilution (0.0 M). For comparison, we also show the optimized

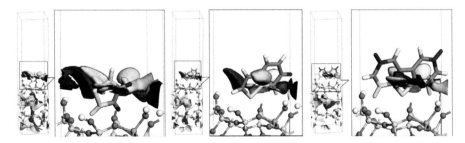

FIGURE 14.6 Solvent distribution function isosurfaces of thiophene (left), benzothiophene (middle), and dibenzothiophene (right) dissolved in benzene (0.0 M) around Ag^+-exchanged chabazite surface. Explicit thiophene, benzothiophene, and dibenzothiophene molecules are shown in stick model at positions optimized using $DMol^3$/GGA-PW91/DNP (Figure 14.3) for reference. The united atom model fragments are colored the same as the solvent distribution isosurfaces. $g_S(T) = 4.7$, $g_{C\alpha}(T) = 7.0$, $g_{H\beta}(T) = 2.3$, $g_S(BT) = 4.4$, $g_{C2a}(BT) = 3.0$, $g_{H5}(BT) = 2.0$, $g_S(DBT) = 8.5$, $g_{\alpha CH} = 4.0$ (DBT), $g_{\gamma CH} = 5.5$ (DBT).

orientation of thiophene and dibenzothiophene in the gas phase (Figure 14.3). Note that the COSMO definition of a solvation cavity is phenomenological, whereas 3D-RISM–KH produces the cavity shape—individual for each solvent interaction site γ—"automatically" in terms of the 3D density distribution functions $g_\gamma(\mathbf{r})$ obtained from Equation 14.13. The isosurface maps of $g_\gamma(\mathbf{r})$ suggest a set of statistically probable solvent orientations that are more realistic for a solvent in liquid phase at room temperature than the gas phase geometry.

14.8 CONCLUSION

We apply PBC-DFT for modeling of the preferred Cu^+ and Ag^+ sites in the bulk and slab of chabazite and the enhanced adsorption of thiophenic bitumen fragments on the slab. We found that the preferred sites of Cu^+ exchanged in bulk chabazite are near the center of the six-membered ring and in the eight-membered ring. In addition to these two sites, Ag^+ occupies the center of the hexagonal prism. Both Cu^+ and Ag^+ are stable at the eight-membered ring pore mouth of the chabazite slab. Thiophene adsorbs preferentially to benzene on chabazite slabs that are exchanged with Cu^+ and Ag^+ cations. The adsorption energies of bitumen fragments on the Ag^+-exchanged chabazite slabs increase in the order benzene < thiophene < benzothiophene < dibenzothiophene. The adsorption energies of bitumen fragments on the Cu^+-exchanged chabazite slabs increase in the order benzene < dibenzothiophene < benzothiophene < thiophene. Our results show that Fukui functions can be safely used as a guide to find the preferred adsorption configuration of bitumen fragments on transition metal cation-exchanged chabazite surfaces. However, local steric effects can alter predicted binding and reactivity trends. We intend to develop a systematic method for description of zeolite nanoparticle surface reactivity toward specific bitumen fragments. These results can be used to map zeolite nanoparticle surfaces for selective adsorption of organic macromolecules, including bitumen.

Solvation behavior can be effectively predicted using electronic structure methods coupled with solvation methods, for example, the combination of continuum solvation methods such as COSMO with DFT as implemented in DMol^3 of Accelrys Materials Studio.[58] An attractive alternative is statistical–mechanical 3D-RISM–KH molecular theory of solvation that predicts, from the first principles, the solvation structure and thermodynamics of solvated macromolecules with full molecular detail at the level of molecular simulation. In particular, this is illustrated here on the adsorption of bitumen fragments on zeolite nanoparticles. Furthermore, we have shown that the self-consistent field combinations of the KS-DFT and the OFE method with 3D-RISM–KH can predict electronic and solvation structure, and properties of various macromolecules in solution in a wide range of solvent composition and thermodynamic conditions.[9,10] This includes the electronic structure, geometry optimization, reaction modeling with transition states,[9] spectroscopic properties,[18] adsorption strength and arrangement,[16] supramolecular self-assembly,[11–15,101–104] and other effects for macromolecular systems in pure solvents, solvent mixtures, electrolyte solutions,[105–107] ionic liquids,[108] and simple and complex solvents confined in nanoporous materials.[109] Currently, the self-consistent field KS-DFT/3D-RISM–KH multiscale method is available only in the ADF software.

ACKNOWLEDGMENTS

This work was supported by the Centre for Oil Sands Innovation, University of Alberta, and the National Research Council of Canada. The computations were supported by the Integrated Nanosystems Research Facility at the University of Alberta.

REFERENCES

1. E. P. Newell. Canada's oil sands industry comes of age. *Oil Gas J.* **97**, 1999, 44–53.
2. T. Jiang, G. Hirasaki, C. Miller, K. Moran and M. Fleury. Diluted bitumen water-in-oil emulsion stability and characterization by nuclear magnetic resonance (NMR) measurements. *Energy Fuels* **21**, 2007, 1325–1336.
3. M. Occelli, editor. Fluid catalytic cracking. II. Concepts in catalyst design. ACS Symposium Ser., ACS, Washington, D.C., 1991.
4. S. Ng, Y. Zhu, A. Humphries, L. Zheng, F. Ding, L. Yang and S. Yui. FCC study of Canadian oil-sands derived vacuum gas oils. 1. Feed and catalyst effects on yield structure. 2. Effects of feedstocks and catalysts on distributions of sulfur and nitrogen in liquid products. *Energy Fuels* **16**, 2002, 1196–1221.
5. S. M. Kuznicki, W. C. McCaffrey, J. Bian, E. Wangen, A. Koenig and C. H. Lin. Natural zeolite bitumen cracking and upgrading. *Microporous Mesoporous Mater.* **105**, 2007, 268–272.
6. D. S. Scott, D. Radlein, J. Piskorz, P. Majerski and T. J. W. deBruijn. Upgrading of bitumen in supercritical fluids. *Fuel* **80**, 2001, 1087–1099.
7. A. Kovalenko. Three-dimensional RISM theory for molecular liquids and solid-liquid interfaces. In *Molecular Theory of Solvation*, Fumio Hirata (Ed.) Series: Understanding Chemical Reactivity, Paul G. Mezey (Ed.), vol. 24, Kluwer Academic Publishers, Dordrecht, 2003, 360 p.) pp.169–275.
8. A. Kovalenko and F. Hirata. Self-consistent description of a metal–water interface by the Kohn–Sham density functional theory and the three-dimensional reference interaction site model. *J. Chem. Phys.* **110**, 1999, 10095–10112.
9. S. Gusarov, T. Ziegler and A. Kovalenko. Self-consistent combination of the three-dimensional RISM theory of molecular solvation with analytical gradients and the Amsterdam density functional package. *J. Phys. Chem. A* **110**, 2006, 6083–6090.
10. D. Casanova, S. Gusarov, A. Kovalenko and T. Ziegler. Evaluation of the SCF combination of KS-DFT and 3D-RISM–KH: Solvation effect on conformational equilibria, tautomerization energies, and activation barriers. *J. Chem. Theory. Comput.* **3**, 2007, 458–476.
11. R. Chhabra, J. G. Moralez, J. Raez, T. Yamazaki, J.-Y. Cho, A. J. Myles, A. Kovalenko and H. Fenniri. One-pot nucleation, growth, morphogenesis, and passivation of 1.4 nm Au nanoparticles on self-assembled rosette nanotubes. *J. Am. Chem. Soc.* **132**, 2010, 32–33.
12. T. Yamazaki, H. Fenniri and A. Kovalenko. Structural water drives self-assembly of organic rosette nanotubes and holds host atoms in the channel. *Chem. Phys. Chem.* **11**, 2010, 361–367.
13. T. Yamazaki, N. Blinov, D. Wishart and A. Kovalenko. Hydration effects on the HET-s prion and amyloid-β fibrillous aggregates. Studied with 3D molecular theory of salvation. *Biophys. J.* **95**, 2008, 4540–4548.
14. T. Yamazaki and A. Kovalenko. Spatial decomposition analysis of the thermodynamics of cyclodextrin complexation. *J. Chem. Theory Comput.* **5**, 2009, 1723–1730.
15. Q. Li, S. Gusarov, S. Evoy and A. Kovalenko. Electronic structure, binding energy, and solvation structure of the streptavidin–biotin supramolecular complex: ONIOM and 3D-RISM study. *J. Phys. Chem. C* **113**, 2009, 9958–9967.

16. S. R. Stoyanov, S. Gusarov and A. Kovalenko. Multiscale modeling of asphaltene disaggregation. *Mol. Simul. Spec. Issue Mater. Studio* **34**, 2008, 953–960.

17. T. A. Wesolowski and A. Warshel. Frozen density functional approach for ab initio calculations of solvated molecules. *J. Phys. Chem.* **97**, 1993, 8050–8053.

18. J. W. Kaminski, S. Gusarov, T. Wesolowski and A. Kovalenko. Modeling solvatochromic shifts using the orbital-free embedding potential at statistically mechanically averaged solvent density. *J. Phys. Chem. A* **114**, 2010, 6082–6096.

19. G. Maurin, P. Senet, S. Devautour, P. Gaveau, F. Henn, V. E. Van Doren and J. C. Giuntini. Combining the Monte Carlo technique with 29Si NMR spectroscopy: Simulations of cation locations in zeolites with various Si/Al ratios. *J. Phys. Chem. B* **105**, 2001, 9157–9161.

20. L. Campana, A. Selloni, J. Weber and A. Goursot. Cation siting and dynamical properties of zeolite offretite from first-principles molecular dynamics. *J. Phys. Chem. B* **101**, 1997, 9932–9939.

21. I. I. Zakharov, Z. R. Ismagilov, S. Ph. Ruzankin, V. F. Anufrienko, S. A. Yashnik and O. I. Zakharova. Density functional theory molecular cluster study of copper interaction with nitric oxide dimer in Cu-ZSM-5 catalysts. *J. Phys. Chem. C* **111**, 2007, 3080–3089.

22. N. U. Zhanpeizov, W. S. Ju, M. Matsuoka and M. Anpo. Quantum chemical calculations on the structure and adsorption properties of NO and N_2O on Ag+ and Cu+ ion-exchanged zeolites. *Struct. Chem.* **14**, 2003, 247–255.

23. V. R. Saunders, R. Dovesi, C. Roetti, M. Causà, N. M. Harrison, R. Orlando and C. M. Zicovich-Wilson. *CRYSTAL-98 User's Manual*. Torino, Italy: Università di Torino, 1999.

24. B. Civalleri, A. M. Ferrari, M. Llunell, R. Oralndo, M. Merawa and P. Ugliengo. Cation selectivity in alkali-exchanged chabazite: An ab initio periodic study. *Chem Mater.* **15**, 2003, 3996–4004.

25. X. Solans-Monfort, V. Branchadell, M. Sodupe, C. M. Zicovich-Wilson, E. Gribov, G. Spoto, C. Busco and P. Ugliengo. Can Cu+-exchanged zeolites store molecular hydrogen? An ab-initio periodic study compared with low-temperature FTIR. *J. Phys. Chem. B* **108**, 2004, 8278–8286.

26. G. Kresse and J. Furthmueller. Efficient iterative schemes for ab initio total-energy calculations using a plane-wave basis set. *Phys. Rev. B* **54**, 1996, 11169–11186.

27. G. Kresse and J. Hafner. Ab initio molecular dynamics for open-shell transition metals. *Phys. Rev. B* **48**, 1993, 13115–13118.

28. G. Kresse and J. Hafner. Ab initio molecular-dynamics simulation of the liquid–metal–amorphous–semiconductor transition in germanium. *Phys. Rev. B* **49**, 1994, 14251–14269.

29. G. Kresse and J. Furthmuller. Efficiency of ab-initio total energy calculations for metals and semiconductors using a plane-wave basis set. *Comput. Mater. Sci.* **6**, 1996, 15–50.

30. L. Benco, T. Bučko, J. Hafner and H. Toulhoat. A density functional theory study of molecular and dissociative adsorption of H2 on active sites in mordenite. *J. Phys. Chem. B* **109**, 2005, 22491–22501.

31. R. K. Roy, S. Krishnamurti, P. Geerlings and S. Pal. Local softness and hardness based reactivity descriptors for predicting intra- and intermolecular reactivity sequences: Carbonyl compounds. *J. Phys. Chem. A* **102**, 1998, 3746–3755.

32. R. C. Deka, R. K. Roy and K. Hirao. Local reactivity descriptors to predict the strength of Lewis acid sites in alkali cation-exchanged zeolites. *Chem. Phys. Lett.* **389**, 2004, 186–190.

33. C. Makedonas and C. A. Mitsopoulou. Introduction of modified electronic parameters—Searching for a unified ligand properties scale through the electrophilicity index concept. *Eur. J. Inorg. Chem.* **26**, 2007, 4176–4189.

34. L. J. Bartolotti and P. W. Ayers. An example where orbital relaxation is an important contribution to the Fukui function. *J. Phys. Chem. A* **109**, 2005, 1146–1151.

35. D. Guerra, P. Fuentealba, A. Aizman and R. Contreras. β-Scission of thioimidoyl radicals ($R1$-N-C¼S-$R2$): A theoretical scale of radical leaving group ability. *Chem. Phys. Lett.* **443**, 2007, 383–388.
36. M. Elanany, B.-L. Su and D. P. Vercauteren. Strong templating effect of TEAOH in the hydrothermal genesis of the AlPO4–5 molecular sieve: Experimental and computational investigations. *J. Mol. Catal. A* **270**, 2007, 295–301.
37. A. Chatterjee. A reactivity index study to rationalize the effect of dopants on Brønsted and Lewis acidity occurring in MeAlPOs. *J. Mol. Graphics Modell.* **24**, 2006, 262–270.
38. K. Hemelsoet, D. Lesthaeghe, V. Van Speybroeck and M. Waroquier. Bifunctional acid–base catalyzed reactions in zeolites from the HSAB viewpoint. *Chem. Phys. Lett.* **419**, 2006, 10–15.
39. X. Rozanska, R. A. van Santen and F. Hutschka. A DFT study of the cracking reaction of thiophene activated on zeolite catalysts: The role of the basic Lewis site. *Stud. Surf. Sci. Catal.* **135**, 2001, 2611–2617.
40. H. Soscun, O. Castellano, J. Hernandez and A. Hinchliffe. Theoretical study of the structural, vibrational, and topologic, properties of the charge distribution of the molecular complexes between thiophene and Brønsted acid sites in zeolites. *Int. J. Quantum Chem.* **87**, 2002, 240–253.
41. R. T. Yang, A. J. Hernandez-Maldonado and F. H. Yang. Desulfurization of transportation fuels with zeolites under ambient conditions. *Science* **301**, 2003, 79–81.
42. A. J. Hernandez-Maldonado and R. T. Yang. Desulfurization of liquid fuels by adsorption via π complexation with Cu(I)–Y and Ag–Y zeolites. *Ind. Eng. Chem. Res.* **42**, 2003, 123–129.
43. J. Ambalavanan, F. H. Yang and R. T. Yang. Effects of nitrogen compounds and polyaromatic hydrocarbons on desulfurization of liquid fuels by adsorption via π-complexation with Cu(I)Y zeolite. *Energy Fuels* **20**, 2006, 909–914.
44. F. H. Yang, A. J. Hernandez-Maldonado and R. T. Yang. Selective adsorption of organosulfur compounds from transportation fuels by π-complexation. *Sep. Sci. Technol.* **39**, 2004, 1717–1732.
45. S. R. Stoyanov, S. Gusarov, S. M. Kuznicki and A. Kovalenko. Theoretical modeling of zeolite nanoparticle surface acidity for heavy oil upgrading. *J. Phys. Chem. C* **112**, 2008, 6794–6810.
46. S. R. Stoyanov, C.-X. Yin, M. R. Gray, J. M. Stryker, S. Gusarov and A. Kovalenko. Computational and experimental study of the spectroscopy of nickel(II) and vanadyl porphyrins in crude oils. *J. Phys. Chem. B* **114**, 2010, 2180–2188.
47. S. R. Stoyanov, S. Gusarov and A. Kovalenko. Modeling of bitumen fragment adsorption on Cu⁺ and Ag⁺ exchanged zeolite nanoparticles. *Mol. Simul. Spec. Issue Mater. Studio* **34**, 2008, 943–951.
48. DMol³, release 4.0. Accelrys, Inc., San Diego, CA, 2001.
49. J. P. Perdew, J. A. Chevary, S. H. Vosko, K. A. Jackson, M. R. Pederson, D. J. Singh and C. Fiolhais. Atoms molecules, solids, and surfaces: Applications of the generalized gradient approximation for exchange and correlation. *Phys. Rev. B* **46**, 1992, 6671–6687.
50. C. Lo and B. L. Trout. Density-functional theory characterization of acid sites in chabazite. *J. Catal.* **227**, 2004, 77–89.
51. F. Pascale, P. Ugliengo, B. Civalleri, R. Orlando, P. D'Arco and R. Dovesi. Hydrogarnet defect in chabazite and sodalite zeolites: A periodic Hartree–Fock and B3-LYP study. *J. Chem. Phys.* **117**, 2002, 5337–5346.
52. S. Tsuzuki and H. P. Luthi. Interaction energies of van der Waals and hydrogen bonded systems calculated using density functional theory: Assessing the PW91 model. *J. Chem. Phys.* **114**, 2001, 3949–3957.
53. B. Delley. An all-electron numerical method for solving the local density functional for polyatomic molecules. *J. Chem. Phys.* **92**, 1990, 508–517.

54. B. Delley. From molecules to solids with the DMol3 approach. *J. Chem. Phys.* **113**, 2000, 7756–7764.

55. S. Tosoni, F. Pascale, P. Ugliengo, R. Orlando, V. R. Saunders and R. Dovesi. Quantum mechanical calculation of the OH vibrational frequency in crystalline solids. *Mol. Phys.* **103**, 2005, 2549–2558.

56. M. Calligaris, G. Nardin, L. Randaccio and P. C. Chiaramonti. Cation-site location in a natural chabazite. *Acta Crystallogr. B* **38**, 1982, 602–605.

57. A. Klamt and G. Schüürmann. COSMO: A new approach to dielectric screening in solvents with explicit expressions for the screening energy and its gradient. *J. Chem. Soc., Perkin Trans.* **2**, 1993, 799.

58. B. Delley. The conductor-like screening model for polymers and surfaces. *Mol. Simul.* **32**, 2006, 117–123.

59. D. Chandler and H. C. Andersen. Optimized cluster expansions for classical fluids. II. Theory of molecular liquids. *J. Chem. Phys.* **57**, 1972, 1930–1937.

60. F. Hirata and P. J. Rossky. An extended RISM equation for molecular polar fluids. *Chem. Phys. Lett.* **83**, 1981, 329–334.

61. J. Perkyns and B. M. Pettitt. A dielectrically consistent interaction site theory for solvent–electrolyte mixtures. *Chem. Phys. Lett.* **190**, 1992, 626–630.

62. C. Wohlfarth. In *Static Dielectric Constants of Pure Liquids and Binary Liquid Mixtures.* Landolt–Börnstein New Series, Group IV, Volume 6, edited by O. Madelung. Berlin: Springer-Verlag, 1991.

63. Y. Y. Akhadov. *Dielectric Properties of Binary Solutions. A Data Handbook.* Oxford: Pergamon Press, 1981.

64. A. K. Rappe, C. J. Casewit, K. S. Colwell, W. A. Goddard-III and W. M. Skiff. UFF: A full periodic table force field for molecular mechanics and molecular dynamics simulations. *J. Am. Chem. Soc.* **114**, 1992, 10024–10035.

65. A. E. Kobryn and A. Kovalenko. Molecular theory of hydrodynamic boundary conditions in nanofluidics. *J. Chem. Phys.* **129**, 2008, 134701.

66. Ş. Erkoç. *Annual Reviews of Computational Physics IX*, edited by D. Stauffer. Singapore: World Scientific, 2001, 1–104.

67. A. K. Rappe and W. A. Goddard. Charge equilibration for molecular dynamics simulations. *J. Phys. Chem.* **95**, 1991, 3358–3363.

68. C. Cramer and D. Truhlar. Implicit solvation models: Equilibria, structure, spectra, and dynamics. *Chem. Rev.* **99**, 1999, 2161–2200.

69. E. J. Baerends, P. Ros and D. E. Ellis. Self-consistent molecular Hartree–Fock–Slater calculations. I. The computational procedure. *Chem. Phys.* **2**, 1973, 41–51.

70. G. teVelde, F. Bickelhaupt, S. van Gisbergen, C. Guerra, E. Baerends, J. Snijders and T. Ziegler. Chemistry with ADF. *J. Comput. Chem.* **22**, 2001, 931.

71. C. F. Guerra, J. Snijders, G. teVelde and E. Baerends. Towards an order-N DFT method. *Theor. Chem. Acc.* **99**, 1998, 391–403.

72. S. Genheden, T. Luchko, S. Gusarov, A. Kovalenko and U. Ryde. An MM/3D-RISM approach for ligand binding affinities. *J. Phys. Chem. B* **114**, 2010, 8505–8516.

73. K. Yoshida, T. Yamaguchi, A. Kovalenko and F. Hirata. Structure of *tert*-butyl alcohol–water mixtures studied by the RISM theory. *J. Phys. Chem. B* **106**, 2002, 5042–5049.

74. A. Kovalenko and F. Hirata. Towards a molecular theory for the van der Waals–Maxwell description of fluid phase transitions. *J. Theor. Comput. Chem.* **1**, 2002, 381–406.

75. T. Imai, A. Kovalenko and F. Hirata. Partial molar volume of proteins studied by the three-dimensional reference interaction site model theory. *J. Phys. Chem. B* **109**, 2005, 6658–6665.

76. L. Verslus and T. Ziegler. The determination of molecular structures by density functional theory. The evaluation of analytical energy gradients by numerical integration. *J. Chem. Phys.* **88**, 1988, 322–328.

77. H. Sato, F. Hirata and S. Kato. Analytical energy gradient for the reference interaction site model multiconfigurational self-consistent-field method: Application to 1,2-difluoroethylene in aqueous solution. *J. Chem. Phys.* **105**, 1996, 1546–1551.

78. L. Benco, T. Bučko, J. Hafner and H. Toulhoat. Ab initio simulation of Lewis sites in mordenite and comparative study of the strength of active sites via CO adsorption. *J. Phys. Chem. B* **108**, 2004, 13656–13666.

79. O. Cairon, T. Chevreau and J.-C. Lavalley. Brønsted acidity of extra framework debris in steamed Y zeolites from the FTIR study of CO adsorption. *J. Chem. Soc., Faraday Trans.* **94**, 1998, 3039–3047.

80. P. Gallezot. Preparation of metal clusters in zeolites. *Mol. Sieves* **3**, 2002, 257–305.

81. K. Seff. Cationic zinc clusters with mean formula Zn6.9+5.4 in the sodalite cavities of zeolite Y (FAU). *Microporous Mesoporous Mater.* **85**, 2005, 351–354.

82. M. Calligaris, A. Mezzetti, G. Nardin and L. Randaccio. Cation sites and framework deformations in dehydrated chabazites. Crystal structure of a fully silver-exchanged chabazite. *Zeolites* **4**, 1984, 323–328.

83. I. J. Drake, Y. Zhang, D. Briggs, B. Lim, T. Chau and A. T. Bell. The local environment of Cu^+ in Cu–Y zeolite and its relationship to the synthesis of dimethyl carbonate. *J. Phys. Chem. B* **110**, 2006, 11654–11664.

84. S. R. Stoyanov, J. M. Villegas, A. J. Cruz, L. L. Lockyear, J. H. Reibenspies and D. P. Rillema. Computational and spectroscopic studies of Re(I) bipyridyl complexes containing 2,6-dimethylphenylisocyanide (CNx) ligand. *J. Chem. Theory Comput.* **1**, 2005, 95–106.

85. J. E. Monat, J. H. Rodriguez and J. K. McCusker. Ground- and excited-state electronic structures of the solar cell sensitizer bis(4,4′-dicarboxylato-2,2′-bipyridine)bis(isothiocyanato)ruthenium(II). *J. Phys. Chem. A* **106**, 2002, 7399–7406.

86. R. G. Pearson. Hard and soft acids and bases. *J. Am. Chem. Soc.* **85**, 1963, 3533–3539.

87. R. G. Parr and W. Yang. Density functional approach to the frontier-electron theory of chemical reactivity. *J. Am. Chem. Soc.* **106**, 1984, 4049–4050.

88. P. W. Ayers and R. G. Parr. Variational principles for describing chemical reactions: The Fukui Function and chemical hardness revisited. *J. Am. Chem. Soc.* **122**, 2000, 2010–2018.

89. K. Fukui, T. Yonezawa and C. Nagata. Theory of substitution in conjugated molecules. *Bull. Chem. Soc. Jpn.* **27**, 1954, 423–427.

90. K. Fukui. Role of frontier orbitals in chemical reactions. *Science* **218**, 1982, 747–754.

91. F. L. Hirshfeld. Bonded-atom fragments for describing molecular charge densities. *Theor. Chim. Acta B* **44**, 1977, 129–138.

92. B. Delley. Calculated electron distribution for tetrafluoroterephthalonitrile (TFT). *Chem. Phys. Lett.* **110**, 1986, 329–338.

93. R. S. Mulliken. Electronic population analysis on LCAO–MO [linear combination of atomic orbital–molecular orbital] molecular wave functions. I. *J. Chem. Phys.* **23**, 1955, 1833–1840.

94. R. S. Mulliken. Electronic population analysis on LCAO–MO [linear combination of atomic orbital–molecular orbital] molecular wave functions. II. Overlap populations, bond orders, and covalent bond energies. *J. Chem. Phys.* **23**, 1955, 1841–1846.

95. R. K. Roy, K. Hirao, S. Krishnamurthy and S. Pal. Mulliken population analysis based evaluation of condensed Fukui function indices using fractional molecular charge. *J. Chem. Phys.* **115**, 2001, 2901–2907.

96. V. V. Mihaleva, R. van Santen and A. P. J. Jansen. Quantum chemical calculation of infrared spectra of acidic groups in chabazite in the presence of water. *J. Chem. Phys.* **119**, 2004, 13053–13060.

97. S. P. Yuan, J. G. Wang, Y. W. Li and S. Y. Peng. Theoretical studies on the properties of acid site in isomorphously substituted ZSM-5. *J. Mol. Catal. A* **178**, 2002, 267–274.

98. A. L. Sargent and E. P. Titus. C–S and C–H bond activation of thiophene by Cp*Rh(PMe3): A DFT theoretical investigation. *Organometallics* **17**, 1998, 65.

99. F. Li, L. Song, L. Duan, X. Li and Z. Sun. A frequency response study of thiophene adsorption in zeolite catalysts. *Appl. Surf. Sci.* **253**, 2007, 8802–8809.

100. P. Mills, S. Korlann, M. E. Bussell, M. A. Reynolds, M. V. Ovchinnikov, R. J. Angelici, C. Stinner, T. Weber and R. Prins. Vibrational study of organometallic complexes with thiophene ligands: Models for adsorbed thiophene on hydrodesulfurization catalysts. *J. Phys. Chem. A* **105**, 2001, 4418–4429.

101. J. G. Moralez, J. Raez, T. Yamazaki, R. K. Motkuri, A. Kovalenko and H. Fenniri. Helical rosette nanotubes with tunable stability and hierarchy. *J. Am. Chem. Soc.* **127**, 2005, 8307–8309.

102. R. S. Johnson, T. Yamazaki, A. Kovalenko and H. Fenniri. Molecular basis for water-promoted supramolecular chirality inversion in helical rosette nanotubes. *J. Am. Chem. Soc.* **129**, 2007, 5735–5743.

103. G. Tikhomirov, T. Yamazaki, A. Kovalenko and H. Fenniri. Hierarchical self-assembly of organic prolatenanospheroids from hydrophobic rosette nanotubes. *Langmuir* **24**, 2008, 4447–4450.

104. T. Yamazaki and A. Kovalenko. Spatial decomposition of solvation free energy based on the 3D integral equation theory of molecular liquid: Application to mini proteins. *J. Phys. Chem. B* **115**, 2011, 310–318.

105. A. Kovalenko and F. Hirata. Potentials of mean force of simple ions in ambient aqueous solution. I. Three-dimensional reference interaction site model approach. *J. Chem. Phys.* **112**, 2000, 10391–10402.

106. A. Kovalenko and F. Hirata. Potentials of mean force of simple ions in ambient aqueous solution. II. Solvation structure from the three-dimensional reference interaction site model approach and comparison with simulations. *J. Chem. Phys.* **112**, 2000, 10403–10417.

107. A. Tanimura, A. Kovalenko and F. Hirata. Structure of electrolyte solutions sorbed in carbon nanospaces, studied by the replica RISM theory. *Langmuir* **23**, 2007, 1507–1517.

108. M. Malvaldi, S. Bruzzone, C. Chiappe, S. Gusarov and A. Kovalenko. Ab initio study of ionic liquids by KS-DFT/3D-RISM–KH theory. *J. Phys. Chem. B* **113**, 2009, 3536–3542.

109. A. Kovalenko. Molecular description of electrosorption in a nanoporous carbon electrode. *J. Comput. Theor. Nanosci.* **1**, 2004, 398–411.

15 Reactive Molecular Dynamics Force Field for the Dissociation of Light Hydrocarbons on Ni(111)

Bin Liu, Mark T. Lusk, James F. Ely,
Adri C.T. van Duin, and William A. Goddard III

CONTENTS

15.1 INTRODUCTION

The C–H bond breaking is the initial and usually also the rate-limiting step in the heterogeneous chemistry involving alkane hydrocarbons. Therefore, it plays a pivotal role in many industrially important processes including steam reforming, cracking, and electrochemical oxidation in solid oxide fuel cells [1–5]. A quantitative understanding of how such hydrocarbons interact with catalyst surfaces continues to be an experimental challenge to chemists and engineers. Nowadays, these research efforts are increasingly relying on the computational interrogations of the elementary steps of such chemistry at atomistic or mesoscale levels.

Of particular interest in this work is the use of *ab initio* data sets to fit empirical potential parameters to be used in molecular dynamics (MD) simulations. ReaxFF is a transferable reactive force field designed by van Duin et al. [6] to cover a wide range of elements in the periodic table including very successful predictions of a variety of hydrocarbon molecular structures. ReaxFF was also developed for boron/nitrogen/hydrogen [7], silicon/silica systems [8], and transition metal complexes [9] systems. So far, the force field has been applied to simulations of many reactive systems [10–14].

The goal of this work is to develop a set of ReaxFF parameters for H/C/Ni systems which are relevant in many industrial processes. It proceeds by separately fitting to *ab initio* data that consist of hydrocarbon molecules [6], atomic Ni–C systems [15], and binding energies of hydrocarbons on small Ni clusters. However, because the predictive accuracy of empirical force fields tends to diminish as the structures diverge from those used to fit the parameters, there is a utility in creating force fields that are optimized to particular classes of interactions.

The physical setting that motivated such an optimized set of MD parameters is methane decomposition on Ni(111). The generic force field was used to quantify the rate of dissociation under conditions of constant volume and species number with the temperature fixed at 1250 K to mimic the fuel cell operating conditions [16]. Although this temperature is somewhat higher than that in real fuel cell operation conditions, which motivated this work, it allowed better statistics to be gathered on hydrocarbon decompositions. The energy barrier for methane dissociative adsorption was first estimated to be 205 kJ/mol by carrying out a single constrained MD dissociation event in which hydrocarbon dissociation was forced through a prescribed reaction path. As detailed below, such a high barrier indicates that dissociation would occur only at a very slow pace. A finite temperature simulation was subsequently carried out on a system consisting of 30 methane molecules above a periodic five-layer 5 × 6 Ni slab for 400 ps, and no dissociation of CH_4 was observed—consistent with the high reaction barrier estimate.

The results of the generic ReaxFF simulations were then compared with theoretical predictions using classical reaction rate theory as governed by

$$\frac{d[CH_4]_{ads}}{dt} = k_{dis.ads.}[CH_4], \tag{15.1}$$

where $[CH_4]_{ads}$ is the surface density of the adsorbed CH_4 molecule (in molecule/m^2), that is, the number of CH_4 molecules per unit surface; $[CH_4]$ is the gas phase density (in molecule/m^3), which can be estimated by dividing the number of CH_4 molecules in the simulation by the volume (calculated using the simulation cell dimensions); and $k_{dis.ads.}$ is the effective rate constant (m/s). According to the kinetics parameters reported by Burghgraef et al. [17], the rate constant for methane dissociation on nickel is 1.05×10^{-6} m/s at 1250 K. This implies that the dissociation probability should be 4.2×10^{-6} during each 400-ps interval at 1250 K. However, this rate estimate is almost certainly too low because the activation energy (200.2 kJ/mol), which was adopted by Burghgraef et al., is much greater than the experimental value of 52.7 kJ/mol ± 5.0 kJ/mol [18]. It therefore makes sense that the dissociation rate of predicted energy barriers using the generic ReaxFF parameter set is too low because it has essentially the same reaction barrier.

To improve the theoretical estimate for the rate constant, density functional theory (DFT) was used to obtain the C–H bond breaking energy barrier, which is 81.9 kJ/mol, although still higher than the experimental measurement, but much lower than the value reported by Burghgraef et al. Using this DFT energy barrier, we obtained an improved estimate of 0.3 dissociation event per 400 ps.

To gain additional confidence in the rate of dissociation, Deutschmann's micro-kinetic modeling approach [19,20] was used to make a second estimate for the dissociation rate of CH_4. It is assumed that the rate constant is the product of the rate of surface collisions and the probability that such collisions result in a reaction, that is, the sticking coefficient. The sticking coefficient is 8×10^{-3} using Deutschmann's mechanism [19]. Assuming that the rate of CH_4 dissociation on surfaces is much greater than that for adsorption, the overall dissociation rate constant for the gas phase is then given by

$$k = \frac{\gamma}{(\Gamma_{tot})^m} \sqrt{\frac{RT}{2\pi W}}, \qquad (15.2)$$

where γ is the sticking coefficient, Γ_{tot} is the adsorption site density (2.6×10^{-9} mol/cm^2), and W is the molecular weight (0.016 kg/mol). Consistent with the reactant being in the gas phase, the exponent parameter m is taken to be zero. The resulting value of rate constant k is then 2.57 m/s at 1250 K, which corresponds to 10 dissociation events every 400 ps.

The two theoretical estimates for the rate of methane decomposition are quite different (0.3 m/s and 2.57 m/s) but both suggest that the generic ReaxFF force field set underestimates the reaction rate.

This, combined with the high value of dissociation barrier measured, has motivated the use of DFT to construct a data set specifically targeting the interactions of light hydrocarbons with Ni(111) surfaces. The information was applied to generate a force field intended to be good at predicting hydrocarbon decomposition on nickel surfaces. Specifically, a training set was populated with the dissociation energies and barriers for H_2, CH_4, C_2H_6, and C_3H_8 on Ni(111). Reactive MD simulations were then performed to consider methane dissociation rates at a finite temperature. The new force field is intended for the use of investigating the dissociation of a range of structurally similar hydrocarbons. To assess this capability, the force field was fitted only to the H_2 and CH_4 data, and then tested by taking advantage of the data of the two other heavier hydrocarbons (i.e., C_2H_6 and C_3H_8). A comparison was then made using the energy barriers for these molecules as estimated by DFT and predicted by ReaxFF.

15.2 METHODS

The DMol^3 DFT code provided in Materials Studio® 4.0 was used to perform all the electronic structure calculations [21,22]. A norm conserving, spin unrestricted, semicore pseudopotential [23] approach was used with electron exchange and correlation accounted for using the Perdew–Wang generalized gradient approximation (GGA) [24].

A five-layer, 3×3, periodic slab was chosen to represent Ni(111) for adsorbates with the bottom two layers frozen with bulk lattice spacing. A 9-Å vacuum space

was placed above the surface to avoid interference from neighboring images. A 2 ×
2 × 1 Monkhorst–Pack k-point mesh [25] was used to sample the Brillouin zone of
the unit cell. Linear/quadratic synchronous transit (LST/QST) analysis was used to
identify transition states [26].

Target parameters in the ReaxFF force field were fit to the training set based
on a successive one-parameter search to reproduce the energy corresponding to
each optimized configuration obtained from DFT calculations [27]. The training
proceeded iteratively by minimizing the respective ReaxFF/DFT differences until
convergence was achieved.

Constant temperature (NVT) MD was used in the validation of the force field.
The velocity Verlet [28] algorithm with a time step of 0.25 fs was used throughout
the simulations. A Berendsen [29] thermostat with a temperature-damping constant
of 250 fs was used to maintain the system at the preset temperature. Constrained
MD was used in the cases to obtain comparative adsorption/desorption barriers and
reaction energies for single CH_4, C_2H_6, and C_3H_8 molecules.

15.3 RESULTS AND DISCUSSION

15.3.1 QUANTUM CHEMISTRY CALCULATIONS

Binding energies of H, CH_3, C_2H_5, and C_3H_7 on planar Ni(111) surfaces were calculated for this work and are listed in Table 15.1, where t_1, b_2, and c_3 represent atop,
bridge, and threefold sites, respectively. Both H and CH_3 favor the high-symmetry
threefold site, whereas the bulkier C_2H_5 and C_3H_7 fragments slightly favor the t_1 site
instead. The dissociation energies and dissociation barriers for H_2, CH_4, C_2H_6, and
C_3H_8 were calculated using the LST/QST algorithm and are shown in Table 15.2.

15.3.2 FORCE FIELD DEVELOPMENT

With the intent of developing force fields tailored to hydrocarbon dissociation on
Ni(111), only H_2 and CH_4 data were used to fit the force fields to test the transferability of the force field to describe the dissociation of the heavier C_2H_6 and C_3H_8
molecules. Force field parameters with respect to the Ni atom, Ni–H/C bonds, and
various H–C–Ni valence angles were chosen to be the fitting target. The training
set consisted of the H atom and CH_3 binding at t_1, b_2, and c_3 sites, the dissociation
energies of H_2 and CH_4, and the dissociation barriers of H_2 and CH_4, as mentioned

---

TABLE 15.1
**Binding Energies (in kJ/mol) for H, CH_3, C_2H_5, and C_3H_7 on
Ni(111) Sites**

Site	H	CH_3	C_2H_5	C_3H_7
t_1	−245	−203	−177	−120
b_2	−273	−199	−157	−120
c_3	−283	−219	−162	−119

TABLE 15.2

Dissociation Energies and Barriers for H$_2$, CH$_4$, C$_2$H$_6$, and C$_3$H$_8$

Adsorbate	Dissociation Energy (kJ/mol)	Dissociation Barrier (kJ/mol)
H$_2$	−82	2
CH$_4$	−122	82
C$_2$H$_6$	−45	84
C$_3$H$_8$	−40	60

Note: Only primary dissociation is considered for the hydrocarbons.

in the previous section. The energies associated with the intermediate geometries generated from the LST/QST algorithm were also included to enforce the prediction reliability of the force field in dealing with the transition states. The fitting results are compared with DFT data in Figure 15.1.

As indicated in Figure 15.1, the generic force field underestimates the H$_2$ dissociation energy and severely overestimates its dissociation barrier. Both problems are corrected in the new fit. The absence of observed CH$_4$, dissociation on Ni(111) can also be explained by Figure 15.1: the dissociation energy and barrier were both overestimated. Both of them have been improved in the new parameter set in this work as well.

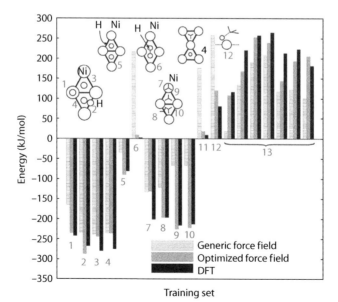

Training set

FIGURE 15.1   (See color insert.) Fitting of ReaxFF parameters against DFT results and comparison with the old force field. 1–4: H binding atop, bridge, threefold hcp, and threefold fcc sites on Ni(111); 5: dissociation energy for H$_2$; 6: dissociation barrier for H$_2$; 7–10: CH$_3$ binding atop, bridge, threefold hcp, and threefold fcc sites on Ni(111); 11: dissociation energy for CH$_4$; 12: dissociation barrier for CH$_4$; 13: the interpolated geometries between 11 and 12.

As a test of the transferability of the optimized force field, the prediction on the adsorption of C_2H_6 and C_3H_8 dissociation is shown in Figure 15.2. A reasonable comparison was obtained, which suggests that this optimized parameter set is transferable to other structurally similar alkane molecules.

15.3.3   Reactive Molecular Dynamics Application

The optimized force field was then applied and tested with the dissociation events at finite temperatures. Figures 15.3a, b, c, d, and e display the energy profiles for the dissociative adsorptions of H_2, CH_4, C_2H_6, and C_3H_8 generated from constrained MD simulations with prescribed reaction pathways. These simulations were performed at a temperature of 5 K because the intent was to reproduce the dissociations kinetics through those *guided* pathways. The inset snapshots in Figure 15.3 illustrate the configurations of some molecular states along the reaction coordinates, for example, predissociation states, the transition states, and the dissociated states. Both the energy profiles and configurations show reasonable behavior for these small hydrocarbons above Ni(111) and are consistent with the *ab initio* data.

An NVT setting was then used to study the rate of CH_4 dissociation on Ni(111). The simulation was initialized by assigning random initial positions and velocities

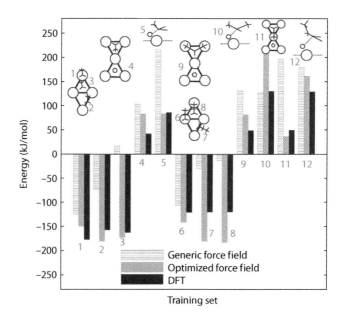

Training set

FIGURE 15.2   (See color insert.) Comparison between the new and old force field on C_2H_6 and C_3H_8 that is not part of training to demonstrate force field transferability. 1–3: C_2H_5 binding atop, bridge, and threefold hcp sites on Ni(111); 4: dissociation energy for C_2H_6; 5: dissociation barrier for C_2H_6; 6–8: C_3H_7 binding atop, bridge, and threefold hcp sites on Ni(111); 9: dissociation for secondary C–H bond of C_3H_8; 10: dissociation barrier for secondary C–H bind of C_3H_8; 11: dissociation energy for primary C–H bond of C_3H_8; 12: dissociation barrier for primary C–H of C_3H_8.

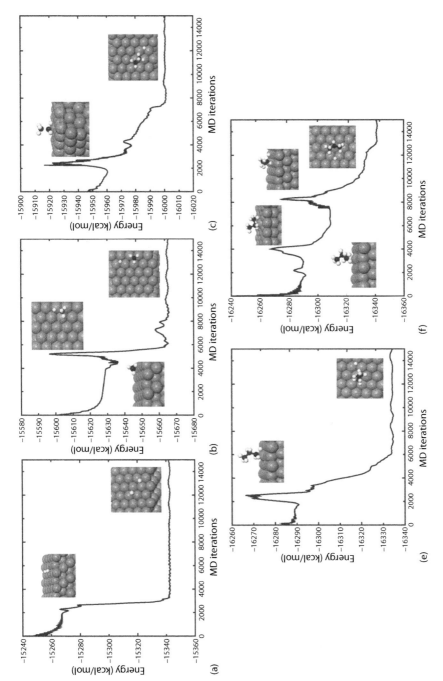

FIGURE 15.3 Energy profiles for dissociation: (a) H_2; (b) CH_4; (c) C_2H_6; (d) C_3H_8 primary C–H bond; (e) C_3H_8 secondary C–H bond.

(scaled to a corresponding temperature of 1250 K) to a single CH_4 molecule followed by measuring the time required for dissociation. The dissociation probability is based on 51 repetitions of such MD simulations, which was fit by a Maxwell–Boltzmann-like distribution expressed in Equation 15.3:

$$f(t) = 3t_{mean}^{-\frac{3}{2}} \sqrt{\frac{3t}{2\pi}} e^{-\frac{3t}{2t_{mean}}}. \tag{15.3}$$

Equation 15.3 was derived by first assuming that the kinetic energy of the molecules exhibits a Maxwell–Boltzmann behavior. A second assumption was made that the dissociation time is proportional to the kinetic energy of the molecule. Figure 15.4 also shows the fit of Equation 15.3 against the simulation data, where the distribution was normalized with the exponent expressed in terms of the mean of the distribution, t_{mean} = 13.07 ps. The distribution was then used to determine a rate constant of k = 270 m/s—higher than the estimates from classical reaction rate theory (0.3 m/s) and the method based on sticking coefficients (2.6 m/s). The single-molecule simulations were stopped immediately following dissociation under the assumption that the initial velocities, positions, and orientations are able to evenly sample the configuration space. The resulting high rate constant prompted a second approach in which a single simulation was run for over 155 ps with 30 methane molecules. A much lower rate constant of 12.9 m/s was obtained and is thought to more accurately reflect the physical setting.

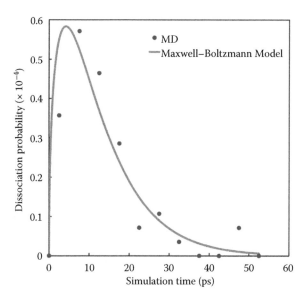

FIGURE 15.4 Fit of simulation time histogram to the Maxwell–Boltzmann model expressed in terms of dissociation time.

15.4 CONCLUSIONS

A new reactive force field set has been developed that is optimized for the dissociation kinetics of small hydrocarbon molecules on Ni(111). The addition of a DFT data set tailored for such interactions resulted in an observed rate of methane decomposition that is higher (12.9 m/s) than either of the estimates derived using experimental data (0.3 m/s and 2.6 m/s). The fitting procedure revealed shortcomings of the generic H/C/Ni force field set—specifically positive dissociation energies (endothermic) and overestimated dissociation barriers. Although the new training was limited to only H_2 and CH_4 dissociation on Ni(111), good reaction barrier agreement between ReaxFF predictions and DFT analysis was obtained for C_2H_6 and C_3H_8 as well—an indication of force field transferability to similarly structured molecules. The newly developed force field still overpredicts the binding energies of H–Ni, C–Ni pairs, and this will be addressed in future optimization work.

ACKNOWLEDGMENTS

This work was partially supported by the U.S. Department of Energy, Office of Science, Grant DE-ER9542165.

REFERENCES

1. E. P. Murray, T. Tsai and S. A. Barnett. A direct-methane fuel cell with a ceria-based anode. *Nature* **400**, 1999, 649–651.
2. S. Park, R. J. Gorte and J. M. Vohs. Applications of heterogeneous catalysis in the direct oxidation of hydrocarbons in a solid-oxide fuel cell. *Appl. Catal, A* **200**, 2000, 55–61.
3. S. D. Park, J. M. Vohs and R. J. Gorte. Direct oxidation of hydrocarbons in a solid-oxide fuel cell. *Nature* **400**, 2000, 265–267.
4. T. V. Choudhary, C. Sivadinarayana, C. C. Chusuei, A. Klinghoffer and D. W. Goodman. Hydrogen production via catalytic decomposition of methane. *J. Catal.* **199**, 2001, 9–18.
5. S. McIntosh and R. J. Gorte. Direct hydrocarbon solid oxide fuel cells. *Chem. Rev.* **104**, 2004, 4845–4865.
6. A. C. T. van Duin, S. Dasgupta, F. Lorant and W. A. Goddard III. ReaxFF: A reactive force field for hydrocarbons. *J. Phys. Chem. A* **105**, 2001, 9396–9409.
7. S. S. Han, J. K. Kang, H. M. Lee, A. C. T. van Duin and W. A. Goddard III. The theoretical study on interaction of hydrogen with single-walled boron nitride nanotubes. I. The reactive force field ReaxFF(HBN) development. *J. Chem. Phys.* **123**, 2005, 114703.
8. A. C. T. van Duin, A. Strachan, S. Stewman, Q. Zhang, X. Xu and W. A. Goddard III. *ReaxFFSiO Reactive Force Field for Silicon and Silicon Oxide Systems*, 2003.
9. W. A. Goddard III, A. C. T. van Duin, K. Chenoweth, M.-J. Cheng, S. Pudar, J. Oxgaard, B. Merinov, Y. H. Jang and P. Persson. Development of the ReaxFF reactive force field for mechanistic studies of catalytic selective oxidation processes on BiMoOx. *Top. Catal.* **38**, 2006, 93–103.
10. D. W. Brenner. Empirical potential for hydrocarbons for use in simulating the chemical vapor deposition of diamond films. *Phys. Rev. B* **42**, 1990, 9458–9471.
11. J. Tersoff. New empirical-model for the structural-properties of silicon. *Phys. Rev. Lett.* **56**, 1986, 632–635.
12. H. S. Johnston and C. Parr. Activation energies from bond energies. I. Hydrogen transfer reactions. *J. Am. Chem. Soc.* **85**, 1963, 2544–2551.

13. H. Sellers. A bond order conservation-Morse potential model of adsorbate–surface interactions: Dissociation of H2, O2, and F2 on the liquid mercury surface. *J. Chem. Phys.* **99**, 1993, 650–655.

14. S. J. Stuart, A. B. Tutein and J. A. Harrison. A reactive potential for hydrocabons with intermolecular interactions. *J. Chem. Phys.* **112**, 2000, 6472–6486.

15. K. D. Nielson, A. C. T. van Duin, J. Oxgaard, W. Q. Deng and W. A. Goddard. Development of the ReaxFF reactive force field for describing transition metal catalyzed reactions, with application to the initial stages of the catalytic formation of carbon nanotubes. *J. Phys. Chem. A* **109**, 2005, 493–499.

16. N. Q. Minh and T. Takahashi. *Science and Technology of Ceramic Fuel Cells.* Amsterdam: Elsevier Science B.V., 1995.

17. H. Burghgraef, A. P. J. Jansen and R. A. van Santen. Electronic structure calculations and dynamics of the chemisorption of methane on a Ni(111) surface. *Chem. Phys.* **177**, 1993, 407–420.

18. J. Thomas, P. Beebe, D. W. Goodman, B. D. Kay and J. T. Yates. Kinetics of the activated dissociative adsorption of methane on the low index planes of nickel single crystal surfaces. *J. Chem. Phys.* **87**, 1987, 2305–2315.

19. E. S. Hecht, G. K. Gupta, H. Y. Zhu, A. M. Dean, R. J. Kee, L. Maier and O. Deutschmann. Methane reforming kinetics within a Ni–YSZ SOFC anode support. *Appl. Catal., A* **295**, 2005, 40–51.

20. V. M. Janardhanan and O. Deutschmann. CFD analysis of a solid oxide fuel cell with internal reforming: Coupled interactions of transport, heterogeneous catalysis and electrochemical processes. *J. Power Sources* **162**, 2006, 1192–1202.

21. B. Delley. An all-electron numerical method for solving the local density functional for polyatomic molecules. *J. Chem. Phys.* **92**, 1990, 508–517.

22. B. Delley. From molecules to solids with the DMol^3 approach. *J. Chem. Phys.* **113**, 2000, 7756–7764.

23. D. R. Hamann, M. Schluter and C. Chiang. Norm-conserving pseudopotentials. *Phys. Rev. Lett.* **43**, 1979, 1494–1497.

24. J. P. Perdew and Y. Wang. Accurate and simple analytic representation of the electron–gas correlation energy. *Phys. Rev. B* **45**, 1992, 13244–13249.

25. H. J. Monkhorst and J. D. Pack. Special points for Brillouin-zone integrations. *Phys. Rev. B* **13**, 1976, 5188–5192.

26. N. Govind, M. Petersen, G. Fitzgerald, D. King-Smith and J. Andzelm. A generalized synchronous transit method for transition state location. *Comput. Mater. Sci.* **28**, 2003, 250–258.

27. A. C. T. van Duin, J. M. A. Baas and B. Vandegraaf. Delft molecular mechanics—A new approach to hydrocarbon force-fields—Inclusion of a geometry-dependent charge calculation. *J. Chem. Soc., Faraday Trans.* **90**, 1994, 2881–2895.

28. L. Verlet. Computer 'experiments' on classical fluids. I. Thermodynamical properties of Lennard–Jones molecules. *Phys. Rev.* **159**, 1967, 98–103.

29. H. J. C. Berendsen, J. P. M. Postma, W. F. van Gunsteren, A. Di Nola and J. R. Haak. Molecular dynamics with coupling to an external bath. *J. Chem. Phys.* **81**, 1984, 3684–3690.

16 Molecular Dynamics Simulations for Drug Dosage Form Development
Thermal and Solubility Characteristics for Hot- Melt Extrusion

Martin Maus, Karl G. Wagner,
Andreas Kornherr, and Gerhard Zifferer

CONTENTS

16.1 INTRODUCTION

In the last decade, the use of molecular dynamics (MD) simulations has become more and more familiar to pharmaceutical developers, for example, by checking on polymorphisms [1,2] and calculating properties of host–guest complexes [3,4]. Furthermore, MD simulations can assist dosage form development by predicting valuable physicochemical properties such as viscosity of solutions [5], diffusivity [6], and water adsorption [7]. However, the above-mentioned parameters are often limited to "small" systems either related to the molecular weight of the drug compound and/or the excipient used. The aim of our study is, therefore, to find *in silico* test models using MD methods for polymers, excipients, drugs, and mixtures thereof exceeding standard MD applications in complexity and, hence, facilitates preformulation even at a stage where drug compound or polymer is in an *in silico* stage of development. Moreover, the risk of unexpected physical or chemical incompatibilities (e.g., immiscibility) during the formulation process, which is time consuming and expensive [8], might be decreased. Fortunately, molecular simulation offers the possibility to calculate many relevant physical properties of the desired excipient and drug molecules as well as of the polymer without the need for costly experiments.

In this respect, the *in silico* prediction of the thermodynamic mixing behavior of different polymer–drug/excipient mixtures is of central interest. A common approach to cope with this problem is the calculation of the solubility parameters according to Hildebrand or Hansen [9–12], which is standard in the development of polymer mixtures [13]. The use of highly developed force fields as the basis of any MD simulation software enables the calculation of solubility parameters with accuracy comparable to those measured experimentally by inverse gas chromatography [14], and an increasing number of other statistical quantitative property relationships between simulated and experimental values are established [15–18].

Focusing on polymer–plasticizer blends, the glass transition temperature (T_g) is an important property to predict the formation of a solid solution. Homogenous polymer–plasticizer blends undergo a second-order–like phase transition at this temperature [19], passing from the rubbery to a glassy state with T_g decreasing with increasing plasticizer concentrations [20,21]. Throughout the literature, several approaches to determine the glass transition temperature of polymers via computer simulations are reported based on the change of various physical properties at T_g. One common approach is to determine the kink in a graph of the specific volume v vs. temperature T [14,19,22–27] originating from the change of the thermal expansion coefficient. Other approaches also use the increase in the potential energy [23–26] at T_g or the temperature dependence of the mean square displacement of polymer chains below and above T_g [26,28]. Normally, all these methods lead to a good prediction of T_g values for *pure* polymers.

Simulations of pharmaceutical relevant *mixtures* of polymer–drug/excipient blends, however, are scarce. Momany and Willett [25] investigated the dependence of T_g of maltodecaose on its hydration (water is known to act as a plasticizer for carbohydrates) and Yoshioka et al. [29] probed T_g values of freeze-dried dextran cakes. Quite recently, we have studied the T_g dependence of an ammonio methacrylate copolymer (Eudragit® RS) as a function of its triethyl citrate (TEC) content [27], which was in good agreement when compared with experimental data.

In this follow-up study, we want to examine the miscibility and the impact on the drug dosage form in respect of T_g from systems including Eudragit RS (a cationic ammonio methacrylate copolymer; type B), three common plasticizers (triethyl citrate, acetyltributyl citrate, and dibutyl sebacate), and two active ingredients: ibuprofen (likely plasticizing impact on Eudragit RS [30]) and theophylline (unlikely plasticizing influence on Eudragit RS [31–34]).

The aim is to assist the developer in the two pivotal development questions of a hot-melt extruded dosage form.

1. What is the best polymer–plasticizer mixture to obtain the lowest possible processing temperature for the extrusion process (a problem that is, of course, closely linked to a low value for T_g)?
2. What is the extent of the drug compound's solubility in the respective polymer–plasticizer blend? The solubility should be rather high for the purpose of forming a solid solution to improve active pharmaceutical ingredient (API) solubility. Moreover, in contrast, it should be very low in case drug dispersion is intended to be formed to extend the API release from the polymer matrix.

To compare the simulated values with the hot-melt extrusion process, the feasible blends are produced through hot-melt extrusion, performed on a lab-scale ram extruder. Employing Eudragit RS as a thermal polymer that is known to form multiple unit-controlled release matrix particles [20,34–36], the purpose in respect of question 2 was to distinguish a system of low solubility for the drug compound.

16.2 MATERIALS AND METHODS

All different substances used in this study are visualized in Figure 16.1 (all chemical structures are not optimized with respect to the conformation due to a better visibility).

16.2.1 MATERIALS

Triethyl citrate (TEC) was purchased from Merck (Darmstadt, Germany). The other substances were kindly donated by various manufacturers: Eudragit RS 100 by Röhm (Darmstadt, Germany), acetyltributyl citrate (ATBC) by Jungbunzlauer Ladenburg (Ladenburg, Germany), dibutyl sebacate (DBS) by Morflex (Greensboro,

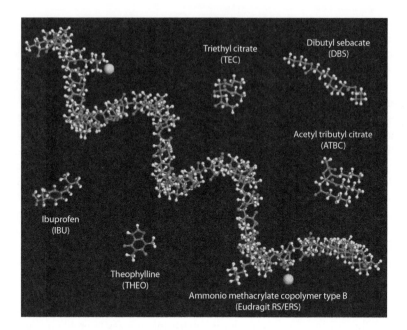

FIGURE 16.1   (See color insert.) Visualization of Eudragit RS polymer chain consisting of 64 monomers and all other molecules (both plasticizers and drugs) used in this study. The chlorine anion is depicted larger for better visibility; the different colors in this and the following figures correspond to different atom types with white reserved for hydrogen, red oxygen, gray carbon, blue nitrogen, and green chlorine. All the graphical displays of molecules were generated with the Materials Visualizer (from Accelrys).

USA), ibuprofen by Knoll Pharmaceuticals (Nottingham, UK), and theophylline anhydrous powder by BASF (Ludwigshafen, Germany).

16.2.2   Numerical Methods

16.2.2.1   Calculation of Solubility Parameters

To calculate the solubility of both plasticizer and drug molecules within a given polymer matrix, it is necessary to compute the so-called cohesive energy E_{coh}. This energy is a measure of the intermolecular forces (both electrostatic as well as dispersive) acting between the molecules of a specific substance. Accordingly, E_{coh} corresponds to the experimental value of the internal heat of vaporization ΔH_{vap} (with R being the real gas constant) [10]

$$E_{\text{coh}} = \Delta H_{\text{vap}} - R \cdot T. \qquad (16.1)$$

In a computer simulation, the energy (and force) between different molecules is directly accessible from the force field. Therefore, it is possible to compute E_{coh}

directly and to transform it into the Hildebrand solubility parameter δ [10] according to

$$\delta = \sqrt{E_{coh}/V},\qquad(16.2)$$

where V is the volume of the phase of which E_{coh} is calculated.

For this numerical procedure, cubic simulation boxes (with periodic boundary conditions in all directions) containing either four polymer chains consisting of 62 monomer units each or 60 plasticizer/drug molecules are constructed. Dependent on the type of the plasticizer, the cubic box size has a side length of about 4 nm (Figure 16.2). For each type of molecule, 10 independently generated simulation boxes are minimized with respect to the total energy. Out of those, the lowest energy system is further relaxed for 2 ns under NPT conditions at ambient conditions, that is, at constant pressure (10^5 Pa) and constant temperature (298 K) to obtain a well-relaxed start structure with the correct density using the Andersen thermostat and barostat [37] with a time step of 1 fs. Afterward, a 200-ps run at constant volume and constant temperature (i.e., NVT conditions) is carried out—100 ps for equilibration and 100 ps for data sampling. Thus, the cohesive energy is averaged over this latter period and the corresponding cohesive energy density E_{coh}/V is calculated by dividing it through the volume of the simulation box.

During minimization as well as during the MD runs, energy summation is performed in direct space, making use of a group-based cutoff for non–bond interactions. Strictly speaking, a cutoff distance of 1.25 nm with a spline switching function is applied

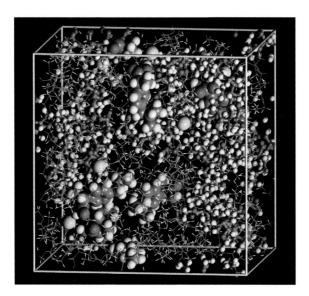

FIGURE 16.2   (See color insert.) Simulation box (3.7 × 3.7 × 3.7 nm) with periodic boundaries showing four Eudragit RS polymer chains and six ibuprofen molecules after a relaxation time of 2 ns at 430 K. One polymer chain and all ibuprofen molecules are visualized in ball-and-stick style to demonstrate the size of the box as well as the distribution of the drug molecules.

for Coulomb and for van der Waals interactions (the latter one being expressed by a 6–9 Lennard–Jones potential) using charge groups to prevent dipoles from being artificially split when one of the atoms is inside and another outside a (atom-based) cutoff.

For the whole simulation procedure, the software package MS Modeling 3.0 (from Accelrys) was applied using the Amorphous Cell tool for construction of the amorphous phase and Discover as the MD code. Atomic charges and interactions between atoms and molecules were accounted for by use of the COMPASS force field [38–41], which is highly optimized for the simulation of condensed phases.

16.2.2.2 Determination of Specific Volume as a Function of Temperature

The numerical procedure is similar to the method briefly described in Section 16.2.2.1 (see also Ref. [27]). However, now, several independent starting structures (generally three to five, to average the obtained final specific volumes to decrease scatter) of the corresponding system are chosen and relaxed for 2 ns under NPT conditions at a temperature approximately 100 K higher than the supposed glass transition temperature. Afterward, a "cooling process" is initiated by lowering the temperature stepwise by 10 K until a temperature ≈100 K lower than T_g is reached. At each temperature, 200-ps NPT ensemble dynamics is carried out—150 ps for equilibration and 50 ps for data sampling (i.e., averaging the specific volume v over this period) using the final configuration of this run as the starting structure for dynamics at the next (10 K lower) temperature.

16.2.3 EXPERIMENTAL METHODS

16.2.3.1 Hot-Melt Extrusion

According to the aim of the study, the mixtures of the polymer and its plasticizing agents were produced by hot-melt extrusion at 140°C. The ram extruder, assembled by an in-house precision mechanic shop, consists of a massive brass cylinder with a 1200-W heat source connected to a self-adapting power controller KS40-1 (PMA Prozess- und Maschinen- Automation, Kassel, Germany) equipped with a PT-100 sensor. The barrel diameter is 10 mm and the length is 280 mm. By connecting a material testing machine DO-FB0.5TS (Zwick, Ulm, Germany) to the plunger, it was possible to control and measure the force and velocity used for the extrusion of the hot melts through the orifice. The material testing machine was controlled by the utilization of programmable logic arrays included in the Software testXpert ver. 2.02 (Zwick).

About 10 g of the physical mixture, containing the polymer and the respective plasticizing substance at quantities from 0% to 25.5% (w/w), was prepared. The mixture was manually poured into the ram extruder and equilibrated for 10 min at 140°C. The stability of the polymer under the chosen conditions has previously been analyzed by thermogravimetric analysis measurements [35,42]. All mixtures were extruded three times at a plunger velocity of 15 mm/min through a 2.0-mm orifice.

16.2.3.2 DSC Measurements

Differential scanning calorimetry (DSC) measurements were performed for the experimental determination of the glass transition temperature T_g. The mean value of the onset and endset temperatures of the transition was calculated, according to DIN 53 765 A20. A Mettler TA 8000 calorimeter with a TAS 811 system and a DSC

820 measuring cell (Mettler-Toledo, Giessen, Germany) was used. Samples containing 20 mg polymer, including the respective plasticizing substance, were accurately weighed and sealed in perforated 40-μl Al pans and further analyzed under a dry nitrogen purge at a flow rate of 50 ml/min.

16.3 RESULTS AND DISCUSSION

This section is divided into three parts.

1. We will discuss the determination of solubility parameters to predict and to explain polymer–plasticizer/drugs miscibility. It is very important—both from a theoretical as well as from an experimental point of view—to ensure that two components are able to build a stable, homogenous mixture.
2. We will apply this knowledge to decide which of the possible polymer–plasticizer/drugs systems are suitable for a calculation of T_g.
3. The calculated T_g values will be compared to experimental values.

16.3.1 NUMERICAL DETERMINATION OF SOLUBILITY PROPERTIES

A computational method in combination with a suitable force field provides an advantage in contrast to the commonly used group-number systems [9,43] for calculating the solubility parameters according to Hildebrand [10] and Hansen [11]. Recent comparisons of simulated solubility parameters with solubility parameters measured using reverse gas chromatography and other experimental methods [44] proved the accurateness of the simulated values.

All plasticizers used in this study are commonly used for the application of polymer coatings as well as for matrix systems [31–33,45–47]. TEC is the standard plasticizer for Eudragit RS with a maximum water solubility of 5%, underlining its hydrophilic character. ATBC was chosen due to its structural similarity compared to TEC, but with an increased lipophilic character. The third plasticizer, the double butylated decanedicarboxy acid (DBS), has a similar lipophilic character as ATBC, but the different molecular structure allows for an additional testing of our computational method.

Ibuprofen (IBU), as a nonsteroidal antiinflammatory, is reported to have plasticizing properties in combination with Eudragit RS [30], thus representing a model drug with severe interactions with the polymer matrix. Its aromatic molecular structure presents a further change in the geometry as well as in the hydrophilic–lipophilic character. In contrast, the antiasthmatic methylxanthine theophylline has proven its inertness within formulations containing Eudragit RS [31–34] and was therefore chosen as the second active ingredient in our study.

The physical states of the substances modeled differ at 298.15 K: DSC measurements (see also Section 16.2.3.2) show that polymer Eudragit RS is amorphous without crystalline sections. In contrast, the plasticizers TEC, ATBC, and DBS are fluids. Furthermore, the drugs ibuprofen and theophylline are crystalline with ibuprofen melting within the hot-melt extrusion process, whereas theophylline stays solid. These differences were taken into account for the calculation of the solubility parameters: using crystallographic data from the Cambridge Structural Database, it was possible to calculate both the solubility of these crystalline drugs as well as the solubility of the corresponding amorphous (i.e., molten) state.

16.3.1.1 Hildebrand Solubility Parameters

The Hildebrand solubility parameters calculated within the MD simulation are presented in Table 16.1. As expected, crystalline drugs have the highest cohesive energy density with the Hildebrand parameter δ (see Equation 16.2), reading 31.64 $(J/cm^3)^{0.5}$ for ibuprofen and 51.43 $(J/cm^3)^{0.5}$ for theophylline, respectively. In contrast, amorphous substances show a reduced energy density due to the absence of the heat of crystallization. With δ reading 17.43 $(J/cm^3)^{0.5}$, the pure polymer has the lowest solubility parameter. The plasticizers and amorphous IBU are in the same range up to a maximum of 20.95 $(J/cm^3)^{0.5}$ for IBU.

16.3.1.2 Hansen Solubility Parameters

Hansen [11] divided the Hildebrand solubility parameter δ into contributions from the nonpolar van der Waals dispersion forces δ_d, the polar (electrostatic) interactions δ_p, and the hydrogen bonding interactions δ_h

$$\delta^2 = \delta_d^2 + \delta_p^2 + \delta_h^2. \tag{16.3}$$

In most modern force fields like COMPASS, however, hydrogen bonding interactions are incorporated in the polar interactions; therefore, discrimination between δ_p and δ_h cannot be made. Accordingly, in our calculations, these parameters are combined to δ_e

$$\delta_e = \sqrt{\delta_p^2 + \delta_h^2}. \tag{16.4}$$

TABLE 16.1
Overview of Densities, Molar Volumes, and Solubility Parameters Calculated via MD Simulations

	ρ (g/cm³)	v_M (cm³)	δ (J/cm³)⁰·⁵	δ_e (J/cm³)⁰·⁵	δ_d (J/cm³)⁰·⁵
TEC	1.114	248.6	20.70	8.62	18.81
ATBC	1.039	387.5	17.96	4.57	17.37
DBS	0.910	345.4	17.63	3.29	17.32
IBU (amorphous)	0.977	211.2	20.95	11.89	17.25
IBU (crystalline)	1.092	188.9	31.64	26.53	17.23
THEO (amorphous)	1.350	133.5	29.48	16.99	24.10
THEO (crystalline)	1.493	120.7	51.43	45.85	23.29
ERS	1.028	6250.1	17.34	11.32	13.13

Note: ρ (g/cm³) = density; v_M (cm³) = molar volume; δ (J/cm³)⁰·⁵ = overall Hildebrand solubility parameter; δ_e (J/cm³)⁰·⁵ = solubility parameter representing the electrostatic interactions, according to Equation 16.4; δ_d (J/cm³)⁰·⁵ = solubility parameter representing the dispersive interactions, according to Hansen.

The Hansen solubility parameters that were calculated according to Equation 16.4 are presented in Table 16.1. With δ_d values of about 17–19 $(J/cm^3)^{0.5}$, all plasticizing molecules were within a narrow range; only the polymer seemed to have a reduced dispersive solubility parameter value reading 13.13 $(J/cm^3)^{0.5}$. A better discrimination was possible by comparison of the different values of the electrostatic interaction parameter δ_e: with a δ_e value of 11.32 $(J/cm^3)^{0.5}$, the pure polymer was in the same range, compared to amorphous ibuprofen [11.89 $(J/cm^3)^{0.5}$] and the hydrophilic plasticizer TEC [8.62 $(J/cm^3)^{0.5}$], thus indicating an excellent miscibility. In contrast, both the values of DBS [3.29 $(J/cm^3)^{0.5}$] and ATBC [4.57 $(J/cm^3)^{0.5}$] showed an extended lipophilic character.

Extremely high values, however, can be observed when considering the crystalline state of the drugs theophylline and ibuprofen with δ_e readings of 45.85 $(J/cm^3)^{0.5}$ and 26.53 $(J/cm^3)^{0.5}$, respectively. In addition, even the—only hypothetical—amorphous state of theophylline showed a δ_e value of 16.99 $(J/cm^3)^{0.5}$ and a δ_d parameter of (24.10 $(J/cm^3)^{0.5}$. Therefore, no miscibility of this drug with the polymer could be expected. For ease of comparison of the different δ_e and δ_d values, a graphical representation is given in Figure 16.3.

16.3.1.3   Flory–Huggins Interaction Parameter

The qualitative considerations of the previous section regarding the solubility of plasticizers/drug/polymer blends can be transferred into a quantitative prediction of miscibility by calculation of the Flory–Huggins interaction parameter χ [48]

$$\chi = \frac{V_{ml} \cdot \left(\delta_1 - \delta_2\right)^2}{R \cdot T},$$

(16.5)

where V_{ml} is the molar volume of plasticizing substance, δ_1 is the solubility parameter of the plasticizer, δ_2 is the solubility parameter of the polymer, $R = 8.314472$ J/(mol·K) (real gas constant), and $T = 298.15$ K (absolute temperature).

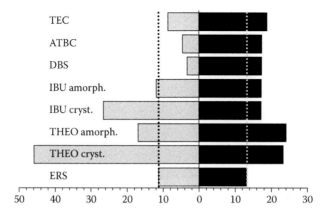

FIGURE 16.3 Visualization of the dispersive δ_d (right side, black) and the electrostatic δ_e (left side, gray) solubility parameters according to Hansen with the upright dashed lines marking the values of the pure Eudragit RS polymer.

An interaction parameter of zero describes an ideal solute–solvent system, where the energetic interactions of two different molecules within a solution are equal to the interactions within the pure substance. The higher the value of χ, the more dissimilar the two substances and, therefore, will not easily dissolve each other.

The calculation of this interaction parameter of different substances and the polymer displayed the differences as well as similarities of the studied blends (see Table 16.2). For all classic plasticizers, the Flory–Huggins parameter χ was equal to or less than 1.132 when mixed with the polymer. For the crystalline drugs ibuprofen and theophylline, χ would read 56.6 and 15.5, respectively, whereas the corresponding amorphous states show an extreme decrease in χ reading 1.1 and 7.9.

16.3.1.4 Gibbs Free Enthalpy Change of Mixing

The Gibbs energy of mixing ΔG_m [48] describes the overall thermodynamic benefit of generating a polymer–plasticizer mixture. Mixing of two substances only occurs in case of negative values of ΔG_m. As a molecular solution of the plasticizer in the polymer matrix is an indispensable requirement for its plasticizing properties, the possibility to calculate the value of ΔG_m before the specific experiment is extremely helpful in practice.

$$\Delta G_m = k \cdot T \cdot \left(\frac{N}{V_{m1}} \cdot v_1 \cdot \ln v_1 + \frac{N}{V_{m2}} \cdot v_2 \cdot \ln v_2 + \chi \cdot \frac{N}{V_{m1}} \cdot v_1 \cdot v_2 \right), \qquad (16.6)$$

where $k = 1.38 \times 10^{-23}$ J/K (Boltzmann constant), $T = 298.15$ K (absolute temperature), $N = 6.02 \times 10^{-23}$ 1/mol (Avogadro constant), V_{m1} is the molar volume of the plasticizing substance, V_{m2} is the molar volume of the polymer chain, $v_1 = 0.1$ (volume fraction of the plasticizer), and $v_2 = 0.9$ (volume fraction of the polymer).

In Table 16.2, the ΔG_m values calculated according to Equation 16.6 for different polymer–plasticizer systems with a typical value for the plasticizer volume fraction of 10% are presented. As expected from the solubility parameters, these values are positive for the crystalline drugs and even for the hypothetical amorphous state of theophylline. Therefore, further simulations (and experiments) considering the plasticizing properties of theophylline made no sense.

TABLE 16.2

Calculated Flory–Huggins Parameter χ and Free Enthalpy of Mixing ΔG_m of 10% Excipient with 90% Polymer at 298 K

	TEC	ATBC	DBS	IBU Amorphous	IBU Crystalline	THEO Amorphous	THEO Crystalline
χ	1.132	0.060	0.012	1.111	15.582	7.939	56.584
ΔG_m (J/cm³)	−1.318	−1.476	−1.682	−1.566	15.343	9.965	99.846

16.3.2 Simulation of Plasticizing Properties

The glass transition temperatures of all substances that show a possible molecular miscibility with the polymer matrix were simulated. As described in Section 16.2.2, a previously reported MD method [27] was applied.

16.3.2.1 Different Polymer–TEC Solid State Solutions

The first check of our improved model (four instead of eight polymer chains, but with a doubled chain length) for the MD simulations was a recalculation of T_g of the previous simulated [27] polymer–TEC systems. Again, a cooling of the new model of the pure polymer from 480 K to 250 K in 10-K steps showed a clearly visible kink in the specific volume v vs. temperature T diagram (Figure 16.4a), thus indicating a second-order–like phase transition from the rubbery to the glassy state. The glass transition temperature (354 K) was indicated by the interception of two linear regression lines from the rubbery and the glassy phases.

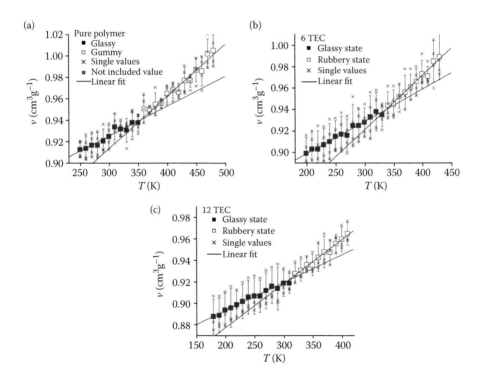

FIGURE 16.4 Plot of the computed specific volume v of pure Eudragit RS polymer (a), Eudragit RS polymer and 6 TEC (b), as well as 12 TEC (c) molecules per simulation box vs. temperature T of the system. The intersection point of the two lines resulting from a linear regression of the data points corresponding to the glassy (■) and the rubbery (□) state determines the simulated T_g values (see Table 16.3). All values are mean values of the specific volumes at the specified temperature of at least three independent systems. Printed error bars represent the standard deviation; circles represent values not used in the analysis.

Analogous to the previous work [27], mixed systems containing four long polymer chains plus 6 TEC or 12 TEC molecules per simulation box were constructed, and the corresponding T_g values were determined (see Figure 16.4b and 16.4c): 332 K for the 6-TEC and 308 K for 12-TEC system. Accordingly, the improved new model for the MD simulation of the TEC containing Eudragit RS systems showed a nearly perfect linear correlation of a decreasing value of T_g at an increasing plasticizer concentration (see Figure 16.5).

Although some deviation from experimental DSC data still occurred, the comparison with the previously reported [27] short-chain system (dotted line in Figure 16.5) clearly highlights the improvement of doubling the polymer chain length in the new model used in this work.

16.3.2.2   Further Solid Polymer–Plasticizer Solutions

Due to the good reproducibility of calculated and experimental T_g values of the plasticizer TEC, the more lipophilic plasticizers ATBC and DBS as well as the model drug ibuprofen were also tested. Amorphous cells containing four polymer chains and six molecules of either ATBC, DBS, or ibuprofen were constructed, and the corresponding T_g values were determined by analyzing the kink in the v vs. T diagrams (Figure 16.6).

All data points within the simulated temperature range of ATBC were included for the subsequent analysis. Due to the distinct kink in the v vs. T graphs, it was easy to separate them into two parts (a rubbery and a glassy part) and fit them to a straight line by linear regression (Figure 16.6a). In contrast, data points of both DBS and ibuprofen showed a larger temperature interval where the glass transition seems to take place. This was indicated by larger deviations from the linear regression lines in the rubbery as well as in the glassy state. Accordingly, some values in this region without a clearly defined physical state (rubbery or glassy) have been left out for analysis (see Figure 16.6b and 16.6c).

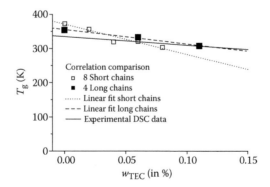

FIGURE 16.5   Comparison of the dependence of the glass transition temperature T_g on the TEC weight proportion w_{TEC} determined by the new MD model (full symbols, dashed line), the MD model used in our previous communication (open symbols, dotted line), and experimental (full line) DSC values. All lines result from a linear regression of the data points.

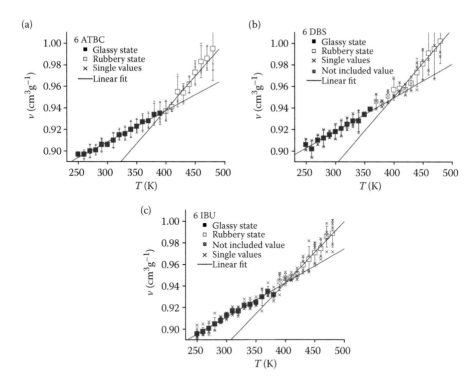

FIGURE 16.6 Plot of the computed specific volume v vs. temperature T of three different polymer–plasticizer systems, each containing four polymer chains and six ATBC (a), six DBS (b), and six ibuprofen (c) molecules per simulation box. Symbols and determination of T_g as in Figure 16.4 (values can be found in Table 16.3).

In Table 16.3, the calculated and the DSC measured glass transition temperatures of all simulated systems are presented. As expected, simulated T_g values were always higher than the experimental values due to the high cooling rate in a computer experiment (i.e., in an *in silico* experiment, one could regard the system as being "shock frozen"). However, the discrepancies were only 4% to 11%, thus indicating the reliability of the presented computational method.

TABLE 16.3

Overview of Extrapolated Experimental and Calculated T_g Values (in Kelvin) of the Different Polymer–Plasticizer Systems with n Symbolizing the Number of Plasticizer Molecules per Simulation Box

	ERS		TEC		IBU		DBS		ATBC	
n	DSC	MD	DSC	MD	DSC	MD	DSC	MD	DSC	MD
0	336.7	354								
6			320.2	332	328.5	349	315.9	348	316.3	353
12			306.5	308						

16.3.3   Experimental Check of the Plasticizing Properties

16.3.3.1   Production of Solid Solutions Using a Hot-Melt Ram Extruder

All physical mixtures were extruded at the conditions stated in Section 16.2.3. By subsequently extruding the mixtures three times, enough shear was applied to achieve well-distributed solid solutions of the plasticizers within the polymer matrix. A comparison of the development of the force effect during the extrusion cycles showed a reduction in the force effect at each step. This trend clearly indicated that the homogeneity increased through the multiple ram extrusions. Furthermore, all three plasticizers, TEC, ATBC, and DBS, showed either extreme stickiness or a too soft consistency of the extruded polymer strands at levels higher than 25.5%. Therefore, a plasticizer fraction w of 25.5% (m/m) indicated the endpoint in the application range.

The crystalline drug ibuprofen melts at 76°C, so the applied extrusion temperature allowed the incorporation of molten drug into the polymer matrix. Therefore, ibuprofen could be homogeneously dissolved up to values of w reading 20% (due to the high stickiness, no higher solid-state solutions could be produced).

In contrast, theophylline is melting at 270°C, so this drug remained crystalline during the entire hot-melt extrusion process and was only suspended in the polymer matrix. The compositions and glass transition temperatures of all samples produced by hot-melt extrusion are shown in Table 16.4.

16.3.3.2   Thermal Analysis of the Extrudates

The subsequent DSC measurements of the produced polymer–TEC, –ATBC, and –DBS extrudates, up to a weight percentage of 19%, showed a clearly visible, strictly amorphous step when the heat flow was plotted against the reference temperature T.

TABLE 16.4

Composition and Experimental T_g Values of All Samples Produced by Hot-Melt Extrusion

Polymer	TEC	ATBC	DBS	IBU	T_g (K)	Standard Deviation
100%					336.7	0.6
89%	11%				305.3	0.9
85%	15%				300.9	1.2
78%	22%				280.4	1.1
90%		10%			310.0	0.6
86%		14%			303.9	1.4
79%		21%			289.7	0.9
91%			9%		304.6	0.5
85%			15%		296.1	0.5
81%			19%		285.0	1.7
90%				10%	318.5	0.3
80%				20%	300.5	0.9

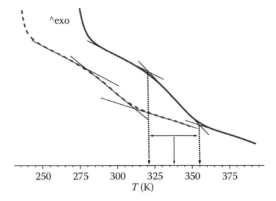

FIGURE 16.7   Plot of the heat flow against temperature T in a DSC experiment of a pure polymer (solid line) as well as a polymer–TEC mixture (dashed line) with $w = 17\%$. Both show a distinct step in the heat flow, marked by the onset/endset point with the midpoint between these two yielding the experimental value of the glass transition temperature T_g.

A representative DSC plot of the transition of the pure polymer as well as a polymer–TEC extrudate at 17% is shown in Figure 16.7.

As it was reported earlier for other polymer–plasticizer blends [27,30,34,42,45], the measured T_g values of the extrudates exhibited a linear dependence on the mass fraction w of plasticizer in the melt. Furthermore, all the plasticizers studied in our work showed a good experimental solubility in the acrylic copolymer Eudragit RS, thus resulting in homogeneous solid solutions by the application of the ram extruder.

However, as one could already mention within the qualitative comparison of the Hansen parameters in Table 16.1 and Figure 16.3, there exists a limit of solubility of plasticizers such as DBS in the polymer matrix (at weight percentage of 19%). This could be proven by DSC measurements of matrices containing 21.5% of DBS (a

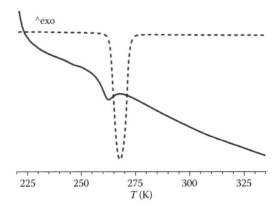

FIGURE 16.8   Plot of the heat flow against temperature T in a DSC experiment of a polymer–DBS mixture (full line) with $w = 21.5\%$ as well as of pure DBS (dashed line). The melting peak of DBS is also visible in the polymer–DBS curve, thus indicating existence of the two different phases.

representative plot is presented in Figure 16.8). In contrast to homogenous polymer–plasticizer blends where an amorphous step in the heat flow vs. temperature plot is visible, a dent—marking the melting point of the plasticizer DBS—occurs on the plot. Therefore, this mixture actually existed at two different, only mechanically mixed, phases: a polymer and a plasticizer phase.

Ibuprofen exhibited a behavior similar to the three plasticizers: again, a linear correlation of the weight proportion of ibuprofen and the T_g of the matrix was found as it was also reported by Kidokoro et al. [34]; its ability to plasticize Eudragit RS was slightly lower compared with the plasticizers.

In contrast, theophylline influenced neither the glass transition temperature of the pure polymer nor the investigated polymer–plasticizer mixtures, which is in good agreement with previous works [30,34].

16.4 CONCLUSION

Our study shows that solubility properties of different polymer–plasticizer/drug blends can be calculated by the application of MD simulations using a proper force field. However, Hildebrand's parameter δ leads to an incomplete picture of the solubility property of a given system, for example, polymer–DBS: although this lipophilic plasticizer exhibited nearly the same solubility parameter δ as the polymer (indicating a nearly ideal mixture), it showed an experimental solubility limit at higher plasticizer fractions. In contrast, solubility parameters according to Hansen deliver a more detailed picture by splitting δ in a dispersive δ_d and an electrostatic δ_e component: now, due to the large difference of δ_e between the polymer and the plasticizer, a solubility limit could be explained at least qualitatively.

However, a final decision on whether a substance has the theoretical potential to build a stable blend with the polymer can only be made by calculating the Gibbs energy. As can be seen in Table 16.2, only TEC, ATBC, DBS, and amorphous ibuprofen showed a theoretical miscibility with the polymer. This was also proven by experiments studying hot-melt extrudates.

Based on these solubility calculations, only miscible polymer–plasticizer/drug systems were further investigated with respect to their plasticizing properties. The correlation of the simulated T_g values with experimental ones reflects the different solubility behavior of the plasticizers studied: less miscible plasticizers like DBS and ATBC show a higher deviation from the experimental T_g values, whereas the good soluble plasticizer TEC, as well as the amorphous drug ibuprofen, exhibits markedly less differences. Accordingly, it seems as if the modeling of the glass transition of polymer-like plasticizers results in a better agreement with experimental values compared to the less miscible substances. With the maximum discrepancies between simulated and experimental values being not more than 11% in the worst case, the validity of our improved computer model is demonstrated. Thus, it is possible to make predictions of the glass transition temperature of these systems with high precision.

It should be noted that the calculation of the glass transition temperature from volumetric traces, that is, specific volume or density as a function of temperature, is not restricted to polymer melts. Furthermore, quantities like total energy and mobility, the latter provided by diffusion coefficients or simply by mean square displacements in NPT simulations, are showing characteristic changes at T_g. These methods have been successfully applied, for example, to various carbohydrates [49,50]—again showing the same trends for T_g as experimental investigations also being systematically higher than experimental ones—to supercooled water [51–53] under high pressure—suggesting that high-density amorphous ices may indeed be proxies of ultraviscous high-density liquids at low temperatures—and recently to aqueous solutions of amides [54]—to predict their cryoprotective ability.

As a concluding remark, it can be stated that our method gives pharmacists a valuable recipe on how to treat unknown polymer–plasticizer/drug systems with respect to their mixing and plasticizing properties. The benefit of the presented method is a direct reduction of time expenditure and costs, in relation to the present trial-and-error principle in dosage form development.

REFERENCES

1. F. J. J. Leusen, S. Wilke, P. Verwer and G. E. Engel. Computational approaches to crystal structure and polymorph prediction, In *Implications of Molecular and Materials Structure for New Technologies*. J.A.K. Howard, F.H. Allen, and G.P. Shields, Eds., NATO Science Series E, Vol. 360, Kluwer Academic: Dordrecht, The Netherlands, 1999, 303–314.
2. P. Verwer and F. J. J. Leusen. Computer simulation to predict possible crystal polymorphs. *Rev. Comput. Chem.* **12**, 1998, 327–365.
3. L. Lawtrakul, H. Viernstein and K. Wolschann. Molecular dynamics simulations of β-cyclodextrin in aqueous solution. *Int. J. Pharm.* **256**, 2003, 33–41.
4. H. Viernstein, P. Weiss-Greiler and K. Wolschann. Quantitative determination of water in y-cyclodextrin by near-infrared spectroscopy. *Pharm. Ind.* **67**, 2005, 122–125.
5. K. Zhang and C. W. Manke. Simulation of polymer solutions by dissipative particle dynamics. *Mol. Simul.* **25** (3–4), 2000, 157–166.
6. M. Meunier. Diffusion coefficients of small gas molecules in amorphous *cis*-1,4-polybutadiene estimated by molecular dynamics simulations. *J. Chem. Phys.* **123** (13), 2005, 134906.
7. Y. Aray, M. Marquez, J. Rodriguez, S. Coll, Y. Simó-Manso, C. Gonzalez and D. A. Weitz. Electrostatics for exploring the nature of water adsorption on the Laponite sheets' surface. *J. Phys. Chem. B* **107**, 2003, 8946–8952.
8. M. Swenson, M. Languell and J. Golden. Modeling and simulation: The return on investment in materials science. IDC White Paper, 2004, 1–24.
9. A. Forster, J. Hempenstall, I. Tucker and T. Rades. Selection of excipients for melt extrusion with two poorly water-soluble drugs by solubility parameter calculation and thermal analysis. *Int. J. Pharm.* **226** (1–2), 2001, 147–16.
10. J. H. Hildebrand. Solubility. *J. Am. Chem. Soc.* **38** (8), 1916, 1452–1473.
11. C. M. Hansen. *Hansen Solubility Parameters, A User's Handbook*. Boca Raton: CRC Press, 2000.
12. J. Burke. Solubility parameters: Theory and application. In *AIC Book and Paper Group Annual* 3, 1984, 13–58. Available online at http://aic.stanford.edu/sg/bpg/annual/v03/bp03-04.html (Accessed 06/27/2006).
13. P. Choi. Molecular dynamics studies of the thermodynamics of HDPE/butene-based LLDPE blends. *Polymer* **41** (24), 2000, 8741–8747.

14. J. Han, R. H. Gee and R. H. Boyd. Glass transition temperatures of polymers from molecular dynamics simulations. *Macromolecules* **27** (26), 1994, 7781–7784.

15. L. Zhao and P. Choi. Study of the correctness of the solubility parameters obtained from indirect methods by molecular dynamics simulation. *Polymer* **45** (4), 2004, 1349–1356.

16. M. Belmares, M. Blanco, W. A. Goddard III, R. B. Ross, G. Caldwell, S. H. Chou, J. Pham, P. M. Olofson and C. Thomas. Hildebrand and Hansen solubility parameters from molecular dynamics with applications to electronic nose polymer sensors. *J. Comput. Chem.* **25** (15), 2004, 1814–1826.

17. P. Choi, T. A. Kavassalis and A. Rudin. Estimation of the three-dimensional solubility parameters of alkyl phenol ethoxylates using molecular dynamics. *J Colloid Interface Sci.* **150** (2), 1992, 386–393.

18. T. A. Kavassalis, P. Choi and A. Rudin. The calculation of 3D solubility parameters using molecular models. *Mol. Simul.* **11** (2–4), 1993, 229–241.

19. H. Yang, L. Ze-Sheng, H. Qian, Y. Yang, X. Zhang and C. Sun. Molecular dynamics simulation studies of binary blend miscibility of poly(3-hydroxybutyrate) and poly(ethylene oxide). *Polymer* **45** (2), 2004, 453–457.

20. Y. Zhu, N. H. Shah, A. W. Malick, M. H. Infeld and J. W. McGinity. Solid-state plasticization of an acrylic polymer with chlorpheniramine maleate and triethyl citrate. *Int. J. Pharm.* **241** (2), 2002, 301–310.

21. R. Gruetzmann and K. G. Wagner. Quantification of the leaching of triethyl citrate/polysorbate 80 mixtures from Eudragit RS films by differential scanning calorimetry. *Eur. J. Pharm. Biopharm.* **60** (1), 2005, 159–162.

22. C. F. Fan, T. Cagin, W. Shi and K. A. Smith. Local chain dynamics of a model polycarbonate near glass transition temperature. A molecular dynamics simulation. *Macromol. Theory Simul.* **6** (1), 1997, 83–102.

23. K. Yu, Z. Li and J. Sun. Polymer structures and glass transition: A molecular dynamics simulation study. *Macromol. Theory Simul.* **10** (6), 2001, 624–633.

24. J. R. Fried and P. Ren. Molecular simulation of the glass transition of polyphosphazenes. *Comput. Theor. Polym. Sci.* **9** (2), 1999, 111–116.

25. F. A. Momany and J. L. Willett. Molecular dynamics calculations on amylose fragments. I. Glass transition temperatures of maltodecaose at 1, 5, 10 and 15.8% hydration. *Biopolymers* **63** (2), 2002, 99–110.

26. R. Roe. MD simulation study of glass transition and short time dynamics in polymer liquids. In *Advances in Polymer Science* 116, edited by Lucien Monnerie and U.W. Suter. Berlin, Heidelberg: Springer, 1994, 111–144.

27. K. G. Wagner, M. Maus, A. Kornherr and G. Zifferer. Glass transition temperature of a cationic polymethacrylate dependent on the plasticizer content—Simulation vs. experiment. *Chem. Phys. Lett.* **406**, 2005, 90–94.

28. Y. Tamai. A practical method to determine glass transition temperature in molecular dynamics simulation of mixed ionic glasses. *Chem. Phys. Lett.* **351** (1–2), 2002, 99–104.

29. S. Yoshioka, Y. Aso and S. Kojima. Prediction of glass transition temperature of freeze-dried formulations by molecular dynamics simulation. *Pharm. Res.* **20** (6), 2003, 873–878.

30. C. Wu and J. W. McGinity. Non-traditional plasticization of polymeric films. *Int. J. Pharm.* **177**, 1999, 15–27.

31. J. Siepmann, F. Lecomte and R. Bodmeier. Diffusion-controlled drug delivery systems: Calculation of the required composition to achieve desired release profiles. *J. Controlled Release* **60** (2–3), 1999, 379–389.

32. H. Rey, K. G. Wagner, P. Wehrle and P. C. Schmidt. Development of matrix-based theophylline sustained-release microtablets. *Drug Dev. Ind. Pharm.* **26** (1), 2000, 21–26.

33. S. Narisawa, M. Nagata, C. Danyoshi, H. Yoshino, K. Murata, Y. Hirakawa and K. Noda. An organic acid-induced sigmoidal release system for oral controlled-release preparations. *Pharm. Res.* **11** (1), 1994, 111–116.

34. M. Kidokoro, N. H. Sha, A. W. Malick, M. H. Infeld and J. W. McGinity. Properties of tablets containing granulations of ibuprofen and an acrylic copolymer prepared by thermal processes. *Pharm. Dev. Technol.* **6** (2), 2001, 263–275.

35. N. Follonier, E. Doelke and E. T. Cole. Evaluation of hot-melt extrusion as a new technique for the production of polymer-based pellets for sustained release capsules containing high loadings of freely soluble drugs. *Drug Dev. Ind. Pharm.* **20** (8), 1994, 1323–1339.

36. Y. Zhu, N. H. Shah, A. W. Malick, M. H. Infeld and J. W. McGinity. Influence of a lipophilic thermal lubricant on the processing conditions and drug release properties of chlorpheniramine maleate tablets prepared by hot-melt extrusion. *J. Drug Delivery Sci. Technol.* **14** (4), 2004, 313–318.

37. H. C. Andersen. Molecular dynamics simulations at constant pressure and/or temperature. *J. Chem. Phys.* **72** (4), 1980, 2384–2393.

38. M. J. Hwang, T. P. Stockfisch and A. T. Hagler. Derivation of class II force fields. 2. Derivation and characterization of a Class II force field, CFF93, for the alkyl functional group and alkane molecules. *J. Am. Chem. Soc.* **116** (6), 1994, 2515–2525.

39. H. Sun. Ab initio calculations and force field development for computer simulation of polysilanes. *Macromolecules* **28** (3), 1995, 701–712.

40. H. Sun. COMPASS: An ab initio force-field optimized for condensed-phase applications—Overview with details on alkane and benzene compounds. *J. Phys. Chem. B* **102** (38), 1998, 7338–7364.

41. S. W. Bunte and H. Sun. Molecular modeling of energetic materials: The parameterization and validation of nitrate esters in the COMPASS force field. *J. Phys. Chem. B* **104** (11), 2000, 2477–2489.

42. Y. Zhu, N. H. Shah, A. W. Malick, M. H. Infeld and J. W. McGinity. Influence of thermal processing on the properties of chlorpheniramine maleate tablets containing an acrylic polymer. *Pharm. Dev. Technol.* **7** (4), 2002, 481–489.

43. N. Schuld and B. A. Wolf. Polymer–solvent interaction parameters. In *Polymer Handbook*, edited by J. Brandrup, E. H. Immergut and E. A. Grulke. New York: John Wiley & Sons, 1999, 675–688.

44. T. A. Kavassalis, P. Choi and A. Rudin. Molecular models and three dimensional solubility parameters of non-ionic surfactants. In *Molecular Simulation and Industrial Applications*, edited by K. E. Gubbins and N. Quirke. Amsterdam: Gordon & Breach, 1996, 315–329.

45. C. Wu and J. W. McGinity. Influence of methylparaben as a solid-state plasticizer on the physicochemical properties of Eudragit RS PO hot-melt extrudates. *Eur. J. Pharm. Biopharm.* **56**, 2003, 95–100.

46. R. Chang and C. Hsiao. Eudragit RL and RS pseudolatices: Properties and performance in pharmaceutical coating as a controlled release membrane for theophylline pellets. *Drug Dev. Ind. Pharm.* **15** (2), 1989, 187–196.

47. J. C. Gutiérrez-Rocca and J. W. McGinity. Influence of water soluble and insoluble plasticizers on the physical and mechanical properties of acrylic resin copolymers. *Int. J. Pharm.* **103** (3), 1994, 293–301.

48. L. H. Sperling. *Introduction to Physical Polymer Science*. New York: John Wiley & Sons, 2001, 70–75.

49. A. Simperler, A. Kornherr, R. Chopra, P. A. Bonneta, W. Jones, W. D. S. Motherwell and G. Zifferer. Glass transition temperature of glucose, sucrose and trehalose: An experimental and in silico study. *J. Phys. Chem. B* **110**, 2006, 19678–19684.

50. A. Simperler, A. Kornherr, R. Chopra, W. Jones, W. D. S. Motherwell and G. Zifferer. The glass transition temperatures of amorphous trehalose–water mixtures and the mobility of water: An experimental and in silico study. *Carbohydr. Res.* **342**, 2007, 1470–1479.

51. N. Giovambattista, C. A. Angell, F. Sciortino and H. E. Stanley. Glass-transition temperature of water: A simulation study. *Phys. Rev. Lett.* **93**, 2004, 047801-1-4.

52. M. Seidl, T. Loerting and G. Zifferer. High density amorphous ice: Molecular dynamics simulations of the glass transition at 0.3 GPa. *J. Chem. Phys.* **131**, 2009, 114502-1-7.

53. M. Seidl, T. Loerting and G. Zifferer. Molecular dynamics simulations on the glass-to-liquid transition in high density amorphous ice. *Z. Phys. Chem.* **223**, 2009, 1047–1062.

54. C. A. Kreck, J. B. Mandumpal and R. L. Mancera. Prediction of the glass transition in aqueous solutions of simple amides by molecular dynamics simulations. *Chem. Phys. Lett.*, **501**, 2011, 273–277.

17 Cobalt Complex Based on Cyclam for Reversible Binding of Nitric Oxide

Olivier Siri, Alain Tabard,
Pluton Pullumbi, and Roger Guilard

CONTENTS

17.1 INTRODUCTION

As nitric oxide (NO) is an endogenous reactive molecule that plays important physiological roles in living organisms [1], its delivery to specific targets receives a great deal of attention through the study of numerous biological processes and therapeutic applications. Only a limited number of exogenous NO donors are available for clinical use, so the development of molecules that can release NO is of considerable interest [2]. The simplest case uses direct inhalation of NO [3], but its clinical use as a gas is technically difficult to administer [4]. Most exogenous NO donors are organic molecules such as nitrates that do not directly release NO owing to the biotransformations that are initially required [5–7]. More recently, the use of metal nitrosyl complexes appeared very attractive because the only direct NO-releasing drug clinically available in the U.S. is sodium nitroprusside (iron complex) for which five toxic cyanide ions are released for every NO molecule [8,9]. Different NO-donor complexes based on ruthenium [10], iron [11–14], manganese [15], and cobalt [11,16,17] are described in the literature. Among the most promising cobalt complexes, nitrosyl-cobinamide, a structural analog of vitamin B12 (*unsaturated* cyclic tetraamine), is an efficient direct NO-releasing agent [16]. To the best of our knowledge, the use of cobalt(II) derivatives based on *saturated* cyclic polyamines such as cyclam **1** has not been reported, whereas their flexibility might allow distinct configurations of the nitrosyl complexes with specific properties [18].

Herein, we report the synthesis and theoretical studies of a cobalt(II) complex based on **1** for reversible binding of NO.

1

17.2   EXPERIMENTAL SECTION

$CoCl_2 \cdot 6H_2O$ (0.15 mmol) under argon atmosphere was added to a solution of 0.15 mmol of **1** in dry methanol. The mixture was stirred at room temperature for 15 min, and NO gas was then bubbled through by means of a gastight glass syringe, connected to a flow meter, equipped with a needle valve for the adjustment of the gas flow rate and calibrated between 5 ml/min and 30 ml/min. The NO gas injection effectuated at room temperature, and an absolute pressure of about 0.9 bar for 60-s time intervals leads to a rapid color change consistent with the formation of the nitrosyl complex. [(**1**)Co(NO)(Cl)]PF_6 was isolated after precipitation with $N(Bu)_4 \cdot PF_6$ as a green solid (89% yield). Its infrared (IR) spectrum (KBr disk) exhibited a strong sharp absorption at 1610 cm^{-1} corresponding to the NO stretching. Calculated for $C_{10}H_{24}N_5CoOCl \cdot PF_6$: C% 25.6; H% 5.2; N% 14.9; Cl% 7.5; F% 24.3; Co% 12.5; P% 6.6. Found: C% 25.5; H% 5.2; N% 14.4; Cl% 7.3; F% 23.5; Co% 11.8; P% 6.2.

17.3   SYNTHESES AND EXPERIMENTAL CHARACTERIZATION

Cyclam **1** is a well-known, macrocyclic tetraamine possessing a 14-membered ring able to form stable complexes with many metal ions [19]. The complex [(**1**)CoCl$_2$] was first synthesized by adding $CoCl_2 \cdot 6H_2O$ to a methanolic solution of ligand **1** under argon atmosphere. This reaction, when carried out with a 1:1 ratio of reagents under reflux for 1 h, led to precipitation of [**1**-Co(III)Cl$_2$]Cl as bright green crystals corresponding to the oxidation of Co(II) in Co(III) as reported in Refs. [20] and [21].

The fresh solution of [(**1**)CoCl$_2$] was then treated with NO gas and the immediate color change indicated the coordination of NO to the metal center. This reaction can be monitored by electron paramagnetic resonance (EPR) spectroscopy because the paramagnetic Co(II) species becomes diamagnetic upon coordination of NO and formation of the *trans*-[(**1**)Co(NO)(Cl)]Cl compound. The nitrosyl complex could be more easily isolated after anion exchange with $N(Bu)_4 \cdot PF_6$, which led to the precipitation of the *trans*-[(**1**)Co(NO)(Cl)]PF_6 salt that was characterized. The *trans* configuration of the complex and the geometry of the CoNO moiety were determined by IR spectroscopy. NO can form two distinct terminal bonding modes that have different geometries, that is, linear and bent [22], and the most distinctive physical property is the IR ν(NO) band for which the frequencies of bent CoNO groups are usually lower than those of linear MNO groups [23]. The IR spectrum of [(**1**)Co(NO)(Cl)]Cl in 1600 cm^{-1}–1900 cm^{-1} showed one intense band at 1610 cm^{-1}, which is consistent with a bent geometry. In addition, the 800–900 cm^{-1} range that gives information about the coordination scheme of the central metal ion [24] revealed three bands in agreement with the *trans* configuration. Complex *trans*-[(**1**)Co(NO)(Cl)]Cl is stable under inert atmosphere in a methanolic solution, that is, argon bubbling did not lead to the release of NO. In contrast, such a compound is unstable in the presence of dioxygen and converted into the Co(III)–nitro complex for which the IR spectrum of the obtained complex did not show the NO vibrations but the presence of two new bands at 1355 cm^{-1} and 828 cm^{-1} characteristic of the coordinated NO$_2$ group.

Interestingly, we noted that related cobalt(II) species can be five-coordinated in solution [23,25] owing to a strong *trans* influence. The *trans* influence of a ligand is defined as the extent to which that ligand weakens the bond *trans* to itself. In the *trans*-[(**1**)Co(NO)(Cl)]Cl complex for which the Co–N–O moiety is bent, NO is formally NO$^-$ reflecting no back-bonding. This assignment is consistent with a strong *trans* labilizing effect as already observed for Co–NO analogs with a bent nitrosyl group. However, in the case of *trans*-[(**1**)Co(NO)(Cl)]Cl, the axial ligand Cl is not *trans* director to NO so that no release of NO was observed.

17.4   COMPUTATIONAL METHODS AND CALCULATIONS

The structures of all of the reactants, complexes, transition states, intermediates, and products were fully optimized using the DMol^3 density functional theory [26,27] code as implemented in the Materials Studio® distribution of Accelrys software package. The nonlocal gradient-corrected functional VWN-BP has been used for all the DMol^3 calculations together with the double numerical plus polarization (DNP) basis sets and the fine numerical integration grid. For all the species, the DMol^3 calculations have been performed using the semicore, pseudopotential option in gas as well as in the solvated phase. The hardness-conserving, semicore pseudopotentials, called density-functional, semicore pseudopotentials, are generated by fitting all-electron relativistic DFT results and have a nonlocal contribution for each channel up to $l = 2$, as well as a nonlocal contribution to account for higher channels. The harmonic vibrational frequencies were calculated at the same level to characterize the nature of the local minima with no imaginary frequency and those of transition states with only one imaginary frequency. The Conductor-like Screening Model (COSMO) [28,29] was used to model the solute molecule forming a cavity within the dielectric continuum with the permittivity representing the solvent, which in our case was methanol. The direct incorporation of the solvent effects within the self-consistent field (SCF) procedure is a major computational advantage of the COSMO scheme. The DMol^3/COSMO orbitals are obtained using the variational scheme enabling the derivation of accurate analytic gradients with respect to the coordinates of the solute atoms. The solvation energies depend on the choice of DMol^3 parameters, such as the type of DFT functional, the basis set, and the integration grid. Recent publications [30] indicate that the DMol^3/COSMO calculations with the VWN-BP functional can predict solvation energies with an accuracy of the order of 2 kcal/mol. Recently, we have reported the use of the semiempiric PM3-tm method, incorporating the parameterization of transition metals to the original PM3 [31,32], as implemented in the SPARTAN package [33] for the study of the structures and stabilities of Fe(1)(NO)(Cl) complexes [18]. In the present study, due to the relatively large size of the [Co(1)]NO complexes, we have applied the same method (PM3-tm) to generate the initial stable geometries that were further optimized using DMol³, and we have obtained the results for the gas-phase isolated complexes as well as for their respective solvated form through DMol³/COSMO calculations. To validate and compare the PM3-tm and DMol³ methods for Co(1) complexes, the *trans*-[Co(1) Cl₂] was optimized with both methods and the obtained structures were compared by overlay [34] to the experimental (x-ray diffraction) data reported for this compound [20]. The similarity between the calculated PM3 and DMol³ and the experimental geometric parameters is quite good, the root mean square deviation being equal to 0.101 Å (PM3) and 0.051 Å (DMol³).

As there could be *a priori* two oxidation states for cobalt in the complex form, and because of the recent interest in model biological systems such as vitamin B12 or isoelectronic systems with Fe(II) complexes, systematic calculations of 1-Co(III) complexes have been carried out in parallel with those of 1-Co(II) ones including all combinations of the axial ligands (NO, NO_2, NO_2^-, and Cl⁻) that could coordinate to both 1-Co(II) or 1-Co(III). These different cases are reported in Table 17.1. The different configurations of the *trans*-[Co(1)] complexes are specified by two numbers, the first one (Roman—in parentheses) stands for the oxidation state of cobalt, whereas the second one indicates the partial charge of the cobalt-containing half-complex. The total charge of the complex is reported in Table 17.1.

The different calculations carried out in this study are reported in Table 17.2. Each complex has been calculated in gas phase (empty) and in solvent (methanol) using the COSMO method to include solvent effects.

In Table 17.3, the stabilities of 1-Co(II) and 1-Co(III) complexes are reported for both gas phase and in the methanol species. It appears clearly from this table that the 1-Co(II) complexes are much more stable than the 1-Co(III) ones as would normally be expected due to the extra electron. However, the difference is reduced when the solvent effect is taken into account. In our further analysis of the chemical reactivity and stability of 1-Cobalt complexes, we will focus only on 1-Co(II) ones.

TABLE 17.1

Total Charge of 1-Co(II) and 1-Co(III) Calculated Complexes with Selected Axial Ligands (NO, NO_2, NO_2^-, and Cl⁻)

Complexes/Ligands	NO	NO_2	NO_2^-	Cl⁻
1-Co(II)–Cl⁺	1	1	0	0
1-Co(III)–Cl²⁺	2	2	1	1
1-Co(II)–NO²⁺	2	2	1	1
1-Co(III)–NO³⁺	3	3	2	2
1-Co(II)–NO_2^+	1	1	0	0
1-Co(III)–NO_2^{2+}	2	2	1	1
1-Co(III)–NO_2^{3+}	3	3	2	2

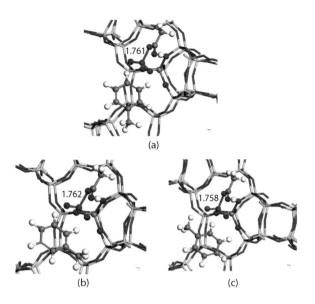

FIGURE 1.7 Reactant structures (acetyl nitrate and toluene) in side cross section of 12-T pore around Brønsted acid site for (a) *para*, (b) *ortho* 1, and (c) *ortho* 2 attack orientations of toluene. Distance and bond lengths are in angstroms.

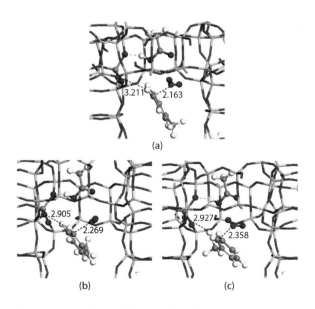

FIGURE 1.10 The π-complex for the (a) *para*, (b) *ortho* 1, and (c) *ortho* 2 attack. Distances and bond lengths are in angstroms.

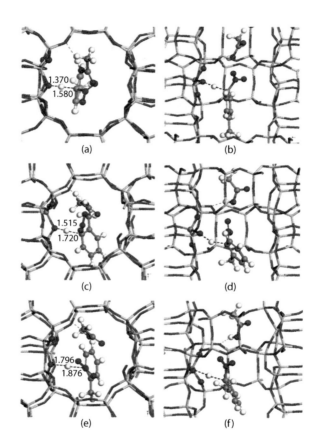

FIGURE 1.13 TS3 for *para* nitration (a) viewing down the 12-T pore and (b) viewing a side cross section of the 12-T pore, *ortho* 1 nitration (c) viewing down the 12-T pore and (d) viewing a side cross section of the 12-T pore, and *ortho* 2 nitration (e) viewing down the 12-T pore and (f) viewing a side cross section of the 12-T pore. Distances shown are in angstroms.

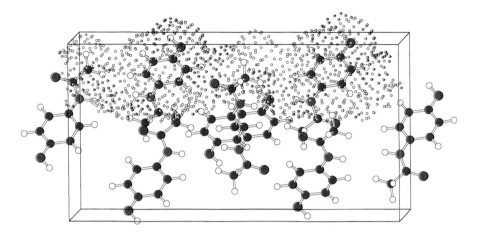

FIGURE 3.1   COSMO charges for the wetted monoclinic (201) paracetamol surface.[26] Gray balls are the carbon atoms, red—nitrogen atoms, blue—oxygen atoms, and dark blue—hydrogen atoms. Color code for COSMO charges: blue, -2 e/nm^2; red, $+2$ e/nm^2; and white, zero. Upper half of the COSMO surface is shown; face of cell indicates atomic surface plane.

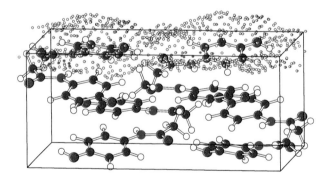

FIGURE 3.2   COSMO charges for the wetted orthorhombic (001) paracetamol surface. Gray balls are the carbon atoms, red—nitrogen atoms, blue—oxygen atoms, and dark blue—hydrogen atoms. Color code for COSMO charges: blue, -2 e/nm^2; red, $+2$ e/nm^2; and white, zero. Upper half of the COSMO surface is shown; face of cell indicates atomic surface plane.

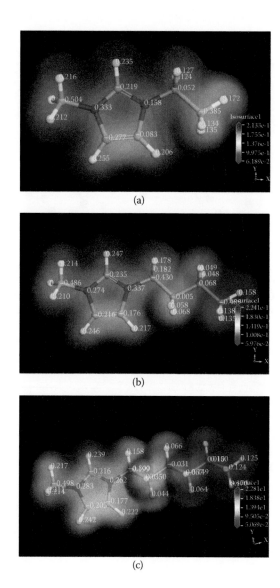

FIGURE 5.1 Electrostatic potential field around various cations as calculated at the PW91/ DNP level. (a) 1-Ethyl-3-methylimidazolium (EMIM), (b) 1-butyl-3-methylimidazolium (BMIM), and (c) 1-hexyl-3-methylimidazolium (HMIM).

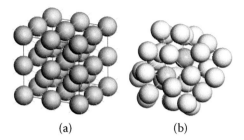

(a)   (b)

FIGURE 6.1 Mo_{35} nanocluster. (a) Initial configuration generated from $2 \times 2 \times 2$ b.c.c. supercell, and (b) the resulting structure optimized at the RHF/AM1* level without periodic boundaries or symmetry constraints. Mo atoms in (b) are colored according to their Mulliken charges: red, negative; blue, positive.

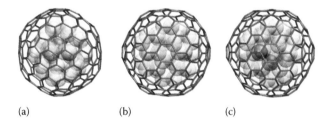

(a)   (b)   (c)

FIGURE 6.7 Endohedral fullerene Mo_n clusters: (a) $Mo_{35}@C_{180}$, (b) $Mo_{36}@C_{180}$, and (c) $Mo_{37}@C_{180}$. Mo atoms are represented by semitransparent spheres, colored according to their Mulliken charges (red, negative; blue, positive) with ranges set the same for each image.

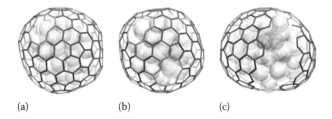

(a)   (b)   (c)

FIGURE 6.9 RHF/AM1* optimized geometries of endohedral fullerene Mo_n clusters: (a) $Mo_{44}@C_{180}$, (b) $Mo_{51}@C_{180}$, and (c) $Mo_{53}@C_{180}$. Mo atoms are represented by semitranspar-ent spheres, colored according to their Mulliken charges (red, negative; blue, positive) with ranges set the same for each image. Fullerene defects in (a) and (b) are highlighted in green and yellow, and full opening of the fullerene cage is observed in (c), with bonding to Mo atoms (which terminate the dangling bonds) not shown for clarity.

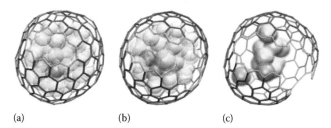

(a)   (b)   (c)

FIGURE 6.12   RHF/AM1* optimized geometries of endohedral fullerene $Mo_{(52-x)}S_x@C_{180}$ clusters: (a) $Mo_{52}@C_{180}$, (b) $Mo_{51}S@C_{180}$, and (c) $Mo_{42}S_{10}@C_{180}$. In (a) and (b), Mo atoms are represented by semitransparent spheres, colored according to their Mulliken charges (red, negative; blue, positive) with ranges set the same for each image. In (b) and (c), sulfur atoms are highlighted by overlaying a semitransparent yellow sphere. The fullerene defects in all three structures are highlighted in green, with full opening of the fullerene cage observed in (c). As in Figure 6.9, bonding to Mo atoms (which terminate the dangling bonds) is not shown for clarity, and Mo atoms are also hidden in (c).

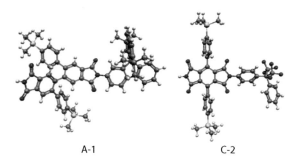

A-1   C-2

FIGURE 11.3   Optimized geometrical structures of A-1 and C-2. A-1 has a very bulky structure, whereas C-2, especially looking at the dianhydride, is quite flat.

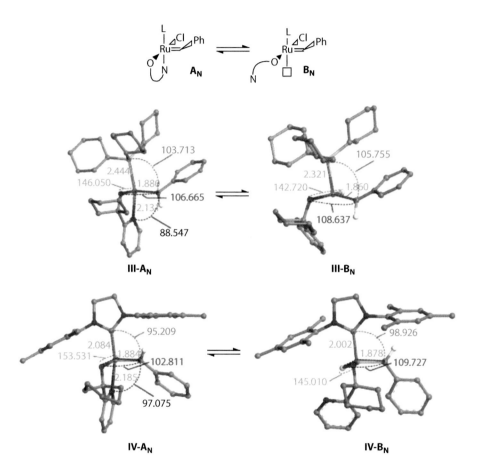

SCHEME 12.6 Dissociation ($\mathbf{A_N}$ to $\mathbf{B_N}$) of labile N-atom of O,N-ligand along with the optimized geometries for the lowest energy catalyst precursor complexes. The hydrogen atoms on the ligands are omitted for clarity and the unit of the indicated bond distances is angstrom (Å).

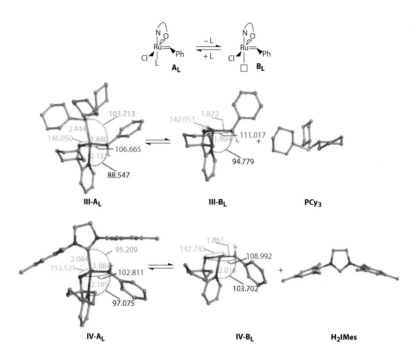

SCHEME 12.7 Dissociation ($\mathbf{A_L}$ to $\mathbf{B_L}$) of ligand L along with the optimized geometries for the lowest energy catalyst precursor complexes. The hydrogen atoms on the ligands are omitted for clarity and the unit of the indicated bond distances is angstrom (Å).

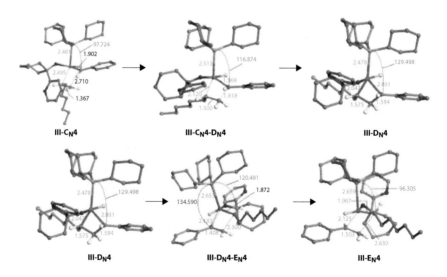

SCHEME 12.9 Optimized geometries of the lowest energy equilibrium and transition state structures involved in the metathesis reaction with **III** for activation step 4 ($\mathbf{C_N}$ to $\mathbf{E_N}$) in the $\mathbf{C_{II}}$ mode when the labile N-atom is dissociated. The hydrogen atoms on the ligands are omitted for clarity and the unit of the indicated bond distances is angstrom (Å).

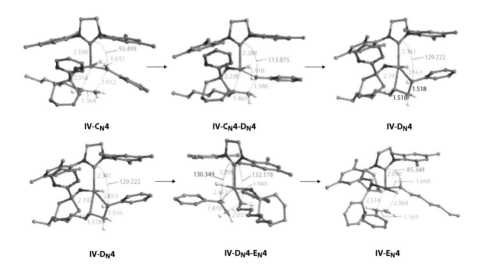

SCHEME 12.10 Optimized geometries of the lowest energy equilibrium and transition state structures involved in the metathesis reaction with **IV** for activation step 4 (C_N to E_N) in the C_{II} mode when the labile N-atom is dissociated. The hydrogen atoms on the ligands are omitted for clarity and the unit of the indicated bond distances is angstrom (Å).

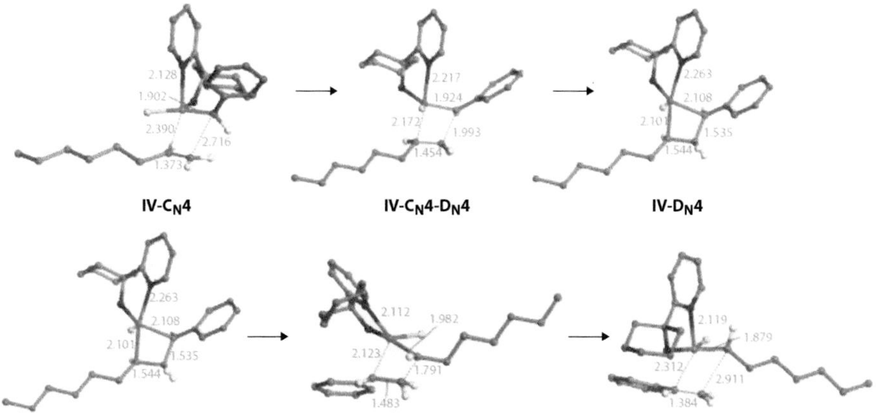

SCHEME 12.11 Optimized geometries of the lowest energy equilibrium and transition state structures involved in the metathesis reaction with **IV** for activation step 4 (C_L to E_L) in the C_{II} mode when L is dissociated. The hydrogen atoms on the ligands are omitted for clarity and the unit of the indicated bond distances is angstrom (Å).

FIGURE 13.1   Fluorometholone molecule with the (020) face of fluorometholone.

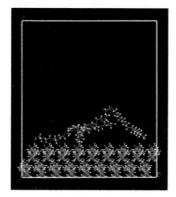

FIGURE 13.2   Polysorbate 80 with the (020) face of fluorometholone.

FIGURE 13.4   Polysorbate 80 with the nonpolar (110) face of carbamazepine.

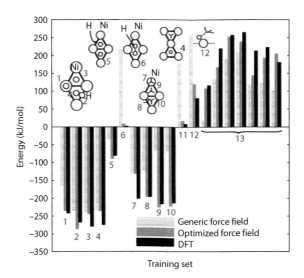

FIGURE 15.1 Fitting of ReaxFF parameters against DFT results and comparison with the old force field. 1–4: H binding atop, bridge, threefold hcp, and threefold fcc sites on Ni(111); 5: dissociation energy for H_2; 6: dissociation barrier for H_2; 7–10: CH_3 binding atop, bridge, threefold hcp, and threefold fcc sites on Ni(111); 11: dissociation energy for CH_4; 12: dissociation barrier for CH_4; 13: the interpolated geometries between 11 and 12.

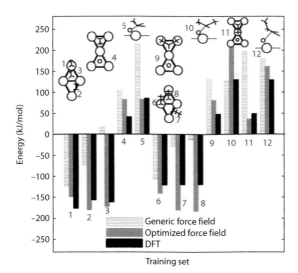

FIGURE 15.2 Comparison between the new and old force field on C_2H_6 and C_3H_8 that is not part of training to demonstrate force field transferability. 1–3: C_2H_5 binding atop, bridge, and threefold hcp sites on Ni(111); 4: dissociation energy for C_2H_6; 5: dissociation barrier for C_2H_6; 6–8: C_3H_7 binding atop, bridge, and threefold hcp sites on Ni(111); 9: dissociation for secondary C–H bond of C_3H_8; 10: dissociation barrier for secondary C–H bind of C_3H_8; 11: dissociation energy for primary C–H bond of C_3H_8; 12: dissociation barrier for primary C–H of C_3H_8.

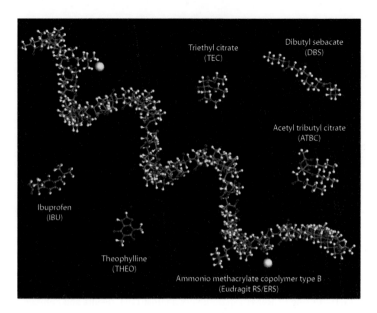

FIGURE 16.1 Visualization of Eudragit RS polymer chain consisting of 64 monomers and all other molecules (both plasticizers and drugs) used in this study. The chlorine anion is depicted larger for better visibility; the different colors in this and the following figures correspond to different atom types with white reserved for hydrogen, red oxygen, grey carbon, blue nitrogen, and green chlorine. All the graphical displays of molecules were generated with the Materials Visualizer (from Accelrys).

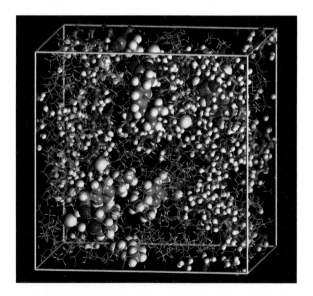

FIGURE 16.2 Simulation box (3.7 × 3.7 × 3.7 nm) with periodic boundaries showing four Eudragit RS polymer chains and six ibuprofen molecules after a relaxation time of 2 ns at 430 K. One polymer chain and all ibuprofen molecules are visualized in ball-and-stick style to demonstrate the size of the box as well as the distribution of the drug molecules.

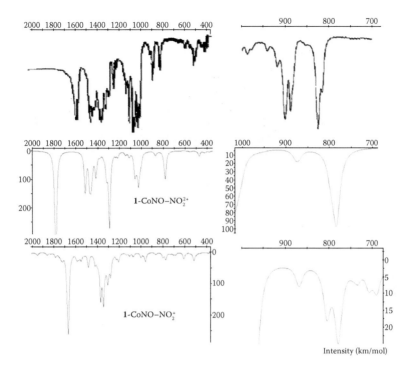

FIGURE 17.4 Experimental and simulated IR spectra of $[\mathbf{1}\text{-Co(II)NONO}_2]^{2+}$ and $[\mathbf{1}\text{-Co(II)}$ NONO$_2]^+$ complexes.

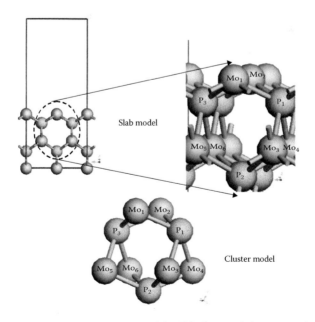

FIGURE 20.1 Comparison between MoP slab (100) face and the Mo_6P_3 cluster model of the present study.

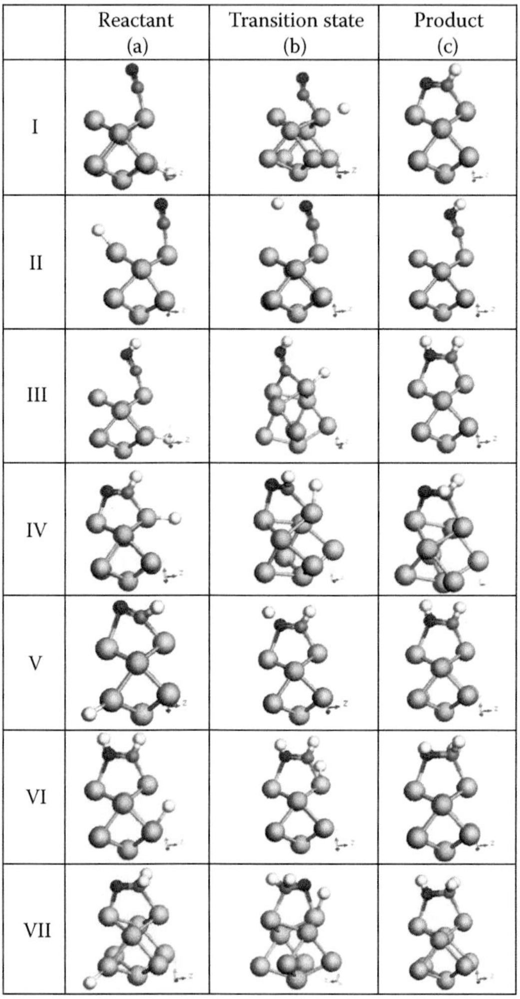

FIGURE 20.4 Methane and methanol formation reaction steps on Mo_6P_3 cluster. Blue = molybdenum; lavender = phosphorus; gray = carbon; red = oxygen; white = hydrogen.

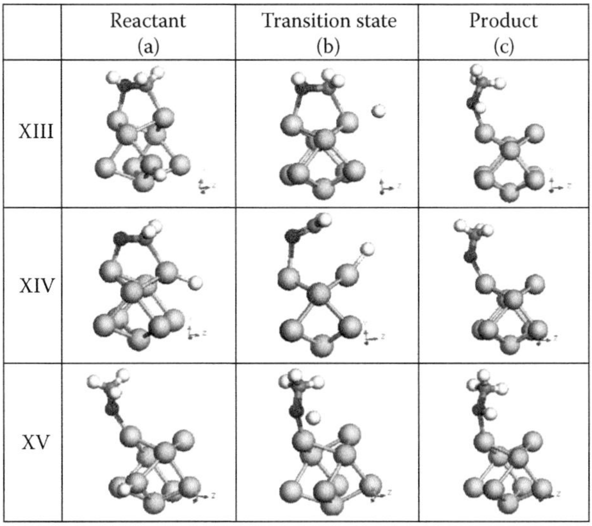

	Reactant (a)	Transition state (b)	Product (c)
XIII			
XIV			
XV			

FIGURE 20.6 Methanol formation reaction steps on Mo_6P_3 cluster. Blue = molybdenum; lavender = phosphorus; gray = carbon; red = oxygen; white = hydrogen.

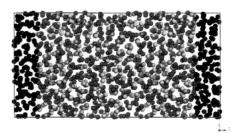

FIGURE 21.3 Model of glass for confined shear. Oxygen atoms—red, silicon—yellow, boron—brown, the alkali ions—purple (lithium smaller than sodium).

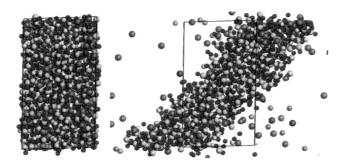

FIGURE 21.4 Shearing of borosilicate glass using confined shear method. Oxygen atoms—red, silicon—yellow, boron—brown, the alkali ions—purple (lithium smaller than sodium).

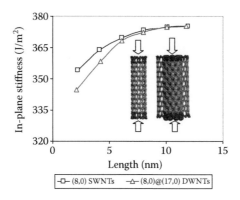

FIGURE 22.1 Comparison of in-plane stiffness between zigzag DWNTs and SWNTs.

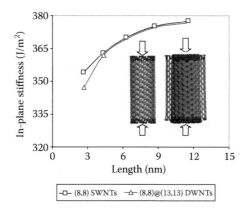

FIGURE 22.2 Comparison of in-plane stiffness between armchair DWNTs and SWNTs.

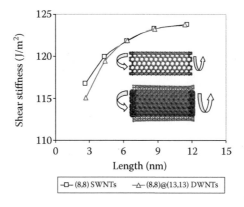

FIGURE 22.3 Comparison of shear modulus between armchair DWNTs and SWNTs.

TABLE 17.2

Energies for the Gas Phase and Methanol Complexes of 1-Co(II) or 1-Co(III) with Different Ligands

1-Co(II)–Cl⁺		NO	NO$_2$	NO$_2^{-1}$	Cl^{-1}
Complex charges	1	1	1	0	0
In gas phase (au)	−1241.893521	−1371.921846	−1447.17903	−1447.338003	−1702.38107
In methanol (au)	−1242.013279	−1372.001621	−1447.259219	−1447.376739	−1702.43868
Solvent effect (kcal/mol)	−75.15	−50.06	−50.32	−24.31	−36.15
HOMO (eV) gas phase	−7.78	−7.85	−9.38	−2.61	−1.36
LUMO (eV) gas phase	−6.08	−6.67	−6.15	−1.56	−0.79
HOMO–LUMO (eV) gas phase	1.70	1.18	3.23	1.05	0.57
HOMO (eV) in methanol	−4.11	−5.05	−6.64	−3.50	−3.50
LUMO (eV) in methanol	−1.92	−3.74	−3.23	−2.08	−1.95
HOMO–LUMO (eV) in methanol	2.19	1.31	3.40	1.42	1.56

1-Co(III)–Cl²⁺		NO	NO$_2$	NO$_2^{-1}$	Cl^{-1}
Charges	2	2	2	1	1
In gas phase (au)	−1241.56782	−1371.552237	−1446.754605	−1447.17903	−1702.26157
In methanol (au)	−1241.854823	−1371.802123	−1447.002374	−1447.259219	−1702.31577
Solvent effect (kcal/mol)	−180.10	−156.81	−155.48	−50.32	−34.01
HOMO (eV) gas phase	−13.62	−13.20	−14.04	−9.38	−8.65
LUMO (eV) gas phase	−11.91	−12.14	−13.59	−6.15	−5.83
HOMO–LUMO (eV) gas phase	1.71	1.05	0.45	3.23	2.82
HOMO (eV) in methanol	−7.14	−6.89	−7.87	−6.64	−6.41
LUMO (eV) in methanol	−4.66	−5.83	−7.42	−3.23	−3.50
HOMO–LUMO (eV) in methanol	2.48	1.07	0.45	3.40	2.91

1-Co(II)–NO²⁺		NO	NO$_2$	NO$_2^{-1}$	Cl^{-1}
Charges	2	2	2	1	1
In gas phase (au)	−911.295746	−1041.256722	−1116.490312	−1116.853893	−1371.92185

(continued)

TABLE 17.2 (Continued)

Energies for the Gas Phase and Methanol Complexes of 1-Co(II) or 1-Co(III) with Different Ligands

1-Co(II)–Cl+		NO	NO$_2$	NO$_2^{-1}$	Cl^{-1}
In methanol (au)	−911.556231	−1041.512176	−1116.742836	−1116.938316	−1372.00162
Solvent effect (kcal/mol)	−163.46	−160.30	−158.46	−52.98	−50.06
HOMO (eV) gas phase	−12.65	−11.90	−12.95	−7.72	−7.85
LUMO (eV) gas phase	−10.87	−11.31	−12.09	−6.84	−6.67
HOMO–LUMO (eV) gas phase	1.78	0.60	0.86	0.88	1.18
HOMO (eV) in methanol	−5.89	−6.64	−6.74	−4.91	−5.05
LUMO (eV) in methanol	−4.14	−5.10	−5.78	−3.84	−3.74
HOMO–LUMO (eV) in methanol	1.75	1.55	0.97	1.07	1.31

1-Co(III)–NO^{3+}		NO	NO$_2$	NO$_2^{-1}$	Cl^{-1}
Charges	3	3	3	2	2
In gas phase (au)	−910.747861	−1040.732515	−1115.941179	−1116.490312	−1371.53799
In methanol (au)	−911.316027	−1041.27666	−1116.489015	−1116.742836	−1371.80212
Solvent effect (kcal/mol)	−356.53	−341.46	−343.77	−158.46	−165.75
HOMO (eV) gas phase	−18.28	−17.69	−18.16	−12.95	−13.70
LUMO (eV) gas phase	−16.79	−16.52	−16.80	−12.09	−13.02
HOMO–LUMO (eV) gas phase	1.49	1.17	1.36	0.86	0.68
HOMO (eV) in methanol	−8.17	−8.02	−8.76	−6.74	−6.89
LUMO (eV) in methanol	−6.99	−6.80	−7.20	−5.78	−5.83
HOMO–LUMO (eV) in methanol	1.56	1.56	1.56	0.97	1.07

1-Co(II)–NO$_2^+$		NO	NO$_2$	NO$_2^{-1}$	Cl^{-1}
Charges	1	1	1	0	0
In gas phase (au)	−986.854343	−1116.853893	−1192.117753	−1192.287363	−1447.338
In methanol (au)	−986.947275	−1116.938316	−1192.202831	−1192.32393	−1447.37674
Solvent effect (kcal/mol)	−58.32	−52.98	−53.39	−22.94	−24.31

TABLE 17.2 (Continued)
Energies for the Gas Phase and Methanol Complexes of 1-Co(II) or 1-Co(III) with Different Ligands

1-Co(II)–Cl⁺		NO	NO$_2$	NO$_2^{-1}$	Cl^{-1}
HOMO (eV) gas phase	−6.80	−7.72	−9.05	−4.62	−2.61
LUMO (eV) gas phase	−5.34	−6.84	−6.21	−1.73	−1.56
HOMO–LUMO (eV) gas phase	1.46	0.88	2.84	2.89	1.05
HOMO (eV) in methanol	−3.80	−4.91	−6.40	−5.2	−3.50
LUMO (eV) in methanol	−2.28	−3.84	−3.21	−2.2	−2.08
HOMO–LUMO (eV) in methanol	1.53	1.07	3.19	2.98	1.42

1-Co(III)–NO$_2^{2+}$		NO	NO$_2$	NO$_2^{-1}$	Cl^{-1}
Charges	2	2	2	1	1
In gas phase (au)	−986.520493	−1116.490312	−1191.70322	−1192.117753	−1447.17903
In methanol (au)	−986.798345	−1116.742836	−1191.950985	−1192.202831	−1447.25922
Solvent effect (kcal/mol)	−174.35	−158.46	−155.48	−53.39	−50.32
HOMO (eV) gas phase	−13.98	−12.95	−14.12	−9.05	−9.38
LUMO (eV) gas phase	−11.33	−12.09	−13.48	−6.21	−6.15
HOMO–LUMO (eV) gas phase	2.65	0.86	0.64	2.84	3.23
HOMO (eV) in methanol	−7.51	−6.74	−7.98	−6.40	−6.64
LUMO (eV) in methanol	−4.44	−5.78	−7.41	−3.21	−3.23
HOMO–LUMO (eV) in methanol	3.07	0.97	0.58	3.19	3.40

1-Co(III)– NO$_2^{3+}$		NO	NO$_2$	NO$_2^{-1}$	Cl^{-1}
Charges	3	3	3	2	2
In gas phase (au)	−985.932523	−1115.941179	−1191.111532	−1191.70322	−1446.75461
In methanol (au)	−986.503361	−1116.489015	−1191.65726	−1191.950985	−1447.00237
Solvent effect (kcal/mol)	−358.21	−343.77	−342.45	−155.48	−155.48

(*continued*)

TABLE 17.2 (Continued)

Energies for the Gas Phase and Methanol Complexes of 1-Co(II) or 1-Co(III) with Different Ligands

1-Co(II)–Cl+		NO	NO$_2$	NO$_2^{-1}$	Cl^{-1}
HOMO (eV) gas phase	−18.56	−18.16	−19.31	−14.12	−14.04
LUMO (eV) gas phase	−17.64	−16.80	−18.20	−13.48	−13.59
HOMO–LUMO (eV) gas phase	0.92	1.36	1.16	0.64	0.45
HOMO (eV) in methanol	−8.87	−8.76	−9.97	−7.98	−7.87
LUMO (eV) in methanol	−8.13	−7.20	−8.82	−7.41	−7.42
HOMO–LUMO (eV) in methanol	0.74	1.56	1.17	0.58	0.45

1-Co(III)$^{3+}$		NO	NO$_2$	NO$_2^{-1}$	Cl^{-1}
Charges	3	3	3	2	2
In gas phase (au)	−780.705952	−910.747861	−985.932523	−986.520493	−1241.56782
In methanol (au)	−781.356894	−911.316027	−986.503361	−986.798345	−1241.85482
Solvent effect (kcal/mol)	−408.47	−356.53	−358.21	−174.35	−180.10
HOMO (eV) gas phase	−18.82	−18.28	−18.56	−13.98	−13.62
LUMO (eV) gas phase	−18.43	−16.79	−17.64	−11.33	−11.91
HOMO–LUMO (eV) gas phase	0.39	1.49	0.92	2.65	1.71
HOMO (eV) in methanol	−8.49	−8.17	−8.87	−7.51	−7.14
LUMO (eV) in methanol	−6.71	−6.99	−8.13	−4.44	−4.66
HOMO–LUMO (eV) in methanol	1.78	1.18	0.74	3.07	2.48

1-Co(II)$^{2+}$		NO	NO$_2$	NO$_2^{-1}$	Cl^{-1}
Charges	2	2	2	1	1
In gas phase (au)	−781.294661	−911.295746	−986.520493	−986.854343	−1241.89352
In methanol (au)	−781.57347	−911.556231	−986.798345	−986.947275	−1242.01328
Solvent effect (kcal/mol)	−174.96	−163.46	−174.35	−58.32	−75.15

TABLE 17.2 (Continued)
Energies for the Gas Phase and Methanol Complexes of 1-Co(II) or 1-Co(III) with Different Ligands

1-Co(II)–Cl+		NO	NO$_2$	NO$_2^{-1}$	Cl^{-1}
HOMO (eV) gas phase	−13.46	−12.65	−13.98	−6.80	−7.78
LUMO (eV) gas phase	−11.41	−10.87	−11.33	−5.34	−6.08
HOMO–LUMO (eV) gas phase	2.06	1.78	2.65	1.46	1.70
HOMO (eV) in methanol	−5.04	−5.89	−7.51	−3.80	−4.11
LUMO (eV) in methanol	−2.53	−4.14	−4.44	−2.28	−1.92
HOMO–LUMO (eV) in methanol	2.51	1.75	3.07	1.53	2.19

Note: (au) = .

The highest occupied molecular orbital (HOMO) and the lowest unoccupied molecular orbital (LUMO) of a species are known as frontier molecular orbitals after Fukui's work [35] on their role in predicting chemical reactivity. It has been shown that these orbitals play a major role in governing many chemical reactions and determining electronic band gaps in organometallic complexes and solids; they are also responsible for the formation of many charge transfer complexes [36]. According to the frontier molecular orbital theory of chemical reactivity, the formation of a transition state is due to the interaction between the frontier orbitals (HOMO and

TABLE 17.3
Relative Stability (kcal/mol) of 1-[Co(II)/Co(III)] Complexes in Gas Phase and Methanol

Co(II)–Co(III) Complexes		NO	NO$_2$	NO$_2^{-1}$	Cl^{-1}
[1-Co(II)–1-CO(III)]Cl	In gas phase	−231.93	−266.33	−99.76	−74.99
	In methanol	−125.19	−161.17	−73.74	−77.13
[1-Co(II)–1-CO(III)NO	In gas phase	−328.95	−344.59	−228.15	−240.88
	In methanol	−147.79	−159.28	−122.67	−125.19
[1-Co(II)–1-CO(III)]NO$_2$_a	In gas phase	−228.15	−260.12	−106.43	−99.76
	In methanol	−122.67	−158.04	−75.99	−73.74
[1-Co(II)–1-CO(III)]NO$_2$_b	In gas phase	−572.74	−631.41	−366.56	−366.09
	In methanol	−281.94	−342.35	−234.03	−234.92
[1-Co(II)–1-CO(III)]	In gas phase	−343.80	−368.96	−209.49	−204.38
	In methanol	−150.73	−185.11	−93.46	−99.43

LUMO) of the reacting species [35,36]. The energy of the HOMO is directly related to the ionization potential and characterizes the reactivity of the molecule toward an electrophilic reaction. The energy of the LUMO is directly linked to the electron affinity and characterizes the susceptibility of the molecule toward an attack by nucleophiles. Moreover, a large HOMO–LUMO gap corresponds to a high stability for the molecule but to a lower reactivity [37,38].

Another useful approach in predicting the reactivity of species, involving a single pair of frontier orbitals, is the hard–soft acid–base principle introduced by Pearson [39–42] and developed in the framework of DFT by Parr and coworkers [43–45]. Following the definition of Pearson [39–41] of global hardness (η) and global softness (S) based on the finite difference approximation, we have calculated the global hardness as $\eta = 1/2 \, (\varepsilon_{LUMO} - \varepsilon_{HOMO})$ and the global softness as $S = 1/\eta$. In Table 17.4, we have reported the values of global hardness (GH) and global softness (GS) for all [1-Co(II)]–L_iL_j, where L_iL_j are two ligands from (NO, NO_2, NO_2^-, and Cl^-). GS has been formally connected with molecular polarizability.

In Table 17.5, we report the calculations carried out for the isolated ligands (Cl^-, NO, NO_2, and NO_2^-) in gas phase and in methanol, including NO stretching vibration frequencies that compare well with experimental observations. In Table 17.6, we report the strength of [L_i-1-Co(II)]–L_j bond, where L_iL_j are two ligands from (NO, NO_2, NO_2^-, and Cl^-) in gas phase and in methanol solvent.

In Table 17.7, we report the calculated NO vibrational stretching frequencies (cm^{-1}) for different [L_i-1-Co(II)]–L_j complexes, where L_i,L_j are two ligands from (NO, NO_2, NO_2^-, and Cl^-). The accuracy of the vibrational frequency calculations is

TABLE 17.4

Global Hardness (GH) and Global Softness (GS) for 1-Co(II) Complexes in Gas Phase and Methanol

GH in (eV) and GS in (1/eV)			NO	NO_2	NO_2^{-1}	Cl^{-1}
1-Co(II)–Cl^+	In gas phase (GH)	0.85	0.59	1.61	0.53	0.29
	In gas phase (GS)	1.18	1.69	0.62	1.90	3.50
	In methanol (GH)	1.10	0.66	1.70	0.71	0.78
	In methanol (GS)	0.91	1.53	0.59	1.41	0.97
1-Co(II)–NO^{2+}	In gas phase (GH)	0.89	0.30	0.43	0.44	0.59
	In gas phase (GS)	1.12	3.36	2.34	2.28	1.69
	In methanol (GH)	0.87	0.77	0.48	0.54	0.66
	In methanol (GS)	1.15	1.29	2.07	1.86	1.53
1-Co(II)–NO^{2+}	In gas phase (GH)	0.73	0.44	1.42	1.44	0.53
	In gas phase (GS)	1.37	2.28	0.70	0.69	1.90
	In methanol (GH)	0.76	0.54	1.60	1.49	0.71
	In methanol (GS)	1.31	1.86	0.63	0.67	1.41
1-Co(II)$^{2+}$	In gas phase (GH)	1.03	0.89	1.33	0.73	0.85
	In gas phase (GS)	0.97	1.12	0.75	1.37	1.18
	In methanol (GH)	3.79	5.02	5.97	3.04	3.02
	In methanol (GS)	0.26	0.20	0.17	0.33	0.33

TABLE 17.5
Calculated Properties of Ligands in Gas Phase and Methanol

Ligands	NO	NO_2	NO_2^{-1}	Cl^{-1}
Charge	0	0	−1	−1
Energy (au) in gas phase	−129.942283	−205.179536	−205.23013	−460.27953
Energy (au) in methanol	−129.942667	−205.181183	−205.30483	−460.400715
Solvent effect (kcal/mol)	−0.24	−1.03	−46.87	−76.04
HOMO gas phase (eV)	−12.13	−9.25	1.44	1.93
LUMO gas phase (eV)	−5.43	−7.4	4.18	16.32
HOMO–LUMO gap_g	6.7	1.85	2.74	14.39
GH gas phase (eV)	3.35	0.93	1.37	7.2
GS gas phase (1/eV)	0.3	1.08	0.73	0.14
HOMO methanol (eV)	−10.97	−6.36	−1.93	−4.72
LUMO methanol (eV)	−4.29	−4.85	0.15	9.7
HOMO–LUMO gap_m	6.68	1.51	2.08	14.42
GH methanol (eV)	3.34	0.75	1.04	7.21
GS methanol (1/eV)	0.3	1.33	0.96	0.14
NO stretching (cm^{-1}) asym	1893	1628	1207	
NO stretching (cm^{-1}) sym		1329	1307	
NO stretching exp (cm^{-1}) asym	1904	1618		
NO stretching exp (cm^{-1}) sym		1318		

a function of the theoretical method used. In many cases, the experimentally measured values differ significantly from those calculated theoretically. One source of discrepancy is that the experimental values are often determined in solution or in a solid matrix, whereas the calculated values refer to the gas phase. Quite good results can be obtained from density functional calculations using gradient-corrected functionals.

TABLE 17.6
Complex–Ligand Bond Energies (kcal/mol) in Gas Phase and Methanol

Complex–Ligand		NO	NO_2	NO_2^{-1}	Cl^{-1}
1-Co(II)–Cl+	In gas phase	−54.0	−66.5	−134.5	−130.5
	In methanol	−28.7	−40.6	−36.8	−15.5
1-Co(II)–NO²⁺	In gas phase	−11.7	−9.4	−205.8	−217.5
	In methanol	−8.3	−3.4	−48.5	−28.0
1-Co(II)–NO⁺₂	In gas phase	−35.9	−52.6	−127.3	−128.1
	In methanol	−30.4	−46.7	−45.1	−18.0
1-Co(II)²⁺	In gas phase	−36.9	−29.1	−206.8	−200.4
	In methanol	−25.2	−27.4	−43.3	−24.5

TABLE 17.7

Calculated NO Stretching Frequencies of Different Complexes

Calculated N–O Vibrational Frequency (cm⁻¹)		NO	NO_2	NO_2^{-1}	Cl^{-1}
1-Co(II)–Cl⁺	NO stretching (cm⁻¹) asym	1593	1432		
	NO stretching (cm⁻¹) asym		1441	1527	
	NO stretching (cm⁻¹) sym		1309	1503	
1-Co(II)–NO²⁺	NO stretching (cm⁻¹) asym	1650	1786	1671	1594
	NO stretching (cm⁻¹) asym	1738	1517	1489	
	NO stretching (cm⁻¹) sym		1293	1376	
1-Co(II)–NO₂⁺	NO stretching (cm⁻¹) asym	1671	1569	1516	1527
	NO stretching (cm⁻¹) asym	1489	1586	1518	
	NO stretching (cm⁻¹) sym	1376	1493	1263	1503
1-Co(II)²⁺	NO stretching (cm⁻¹) asym	1778			
	NO stretching (cm⁻¹) asym		1618	1389	
	NO stretching (cm⁻¹) sym		1444	1237	

17.5 DISCUSSION

In our combined experimental–theoretical approach, DFT calculations were used to help understand NO competitive coordination to **1**-Cobalt complexes in methanol solution in the presence of other ligands (O_2, NO_2, NO_2^-, and Cl^-). Systematic calculations of all possible combinations of cobalt oxidation state and the two *trans*-coordinating ligands have been undertaken for the isolated structures (35 [**1**-Co(II)]–L_iL_j structures, where L_iL_j are two ligands from [NO, NO_2, NO_2^-, and Cl^-]) and their analogs in methanol counterparts. The results of such calculations reported in Table 17.2 indicate that the solvent effect has an important role on the stability of these complexes. The comparative analysis of the stability of Co(II) complexes with their Co(III) analog structures, reported in Table 17.3, clearly indicates that Co(II) complexes are more stable due to the electron imbalance of the Co(III) species than that which is observed experimentally when compensated by an anion species. The electrostatic interaction between the two charged species would normally stabilize the Co(III) complex. When the effect of the solvent is taken into account, the difference in stability between calculated **1**-Co(II) and **1**-Co(III) complexes is reduced.

   In Table 17.4, the global stability (GH) and reactivity (GS) indices are reported for **1**-Co(II)–L_iL_j complexes, where L_iL_j are two ligands from (NO, NO_2, NO_2^-, or Cl^-). According to the principle of maximum hardness [43], the more stable complexes should have a maximum hardness value. The calculated GH values for each complex in the gas phase or methanol, reported in Table 17.6, indicate that between the calculated complexes **1**-Co(II)–L_iL_j, the [**1**-Co(II)–ClNO_2]⁺, [**1**-Co(II)–(NO_2)₂]⁺, and [**1**-Co(II)–(NO_2)₂]⁺ are the most stable ones. The reported values of GS for the same complexes also indicate that the most reactive complexes are [**1**-Co(II)–ClNO]⁺, [**1**-Co(II)–NONO₂]²⁺, and [**1**-Co(II)–NONO₂]⁺. Table 17.5 shows the calculated properties of the considered ligands (NO, NO_2, NO_2^-, and Cl^-). It appears clearly that the most stable ligand is the anion Cl^- and the most reactive one is the NO_2 radical.

In this table, we also report the vibrational frequencies of the isolated ligands that will be further discussed when the IR spectra of the complexes are compared with experimental ones.

In Figure 17.1, we report the optimized structures of all stable complexes [1-Co(II)]–L_iL_j, where L_iL_j are two ligands from (NO, NO_2, NO_2^-, and Cl–) reported in Table 17.6. The analysis of the interaction of 1-Co(II)–Cl_2 with the NO reported in this table confirms the experimental finding that the formation of the *trans* Cl–Co–NO cyclam complex is observed at low NO concentrations and in the absence of oxygen. As a matter of fact, the calculated Co–NO bond energy (28.7 kcal/mol) is nearly double of the Co–Cl one (15.5 kcal/mol). It should be noted, however, that in this analysis, only the calculations that take into account the solvent effects should be considered. These results are well correlated with the GH data reported in Table 17.4. If we consider the replacement of the second Cl– by an NO ligand and the formation of a *trans* NO–Co–NO cyclam complex, the predicted binding energy of the second NO to Co is only 8.3 kcal/mol, which is nearly half of the corresponding Cl–Co binding energy (15.5 kcal/mol). This prediction perfectly fits with the experimental observations. The formation of the *trans* NO–Co–NO cyclam complex has not been observed experimentally. Following the analysis of the stability of the complexes reported in Table 17.6, we can say that the binding of the [1-Co(II)]$^{2+}$ complex with the NO ligand in methanol (25.2 kcal/mol) is nearly same as its binding with the Cl– ligand (24.5 kcal/mol), and adding a *trans*-second ligand different from the first one is energetically favored. A [1-Co(II)]–L_i complex will preferably form a [1-Co(II)]–L_iL_j structure, where L_iL_j are two ligands from (NO, Cl–) and L_i is different from L_j.

The addition of a neutral NO_2 ligand to a methanol solution containing 1-Co(II)–Cl_2 should favor the *trans* 1-Co(II)–$ClNO_2$. As reported in Table 17.6, the binding energy of NO_2 on 1-Co(II)–Cl complex is about 40.6 kcal/mol, which is sensibly higher than the NO binding on the same complex. At the same time, the NO_2 binding energy to the ligand-free [1-Co(II)]$^{2+}$ complex in methanol (27.4 kcal/mol) is very similar to the bond energy of the NO and Cl– ligands (25.2 kcal/mol and

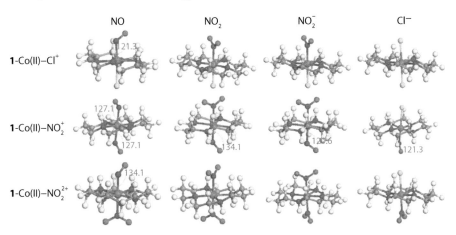

FIGURE 17.1 Optimized structures of 1-Co(II) *trans* L_i,L_j complexes. L_i,L_j are ligands from (NO, NO_2, NO_2^-, and Cl–).

24.5 kcal/mol, respectively), but once the [**1**-Co(II)]NO$_2$ half-complex is formed, binding of a second NO$_2$ ligand is favored (46 kcal/mol) over NO (35.9 kcal/mol) or Cl$^-$ (18 kcal/mol). This trend is in good agreement with the stability estimation from the GH values in Table 17.4. In the experiments in which dioxygen was introduced to the methanol solution containing the *trans* **1**-Co(II)–ClNO complex, new IR bands were observed corresponding to a coordinated NO$_2$ group to the cobalt center. This result confirms the theoretical predictions related to the stability of [**1**-Co(II)]L$_i$NO$_2$ complexes, where L$_i$ is a ligand from (NO, NO$_2$, NO$_2^-$, and Cl$^-$). The predicted NO vibrational spectra of the different stable **1**-Co(II)–L$_i$L$_j$ complexes, where L$_i$L$_j$ are two ligands from (NO, NO$_2$, NO$_2^-$, and Cl$^-$), are reported in Table 17.7. These predictions were confirmed by the experimental analysis of the IR spectra evolution during sequential treatment of the solution containing **1**-Co(II)Cl$_2$ with NO, Ar, and O$_2$ gas streams and reported in Figures 17.2 through 17.5. In Figure 17.2, the observed intense band at 1610 cm^{-1} for the **1**-Co(II)NOCl complex has been predicted in quite good agreement at 1593 cm^{-1}, and the overall shape of the IR spectrum is very well reproduced apart from the region around 3500 cm^{-1} corresponding to the H–O stretching of the solvent (methanol) molecules. The introduction of the oxygen in the methanol solution containing the **1**-Co(II)NOCl was followed by the presence of new NO stretching bands around 1355 cm^{-1} characteristic of coordination of NO$_2$. According to our calculations, these bands can be attributed either to **1**-Co(II)ClNO$_2$ (calculated around 1309 cm^{-1}) or **1**-Co(II)NONO$_2$ (predicted at 1376 cm^{-1}) or **1**-Co(II) NO$_2$NO$_2$ (predicted at 1263 cm^{-1}). In Figure 17.3, we reported the experimental and predicted IR spectra of [**1**-Co(II)ClNO$_2$]$^+$ and [**1**-Co(II)ClNO$_2$] species corresponding to two different (NO$_2$ and NO$_2^-$) species. From the comparison of the overall shape of the predicted spectra, it is clear that the one observed corresponds well to the [**1**-Co(II)ClNO$_2$]$^+$ complex. This result is in perfect agreement with the predicted bond strength of [**1**-Co(II)ClNO$_2$]$^+$ complex corresponding to the coordination of NO$_2$ ligand on the metal center of the [**1**-Co(II)Cl]$^{2+}$ complex.

By varying the flow rate of NO as well as the quantity and duration of dioxygen bubbling into the solution flask, a drastic evolution of the experimental spectra was observed; however, it is not evident to experimentally attribute the vibrational bands to different species due to the evolution of the axial ligands in solution.

To gain some insight into the evolution of the complexes, in Figures 17.4 and 17.5, the same experimental spectrum has been compared to the simulated ones of **1**-Co(II)NONO$_2$ and **1**-Co(II)NO$_2$NO$_2$.

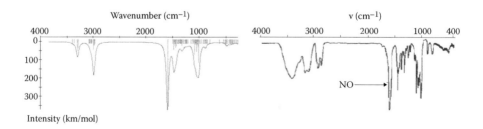

FIGURE 17.2 Experimental and simulated IR spectra of **1**-Co(II)ClNO complex.

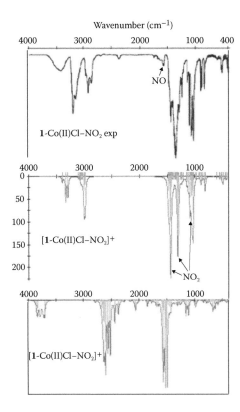

FIGURE 17.3 Experimental and simulated IR spectra of [1-Co(II)ClNO₂]⁺ and [1-Co(II)ClNO₂] complexes.

From Figure 17.4, it is clear that the attribution of the observed bands is less evident than in the case of the initial complexes. This is also due to the superposition of spectra of different species present in the solution. However, the trend corresponding to the similarity of the bands of experimental and simulated spectra allows us to further correlate the observed spectrum to the one corresponding to the [1-Co(II) NONO₂]⁺ complex. This is supported also by the fact that the latter is much more stable in solution, as reported in Table 17.6.

In Figure 17.5, we report the comparison of the same experimental spectrum reported in Figure 17.4, together with the corresponding ones to [1-Co(II)(NO₂)₂]²⁺ and [1-Co(II)(NO₂)₂]⁺. From a rapid comparison of the aligned spectra, it is evident that the experimental one can be considered as a combination of the simulated ones with some contribution from the spectrum of [1-Co(II)NONO₂]⁺ reported in Figure 17.4. As predicted in Table 17.6, all three species have similar stabilities. The analysis of the bond strength of these three complexes also suggests that 1-Co(II)NO₂NO₂ (both species) are very stable, but no direct experimental observation has been possible. Following this study, we propose a modification of the experimental conditions that would favor the isolation of these complexes.

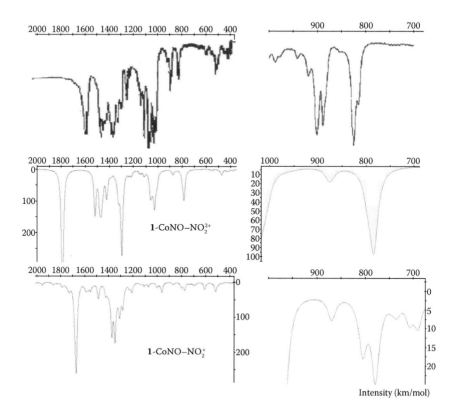

FIGURE 17.4   (See color insert.) Experimental and simulated IR spectra of [1-Co(II)
NONO₂]²⁺ and [1-Co(II)NONO₂]⁺ complexes.

In the following section, we discuss the mechanism of the formation of different
predicted complexes in methanol solution starting with the oxidation of NO in the
presence of dioxygen to give nitrogen dioxide; this is an important reaction that has
been extremely well studied in the gas phase in relation to air pollution caused by different nitrogen oxides but whose outcome in water solution gives exclusively nitrite
ions, with a few nitrate ions. The mechanism of NO oxidation by dioxygen to NO_2 in
gas phase has been recently [46,47] explained by the formation of the intermediate
ONOONO species that undergoes homolytic O–O bond breaking. The aqueous reaction of NO with dioxygen occurs with the following overall stoichiometry:

$$4NO + O_2 + 2H_2O = 4H^+ + 4NO_2^-.$$

As a result of these reactions in methanol solution, in the presence of NO and O_2,
one also finds NO_2 and NO_2^-. To respond to the question—whether the oxidation of
the NO coordinated to **1**-Co(II), to NO_2, is catalyzed by the Co(II) center or whether
NO is first oxidized in solution and is then exchanged with the co-coordinated NO—
we have undertaken calculations of the oxidations of NO in both these cases in gas

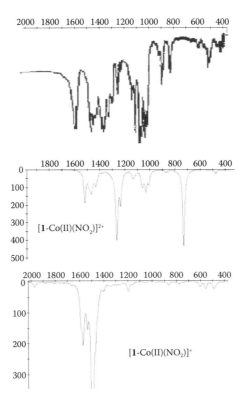

FIGURE 17.5   Experimental and simulated IR spectra of [**1**-Co(II)(NO₂)₂]²⁺ and [**1**-Co(II) (NO₂)₂]⁺ complexes.

phase as well as in methanol solutions. We found that, as previously reported [46,47], the first stage of NO oxidation corresponds to the formation of the ON–OO interme- diate with a small barrier of 13.2 (14.5) kcal/mol in gas phase (in solution), whereas for the NO coordinated to the cyclam–Co(II) complex, the formation of 1-Co(II) NO–OO showed a barrier of –0.4 (5.0) kcal/mol in gas phase (in methanol solution). This indicates that the cobalt acts as a catalyst for NO oxidation. The second step corresponds to the coordination of a second NO to the ON–OO species followed by the homolytic breaking of the O–O bond and formation of two NO₂ species. The calculation for the simple and Co-coordinated NO–OO–NO species indicated that formation of 1-Co(II)NO₂ + NO₂ species in the solution is 38.62 kcal/mol more stable than the cyclam–Co(II)NO–OO–NO complex, whereas this difference for the simple species is only 15.61 kcal/mol. This result clearly indicates that the formation of the 1-Co(II)NO₂NO₂ complex is favored not only energetically but also based on reactivity data.

Several hypotheses have been posed to explain nitrite formation in solutions. The electrophile–nucleophile interaction involves the overlap of the HOMO of the nucleophile with the LUMO of the electrophile to form a couple of new bonding and antibonding orbitals. The closer the energy between the two interacting orbitals, the greater their interaction. Soft–soft interactions are mostly controlled by the frontier orbitals of the interacting systems, and in the case of the hard–hard type of interactions, the contribution of electrostatic interaction becomes more significant.

It should be noticed, however, that the orbital interaction alone does not take into account the charge interaction effects (interaction between charged species) as well as the overlapping of the orbitals that is partially controlled by their respective symmetries. In Table 17.8, the energy difference between the HOMO of the nucleophile [1-Co(II)] complex and the LUMO of the electrophile ligand (Cl⁻, NO, NO_2, and NO_2^-) has been reported. The overall effect of the solvent is to reduce the gap between the HOMO of the nucleophile and the LUMO of the electrophile, in agreement with the experimental observation of new species formation in methanol solutions [1-Co(II)]–L_i in the presence of different ligands (NO, NO_2, NO_2^-, and Cl⁻). The differences in the HOMO–LUMO gap between the reactants reported in Table 17.8 for the charged ligands NO_2^- and Cl⁻ are compensated by the charge interaction between the 1-Co(II) complex that bears a positive charge and the negatively charged ligands.

It is important to note that it is not only the charge interaction between the reactants and the energy difference of their frontier orbitals that are important in explaining the reactivity of these complexes but also the localization of the orbitals on the reacting species. For this reason, we have calculated the Fukui functions corresponding to different electrophilic and nucleophilic attacks and have used these data to complement the reactivity information.

In Figure 17.6, we report the different steps of the NO coordination to the 1-Co(II) complex followed by its oxidation and further transformation of NO into NO_2. In this figure, we represent graphically the Fukui function indices located at the sites of nucleophilic and electrophilic attack for all the intermediaries of the reaction. The Fukui functions provide valuable information on the possible positions of reactivity and confirm the experimental observations for these complexes. As reported in

---

TABLE 17.8

Energy Difference (kcal/mol) between HOMO (nu) and LUMO (el) of Reactants

Complexes/Ligands		NO	NO_2	NO_2^{-1}	Cl^{-1}
1-Co(II)–Cl⁺	In gas phase	−2.36	−0.38	−11.97	−24.11
	In methanol	0.18	0.74	−4.26	−13.81
1-Co(II)–NO²⁺	In gas phase	−7.22	−5.25	−16.83	−28.97
	In methanol	−1.60	−1.04	−6.04	−15.59
1-Co(II)–NO_2^+	In gas phase	−1.38	0.60	−10.99	−23.13
	In methanol	0.49	1.05	−3.95	−13.50
1-Co(II)²⁺	In gas phase	−8.04	−6.06	−17.65	−29.79
	In methanol	−0.76	−0.19	−5.19	−14.74

Note: (nu) = nucleophile; (el) = electrophile.

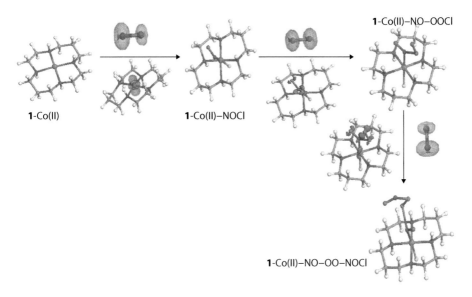

FIGURE 17.6   Oxidation scheme for NO catalyzed by **1**-Co(II) complex.

Figure 17.2, the Fukui functions for NO (electrophile) and **1**-Co(II)Cl (nucleophile) indicate that the maximum interaction is obtained between the metal center and the nitrogen atom of NO. The optimized structure obtained from the interaction of [**1**-Co(II)Cl]$^+$ species and the NO is the [**1**-Co(II)ClNO]$^+$ cation reported in the same figure. The calculation of the Fukui function descriptors for this last complex, as well as the oxygen molecule, gives insight on how these species should be approached to react. As it clearly appears from the figure, the reactive site on the complex is located on the nitrogen atom of NO coordinated to the cobalt center. Following the same reasoning, step by step, it is possible to investigate complicated pathways of reaction with several intermediary steps.

17.6   CONCLUSION

The delivery of NO to specific targets has received a great deal of attention due to the important role of NO in biological systems. Among the possible methods, the use of exogenous NO gas carried by transition metal complexes has remained poorly explored. We have adopted a combined theoretical–experimental approach to determine important parameters that could be used for further tuning the properties of the investigated complex for the target application. The first conclusion of this study is that the solvent effects are extremely important for the realistic description of these complexes and must be taken into account to reproduce experimentally observed trends such as the order of relative stabilities of **1**-Co(II) complexes in solutions as well as their reactivity toward the different ligands present in the solution. We have predicted that the **1**-Co(II)NO$_2$NO$_2$ complex is the most stable one and that it would be possible to observe and isolate it. Moreover, the complex **1**-Co(II)NONO$_2$ turns out to be relatively stable and that it is possible to define experimental conditions in solution to isolate it.

The second conclusion of this work is that calculated vibrational spectra and the Fukui reactivity indices give valuable information on the attribution of the experimental bands and understanding of local reactivity of macrocycle ligands as well as the evolution of reactions involving several steps.

The third conclusion is related to the systematic use of combined theoretical–experimental approaches. We have found that to develop parallel theoretical and experimental investigations is of mutual benefit. More fundamental work on additional, well-characterized transition metal NO complexes is desirable to extend the understanding of the reaction mechanisms and validate the predictive approach to NO-controlled release through designing improved macrocycle ligands. One has to recognize that most of the reported studies dealing with the interaction of NO with organometallic and bioinorganic metal complexes have been confronted with the complex redox solution chemistry of NO and obvious experimental difficulties in studying such reactions [48]. Detailed kinetic and mechanistic studies of the autoxidation reactions of bioinorganic–NO complexes in solution are remarkably scarce. Recent studies have revealed a complex picture in which the coordination of NO becomes associated with drastic changes in the redox nature of the ligand and in the spin state of the metal center. Important complications that may affect the studies on the interaction of NO with metal centers of macrocycle complexes arise also from the easy oxidation of NO by molecular oxygen in solution to produce NO_2 and N_2O_3 as well as nitrite ions in the presence of an aqueous solution. Computational chemistry when appropriately combined with experiment on these complexes enhances the understanding of their stability and reactivity, and complements the interpretation of observed characterizations as, for example, the IR vibrational, nuclear magnetic resonance, or EPR spectra.

REFERENCES

1. L. J. Ignarro, editor. *Nitric Oxide: Biology and Pathobiology*. San Diego: Academic Press, 2000.
2. P. G. Wang, T. B. Cai and N. Taniguchi. *Nitric Oxide Donors*. Germany: Weinheim, 2005.
3. R. Rossaint, K. J. Falke, F. Lopez, K. Slama, U. Pison and W. M. Zapol. Inhaled nitric oxide for the adult respiratory distress syndrome. *N. Engl. J. Med.* **328**, 1993, 399–405.
4. M. Haj Reem, J. E. Cinco and C. D. Mazer. Treatment of pulmonary hypertension with selective pulmonary vasodilators. *Curr. Opin. Anaesthesiol.* **19**, 2006, 88–95.
5. K. Sydow, A. Daiber, M. Oelze, Z. Chen, M. August, M. Wendt, V. Ullrich, A. Muelsch, E. Schulz, J. F. Keaney Jr., J. S. Stamler and T. Muenzel. Central role of mitochondrial aldehyde dehydrogenase and reactive oxygen species in nitroglycerin tolerance and cross-tolerance. *J. Clin. Invest.* **113**, 2004, 482–489.
6. A. Daiber, A. Muelsch, U. Hink, H. Mollnau, A. Warnholtz, M. Oelze and T. Muenzel. The oxidative stress concept of nitrate tolerance and the antioxidant properties of hydralazine. *Am. J. Cardiol.* **96**, 2005, 25i–36i.
7. J. D. Parker. Nitrate tolerance, oxidative stress, and mitochondrial function: Another worrisome chapter on the effects of organic nitrates. *J. Clin. Invest.* **113**, 2004, 352–354.
8. O. R. Leeuwenkamp, W. P. Van Bennekom, E. J. Van der Mark and A. Bult. Nitroprusside, antihypertensive drug and analytical reagent. Review of (photo) stability, pharmacology and analytical properties. *Pharm. Weekbl., Sci. Ed.* **6**, 1984, 129–140.

9. V. Schulz. Clinical pharmacokinetics of nitroprusside, cyanide, thiosulfate and thiocyanate. *Clin. Pharmacokinet.* **9**, 1984, 239–251.

10. M. Halpenny, M. Genevieve, M. Olmstead and P. K. Mascharak. Incorporation of a designed ruthenium nitrosyl in polyhema hydrogel and light-activated delivery of NO to myoglobin. *Inorg. Chem.* **46**, 2007, 6601–6606.

11. A. Franke, F. Roncaroli and R. van Eldik. Mechanistic studies on the activation of NO by iron and cobalt complexes. *Eur. J. Inorg. Chem.* **2007**, 2007, 773–798.

12. C.-Y. Chiang and M. Y. Darensbourg. Iron nitrosyl complexes as models for biological nitric oxide transfer reagents. *J. Biol. Inorg. Chem.* **11**, 2006, 359–370.

13. S. R. Wecksler, A. Mikhailovsky, D. Korystov and P. C. Ford. A two-photon antenna for photochemical delivery of nitric oxide from a water-soluble, dye-derivatized iron nitrosyl complex using NIR light. *J. Am. Chem. Soc.* **128**, 2006, 3831–3837.

14. P. C. Ford and S. Wecksler. Photochemical reactions leading to NO and NOx generation. *Coord. Chem. Rev.* **249**, 2005, 1382–1395.

15. A. A. Eroy-Reveles, Y. Leung and P. K. Mascharak. Release of nitric oxide from a sol–gel hybrid material containing a photoactive manganese nitrosyl upon illumination with visible light. *J. Am. Chem. Soc.* **128**, 2006, 7166–7167.

16. K. E. Broderick, L. Alvarez, M. Balasubramanian, D. D. Belke, A. Makino, A. Chan, V. L. Woods Jr., et al. Nitrosyl-cobinamide, a new and direct nitric oxide-releasing drug effective in vivo. *Exp. Biol. Med.* **232**, 2007, 1432–1440.

17. J. T. Mitchell-Koch, K. M. Padden and A. S. Borovik. Modification of immobilized metal complexes toward the design and synthesis of functional materials for nitric oxide delivery. *J. Polym. Sci., Part A: Polym. Chem.* **44**, 2006, 2282–2292.

18. O. Siri, A. Tabard, P. Pullumbi and R. Guilard. Iron complexes acting as nitric oxide carriers. *Inorg. Chim. Acta* **350**, 2003, 633–640.

19. R. M. Izatt, K. Pawlak, J. S. Bradshaw and R. L. Bruening. Thermodynamic and kinetic data for macrocycle interaction with cations, anions, and neutral molecules. *Chem. Rev.* **95**, 1995, 2529–2586.

20. R. Ivanikova, I. Svoboda, H. Fuess and A. Maslejova. *Trans*-dichloro(1,4,8,11-tetra-azacyclotetradecane)cobalt(iii) chloride. *Acta Crystallogr., Sect. E: Struct. Rep. Online* **E62**, 2006, m1553–m1554.

21. M. E. Sosa-Torres and R. A. Toscano. *Trans*-dichloro(1,4,8,11-tetraazacyclotetradecane) cobalt(iii) chloride tetrahydrate 0.47-hydrochloride. *Acta Crystallogr., Sect. C: Cryst. Struct. Commun.* **C53**, 1997, 1585–1588.

22. J. A. McCleverty. Reactions of nitric oxide coordinated to transition metals. *Chem. Rev. (Washington, DC, United States)* **79**, 1979, 53–76.

23. K. Nakamoto. *Infrared and Raman Spectra of Inorganic and Coordination Compounds*, 3rd edition, edited by K. Nakamoto. New York: John Wiley & Sons, 1978, 448.

24. C. K. Poon. Infrared spectra of some *cis*- and *trans*-isomers of octahedral cobalt(iii) complexes with a cyclic quadridentate secondary amine. *Inorg. Chim. Acta.* **5**, 1971, 322–324.

25. J. A. McCleverty. Chemistry of nitric oxide relevant to biology. *Chem. Rev. (Washington, DC, United States)* **104**, 2004, 403–418.

26. B. Delley. An all-electron numerical method for solving the local density functional for polyatomic molecules. *J. Chem. Phys.* **92**, 1990, 508–517.

27. B. Delley. From molecules to solids with the DMol^3 approach. *J. Chem. Phys.* **113**, 2000, 7756–7764.

28. B. Delley. The conductor-like screening model for polymers and surfaces. *Mol. Simul.* **32**, 2006, 117–123.

29. A. Klamt and G. Schueuermann. COSMO: A new approach to dielectric screening in solvents with explicit expressions for the screening energy and its gradient. *J. Chem. Soc., Perkin Trans. II* 1993, 799–805.

30. J. Andzelm, C. Kolmel and A. Klamt. Incorporation of solvent effects into density functional calculations of molecular energies and geometries. *J. Chem. Phys.* **103**, 1995, 9312–9320.

31. J. J. P. Stewart. Optimization of parameters for semiempirical methods. I. Method. *J. Comput. Chem.* **10**, 1989, 209–220.

32. J. J. P. Stewart. Optimization of parameters for semiempirical methods. II. Applications. *J. Comput. Chem.* **10**, 1989, 221–264.

33. Spartan '02. Wavefunction, Irvine, CA, 2002.

34. CS Chem3D Version 10.0. CambridgeSoft, Cambridge, MA, 2006.

35. K. Fukui. *Reactivity and Structure Concepts in Organic Chemistry, Volume 2: Theory of Orientation and Stereoselection.* New York, USA: Springer-Verlag, 1975.

36. R. Franke. *Theoretical Drug Design Methods.* Amsterdam: Elsevier, 1984.

37. D. F. V. Lewis, C. Ioannides and D. V. Parke. Interaction of a series of nitriles with the alcohol-inducible isoform of P450: Computer analysis of structure–activity relationships. *Xenobiotica* **24**, 1994, 401–408.

38. Z. Zhou and R. G. Parr. Activation hardness: New index for describing the orientation of electrophilic aromatic substitution. *J. Am. Chem. Soc.* **112**, 1990, 5720–5724.

39. R. G. Pearson. Hard and soft acids and bases. *J. Am. Chem. Soc.* **85**, 1963, 3533–3539.

40. R. G. Pearson. Acids and bases. *Science (Washington, DC, United States)* **151**, 1966, 1721–1727.

41. R. G. Pearson. Absolute electronegativity and hardness correlated with molecular orbital theory. *Proc. Natl. Acad. Sci. U.S.A.* **83**, 1986, 8440–8441.

42. R. G. Pearson and J. Songstad. Application of the principle of hard and soft acids and bases to organic chemistry. *J. Am. Chem. Soc.* **89**, 1967, 1827–1836.

43. R. G. Parr and P. K. Chattaraj. Principle of maximum hardness. *J. Am. Chem. Soc.* **113**, 1991, 1854–1855.

44. R. G. Parr and J. L. Gazquez. Hardness functional. *J. Phys. Chem.* **97**, 1993, 3939–3940.

45. R. G. Parr and R. G. Pearson. Absolute hardness: Companion parameter to absolute electronegativity. *J. Am. Chem. Soc.* **105**, 1983, 7512–7516.

46. M. L. McKee. Ab initio study of the N2O4 potential energy surface. Computational evidence for a new N2O4 isomer. *J. Am. Chem. Soc.* **117**, 1995, 1629–1637.

47. L. P. Olson, K. T. Kuwata, M. D. Bartberger and K. N. Houk. Conformation-dependent state selectivity in O–O cleavage of ONOONO: An inorganic cope rearrangement helps explain the observed negative activation energy in the oxidation of nitric oxide by dioxygen. *J. Am. Chem. Soc.* **124**, 2002, 9469–9475.

48. M. Feelisch and J. S. Stamler, editors. *Methods in Nitric Oxide Research.* New York: John Wiley & Sons, 1996.

18 Design of Highly Selective Industrial Performance Chemicals
A Molecular Modeling Approach

Beena Rai and Pradip

CONTENTS

18.1 INTRODUCTION

Design, development, and selection of reagents (surfactants and dispersants) for different industrial applications (mineral processing, ceramics, paints and coatings, cement, and nanoparticles dispersion) remain an art. Most currently available commercial reagents are selected primarily by trial-and-error methods based on rules of thumb and past experience. The time and resources required to come up with an acceptable formulation are, therefore, prohibitively expensive for difficult-to-process systems. Additionally, because of the high cost of empirical search for novel dispersants, the search envelope is severely restricted to a few well-known families of dispersants. A quantitative methodology to screen out/identify the appropriate molecular architectures, based on determination of the relative efficacy of various structures through the results of theoretical computations, is evidently an economically attractive and elegant methodology as compared with the conventional approach. Selecting the most promising molecules from a wide variety of possibilities based on computer-aided design tools for subsequent synthesis, characterization, testing, and pilot plant/plant trials will certainly save enormous costs in time and efforts to arrive at new formulations.

With recent advances in the understanding of molecular-level phenomena governing adsorption of reagents at interfaces, accessibility of application-oriented molecular modeling tools, and availability of relatively inexpensive computing power, it is possible to design reagents customized for specific applications that are based on theoretical computations. Though there have been isolated attempts in the past to study surfactant–surface interactions using molecular modeling tools [1–11], a comprehensive methodology to reagent design has been lacking. We have elucidated the building blocks of this novel paradigm through our recent publications on this topic [12–22]. Two key features of the proposed approach are (1) identification of the molecular recognition mechanisms underlying the adsorption of reagents at the interface and (2) use of advanced molecular modeling techniques for theoretical computations of the relative magnitude of interaction. The molecular modeling thus provides a quantitative search technique for screening and identifying the most promising molecular architectures from a large set of candidates available for a particular application. As a consequence, considerable saving in time and effort needed for developing new formulations is possible.

In this paper, we present a brief overview of this rational design paradigm through a case study drawn from our work on separation among calcium minerals.

18.2 METHODOLOGY

Universal force field (UFF) [23–26] as implemented in Materials Studio® (MS) [27] has been used to model the inorganic surface–reagent interactions. We have successfully demonstrated through our earlier work that UFF can be used to model the mineral–reagent systems with reasonable accuracy [16–19]. A detailed methodology of modeling the mineral–reagent interactions has been reported earlier [16]. The geometry of reagent molecules (surfactant/dispersant/flocculant) was optimized using UFF as implemented in the Forcite module of MS. A surface cell was created from the unit cell of the inorganic crystal at its cleavage plane and optimized with the help of Surface Builder module in MS. The optimized reagent molecule was docked on the mineral surface. The initial geometry of the surface–reagent complex was created physically on the screen with the help of molecular graphics tools, taking into consideration the possible interactions of reagent functional groups with surface atoms. The reagent molecule was then allowed to relax completely on the surface using Forcite Geometry Optimization. Several initial conformations (~20) were assessed so as to locate the minimum energy conformation of the inorganic surface–reagent complex. The partial charges on the atoms were calculated using charge equilibration method [28]. The intramolecular van der Waal interactions were calculated only between atoms that are located at distances greater than fourth nearest neighbors. A modified Ewald summation method [29] was used for calculating the nonbonded Coulomb interactions, whereas for van der Waal interactions, a direct cutoff at a distance less than $r/2$ (where r is the length of the simulation cell) was used. Smart minimizer as implemented in the Forcite module of MS was used for geometry optimization. The optimization was considered to be converged when a gradient of 0.0001 kcal/mol is reached. The interaction energy was calculated for the most likely/favorable conformation using the following equation:

$$\text{Interaction energy } (\Delta E) = E_{complex} - ([E_{reagent} + E_{surface}]), \tag{18.1}$$

where $E_{complex}$, $E_{reagent}$, and $E_{surface}$ are the total energies of the optimized surface–reagent complex, reagent molecule, and surface cluster, respectively.

The structure of the complex obtained through static energy minimization method represents only a local minimum energy structure; it was further optimized to find a global minimum energy structure through molecular dynamics (MD) simulations (both in vacuum and in presence of the solvent) using Forcite Dynamics. MD calculations were run using the constant energy microcanonical ensemble method (i.e., NVE—a system with a fixed number of particles N, a fixed volume V, and a fixed energy E) at 300 K with time step of 1 fs. Total run length was ~300 ps. During the simulations, the temperature was controlled by the velocity scaling method and also atom-based cutoff method was used for calculating both van der Waals and electrostatic forces. The interaction energies were computed as

$$IE_{(in \; solvent)} = \Delta E_{combined} - \sum (\Delta E_{solvent-reagent} + \Delta E_{surface-solvent}) \tag{18.2}$$

$$\Delta E_{combined} = E_{complex,solvent} - \sum (E_{surface} + E_{reagent} + E_{solvent}), \tag{18.3}$$

where $E_{combined}$ is the total interaction energy calculated using Equation 18.3; $E_{complex,solvent}$ is the total energy of the optimized surface–reagent complex in the presence of solvent; $E_{surface}$, $E_{reagent}$, and $E_{solvent}$ are the total energies of free surface, reagent, and solvent molecules computed separately; $\Delta E_{solvent-reagent}$ is the interaction energy computed for the interaction of solvent and reagent molecules; and $\Delta E_{surface-solvent}$ is the contribution due to interaction of solvent molecules with the surface. These energies are subtracted from $\Delta E_{combined}$ to get the final interaction energy [$IE_{(in \; solvent)}$] of the reagent molecule with the inorganic surface (Equation 18.2). It is worth noting that the more negative magnitude of interaction energy indicates more favorable interactions between the reagent and inorganic surface. The magnitude of this quantity is, thus, an excellent measure of the relative intensity/efficiency of interaction among various reagents.

For modeling self-assembled monolayers (SAMs), a monolayer of reagent molecules was placed on the inorganic surface. The reagent molecules were placed as per the most stable conformation obtained through optimized single molecule–surface complex. To find the equilibrium structures of the adsorbed monolayers, the clusters thus created were subjected to geometry optimization followed by MD simulations at 300 K. The surface atoms were kept fixed during the entire simulation run and only adsorbed reagent molecules were allowed to relax.

For simulating wetting behavior of SAMs, a water droplet was placed on the surface and equilibrated using MD simulations. To create a water droplet of experimental density, first a 3D-periodic box containing 512 water molecules was equilibrated at 300 K using MD simulations. A sphere of 20-Å radius (329 water molecules) was cut out of this 3D-periodic box and placed over the SAMs. The system was equilibrated using MD simulations, and during the entire MD run, inorganic surface atoms were kept fixed. The methodology proposed by Fan and Cagin [30] was used to extract the microscopic parameters, namely, the drop volume (V) and the interfacial area (S) of water droplet, which was then put in Equation 18.6 to calculate contact angles.

The height of the droplet (h) from the terminal methyl layer and the radius of the droplet (R) are calculated as follows:

$$h3 + 3S\,h/\pi - 6V/\pi = 0 \tag{18.4}$$

$$R = h/2 + S/2\,\pi\,h. \tag{18.5}$$

The contact angle (θ) is related to the height and the droplet radius as shown below:

$$\cos\theta = 1 - h/R. \tag{18.6}$$

18.3   RESULTS AND DISCUSSIONS

18.3.1   SELECTIVE COLLECTORS (SURFACTANTS) FOR SEPARATION AMONG CALCIUM MINERALS

The separation of sparingly soluble calcium minerals such as fluorite [CaF_2], calcite [$CaCO_3$], fluorapatite [$Ca_{10}(PO_4)_6F_2$], dolomite [$(CaMg)CO_3$], and scheelite [$CaWO_4$] from each other remains a challenging problem without a satisfactory solution to date. The difficulty arises out of their similar surface properties and solubility, same chelating cation (Ca) in their structure, and similar response to various known families of flotation collectors such as fatty acids. Many possible hypotheses have also been suggested in literature to explain this observation [31–35]. We have applied molecular modeling methodology to design more selective collectors for their separation [16,17]. As shown in Figure 18.1, diphosphonic acid-based collectors are found to be more effective than conventional fatty acids [16]. The order of selectivity, as predicted by theoretical computations (interaction energies), compares very well with the experimental flotation response of these reagents. We used this methodology further to design selective depressants for beneficiation of difficult-to-process phosphate ores (separation of calcite/dolomite from fluorapatite) (Indian patent nos. 227449 and 238197).

18.3.2   CONFORMATION OF ADSORBED MOLECULE

Another aspect of these computations was to obtain the most favorable configuration of the adsorbed molecule on a given inorganic surface. The theoretical results were validated with experimental data. For example, as observed in experimental investigations, two most stable fluorite–oleate complexes, namely, unidentate and bidentate, were predicted through molecular modeling calculations as well [36]. As shown in Figure 18.2, the theoretically simulated conformation (as indicated

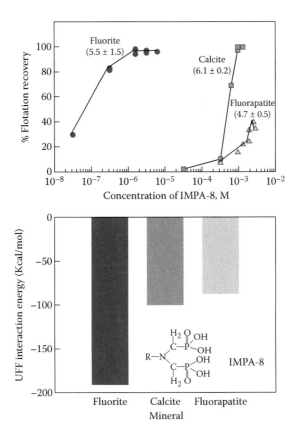

FIGURE 18.1 Comparison of experimental flotation results for the most selective phosphonic acid reagent (octyliminobismethylene phosphonic acid IMPA-8) with its computed interaction energies at different mineral surfaces. (After Pradip et al., *Langmuir* 18, 2002, 932–940.)

through adsorption angles [36]) matched well with the experimental one (obtained through a sophisticated *in situ* infrared external reflection spectroscopy technique [37]).

18.3.3 SELF-ASSEMBLED MONOLAYERS

Interaction energies computed on the basis of single reagent molecule adsorbing on the inorganic surface do provide a very useful quantitative and theoretical measure for assessing the relative affinity of various surface–reagent combinations. Such computations are, therefore, very useful inputs in the design/screening of different molecular architectures for a given separation system. These computations, however, do not capture the associative interactions among the adsorbed molecules. To study the structure of the adsorbed monolayers so as to delineate the more subtle template effects of the substrate in determining the macroscopic behavior of the reagents in

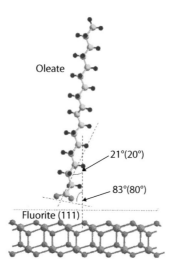

FIGURE 18.2 Conformation of adsorbed oleate molecule on fluorite surface. The numbers in parentheses denote experimental angles obtained through *in situ* infrared external reflection spectroscopy technique. (After Mielczarski, E. et al., *Colloids Surf., A* 205, 73–84, 2002.)

actual application, we have also simulated adsorption of several molecules together adsorbing on the surface, leading to well-defined monolayers. More detailed information on the most likely structure of the adsorbed monolayers was obtained.

We have modeled self-assembled monolayers of a conventional fatty acid (oleate) molecule on calcium mineral surfaces [22]. Different coverages, in terms of site occupancy, leading to different structures were modeled (Figure 18.3). On the fluorite surface, the most favorable conformation was found to be with oleates adsorbed in a hexagonal lattice (67% coverage). The coverage area per molecule (22.2 Å^2) for this conformation matched closely that of the oleate molecular area (20 Å^2),

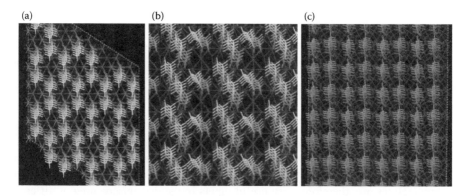

FIGURE 18.3 Structure of adsorbed oleate SAMs on (a) fluorite, (b) fluorapatite, and (c) calcite surfaces. (After Rai, B. and Pradip. *Proceedings of XXII International Mineral Processing Congress (IMPC)*, edited by L. Lorenzen and D. J. Bradshaw. Cape Town, South Africa: South African Institute of Mining & Metallurgy (SAIMM), 1085–1093, 2003.) (Graphical displays were generated using Materials Studio.)

whereas on the calcite surface, 100% coverage was possible with the computed area (20.2 Å^2) matching that of the oleate molecule. On the fluorapatite surface, best fit was obtained with 75% occupancy leading to matching computed area per molecule (21.5 Å^2).

18.3.4 WETTABILITY OF SAMs

Based on the results of adsorbed SAMs, calcite was found to be the best template for oleate adsorption; however, fluorite is the most responsive to experimental flotation with oleate (Figure 18.5). Because the flotation is directly related to wettability of adsorbed SAMs, which is quantified using contact-angle measurements, we also computed the contact angle of the water droplet placed on the adsorbed SAMs at different inorganic surfaces. As shown in Figure 18.4, oleate SAM adsorbed on the fluorite surface seems to be more hydrophobic as compared with those adsorbed on calcite or fluorapatite surfaces. Thus, even though calcite tends to be the best template for adsorption, the wetting characteristics of adsorbed SAMs predict better floatability for fluorite in the presence of oleate. Indeed, it is heartening to note that experimental flotation response follows the same order as predicted by wettability (contact angle) simulations (Figure 18.5).

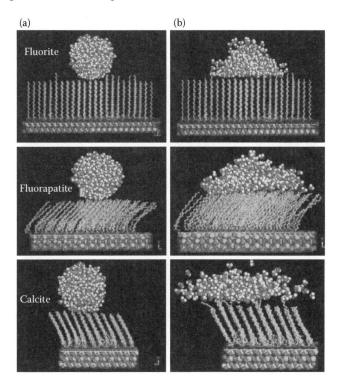

FIGURE 18.4 Snapshots of a water droplet on adsorbed SAMs on fluorite, fluorapatite, and calcite surfaces (a) before equilibration and (b) after 300 ps of MD run. (Graphical displays were generated using Materials Studio.)

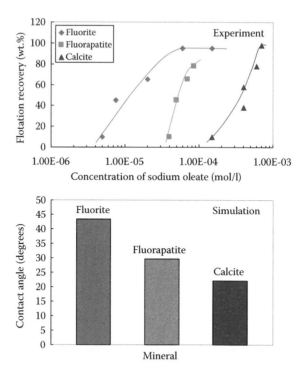

FIGURE 18.5 Comparison of computed contact angles with experimental flotation response of oleate with different calcium minerals. (Experimental flotation data were taken from Sorensen, E. J., *J. Colloids Interface Sci.* 45, 601, 1973.)

18.4 CONCLUDING REMARKS

The utility of currently available molecular modeling tools in the design and screening of different molecular architectures for designing reagents for industrial applications is demonstrated through this communication. Even though this paper deals with reagents for mineral separation system, the proposed molecular modeling-based theoretical approach is very generic in nature and provides a scientifically robust framework for the design/selection of appropriate reagent and/or solvent for a given application. With steady advancements being made in the theory and practice of the molecular modeling approach, it is anticipated that even more complex design problems can be solved, in particular, with respect to the optimization of the molecular architecture to develop performance chemicals tailor-made for the desired application.

REFERENCES

1. W. Aliga and P. Somasundaran. Molecular orbital modeling and UV spectroscopic investigation of adsorption of oxime surfactants. *Langmuir* **3**, 1987, 1103.
2. S. Mann. Molecular recognition in biomineralization. *Nature* **332**, 1988, 119.

3. K. Takahashi. *Proceedings of XVII IMPC*, Dresden, Germany, **2**, 1991, 393.

4. S. N. Black, L. A. Bromley, D. Cottier, R. J. Davey, B. Dobbs and J. E. Rout. Interactions at the organic/inorganic interface: Binding motifs for phosphonates at the surface of barite crystals. *J. Chem. Soc., Faraday Trans.* **87** (20), 1991, 3409.

5. R. J. Davey, S. N. Black, L. A. Bromley, D. Cottier, B. Dobbs and J. E. Rout. Molecular design based on recognition at inorganic surfaces. *Nature* **353**, 1991, 549.

6. F. Grases, A. Gracia-Raso, J. Palou, A. Costa-Bauza and J. G. March. A study of the relationship between chemical structure of some carboxylic acids and their capacity to inhibit the crystal growth of calcium fluoride. *Colloids Surf.* **54**, 1991, 313.

7. I. Addadi, I. Weissbuch, M. Lahav and L. Leiserowitz. Molecular recognition at crystal interfaces. *Science* **253**, 1991, 637.

8. D. Arad, M. A. Kraftory, B. N. Zolotoy, P. Finkelstein and A. Weissman. Molecular modeling for oxidative cross-linking of oleates adsorbed on surfaces of minerals. *Langmuir* **9**, 1993, 1446.

9. P. V. Coveney and W. Humphries. Molecular modeling of the mechanism of phosphonate retarders on hydrating cements. *J. Chem. Soc., Faraday Trans.* **92**, 1996, 831.

10. Y. Numata, K. Takahashi, R. Liang and J. Wakamatser. Adsorption of mercaptobenzothiazole onto pyrite. *Int. J. Miner. Process.* **53**, 1998, 75.

11. N. H. de Leeuw, S. C. Parker and K. Hanumantha Rao. Modeling the competitive adsorption of water and methanoic acid on calcite and fluorite surfaces. *Langmuir* **14**, 1998, 5900.

12. Pradip. The science of reagents design: An historical perspective. *Trans. Indian Inst. Met.* **50**, 1997, 481–494.

13. Pradip. *Reagents for Better Metallurgy*, Chapter 24, edited by P. S. Mulukutla. Denver, CO: SME-AIME, 1994, 245–252.

14. Pradip. Design of crystal structure-specific surfactants based on molecular recognition at mineral surfaces. *Curr. Sci.* **63**, 1992, 180–186.

15. Pradip. On the design of selective reagents for mineral processing applications. *Met., Mater. Processes* **3**, 1991, 15–36.

16. Pradip, B. Rai, T. K. Rao, S. Krishnamurthy, R. Vetrivel, J. Mielczarski and J. M. Cases. Molecular modelling of interactions of diphosphonic acid based surfactants with calcium minerals. *Langmuir* **18**, 2002, 932–940.

17. Pradip, B. Rai, T. K. Rao, S. Krishnamurthy, R. Vetrivel, J. Mielczarski and J. M. Cases. Molecular modelling of interactions of alkyl hydroxamates with calcium minerals. *J. Colloid Interface Sci.* **256**, 2002, 106–113.

18. Pradip and B. Rai. Design of tailor-made surfactants for industrial applications using a molecular modelling approach. *Colloids Surf.* **205**, 2002, 139–148.

19. Pradip and B. Rai. Molecular modelling and rational design of flotation reagents. *Int. J. Miner. Process.* **72**, 2003, 95–110.

20. Pradip, B. Rai and P. Sathish. Rational design of dispersants by molecular modelling for advanced ceramic processing applications. *KONA* **22**, 2004, 151–158.

21. Pradip, B. Rai, P. Sathish and S. Krishnamurty. *Ferroelectrics* **306**, 2004, 195–208.

22. B. Rai and Pradip. *Proceedings of XXII International Mineral Processing Congress (IMPC)*, edited by L. Lorenzen and D. J. Bradshaw. Cape Town, South Africa: South African Institute of Mining & Metallurgy (SAIMM), 2003, 1085–1093.

23. A. K. Rappé, K. S. Colwell and C. J. Casewit. Application of a universal force field to metal complexes. *J. Inorg. Chem.* **32**, 1993, 3438.

24. A. K. Rappé, C. J. Casewit, K. S. Colwell, W. A. Goddard and W. M. Skiff. UFF, a full periodic table force field for molecular mechanics and molecular dynamics simulations. *J. Am. Chem. Soc.* **114**, 1992, 10024.

25. A. K. Rappé, C. J. Casewit and K. S. Colwell. Application of a universal force field to organic molecules. *J. Am. Chem. Soc.* **114**, 1992, 10035.

26. A. K. Rappé, C. J. Casewit and K. S. J. Colwell. Application of a universal force field to main group compounds. *J. Am. Chem. Soc.* **114**, 1992, 10046.

27. Accelrys. Materials Studio Release Notes, Release 4.1. Accelrys Software, San Diego, CA, 2006.

28. A. K. Rappé and W. A. Goddard. Charge equilibration for molecular dynamics simulations. *J. Phys. Chem.* **95**, 1991, 3358.

29. N. Karasawa and W. A. Goddard III. Acceleration of convergence for lattice sums. *J. Phys. Chem.* **93**, 1989, 7320–7327.

30. C. F. Fan and T. Cagin. Wetting of crystalline polymer surfaces–a molecular dynamics simulation. *J. Chem. Phys.* **103**, 1995, 9053.

31. E. J. Sorensen. On the adsorption of some anionic collectors on fluoride minerals. *J. Colloids Interface Sci.* **45**, 1973, 601.

32. R. Pugh and P. Stenius. Solution chemistry studies and flotation behavior of apatite, calcite and fluorite minerals with sodium oleate collector. *Int. J. Miner. Process.* **15**, 1985, 193.

33. N. K. Khosla and A. K. Biswas. Effects of tannin-fatty acid interactions on selectivity of adsoption on calcite and fluorite surfaces. *Trans.—Inst. Min. Metall.* **94**, 1985, C4.

34. N. P. Finkelstein. Review of interactions in flotation of sparingly soluble calcium minerals with anionic collectors. *Trans.—Inst. Min. Metall.* **98**, 1989, C157.

35. D. Arad, M. A. Kraftory, B. N. Zolotoy, P. Finkelstein and A. Weissman. Molecular modeling for oxidative cross-linking of oleates adsorbed on surfaces of minerals. *Langmuir* **9**, 1993, 1446.

36. E. Mielczarski, J. A. Mielczarski, J. M. Cases, B. Rai and Pradip. Influence of solution conditions and mineral surface structure on the formation of oleate adsorption layers on fluorite. *Colloids Surf., A* **205**, 2002, 73–84.

37. E. Mielczarski, J. A. Mielczarski and J. M. Cases. Molecular recognition effect in monolayer formation of oleate on fluorite. *Langmuir* **14**, 1998, 1739.

19 Density Functional Theory Calculations of ^{11}B NMR Parameters in Crystalline Borates

Sabyasachi Sen

CONTENTS

19.1 INTRODUCTION

Borate crystals and glasses are an important class of materials that have received much attention because of their wide-ranging importance in technological processes, including extensive use in the areas of optics, display, and telecommunication [1–3]. The fundamental understanding of these materials and formulation of accurate predictive models for the compositional dependence of physicochemical properties require detailed knowledge of their atomic scale structure and the dynamical phenomena they exhibit, including transport and relaxation. High-resolution ^{11}B nuclear magnetic resonance (NMR) spectroscopy has played an important role in elucidating the short-range structure around B atoms in a wide variety of borate crystals and glasses [4–10]. The natural abundance of the ^{11}B isotope is ~80% and it is a highly NMR-sensitive quadrupolar (nuclear spin $I = 3/2$) nuclide with a relatively large gyromagnetic ratio. Boron atoms in borates are either three- or fourfold coordinated to oxygen atoms, forming BO_3 planar triangles or BO_4 tetrahedra, respectively. The B sites in planar BO_3 triangles are characterized approximately by C_3 symmetry and, consequently, by an NMR line shape dominated by a relatively large quadrupolar coupling constant C_Q ($2.4 \leq C_Q \leq 2.9$ MHz). On the other hand, the B sites in the BO_4 tetrahedra are characterized by higher symmetry than those in BO_3 triangles. Consequently, the NMR line shapes of BO_4 sites are nearly Gaussian, resulting from small C_Q values (typically $0.0 \leq C_Q \leq 0.5$ MHz). Moreover, the ^{11}B isotropic chemical shifts δ_{iso} of the BO_3 and BO_4 sites differ by ~15 ppm to 20 ppm [11]. These differences in C_Q and

δ_{iso} of the BO_3 and BO_4 sites make them easily identifiable in the ^{11}B NMR spectra. Although δ_{iso} for each of these sites vary over a small range, recent detailed ^{11}B NMR studies of a wide variety of borate crystals have shown that at least in the case of BO_3 sites, such variation can be empirically associated with the existence of structural systematics such as the number of bridging vs. nonbridging oxygen nearest neighbors and the sum of cation–oxygen bond strengths [11]. Recently available high-field ^{11}B NMR data have also shown the presence of chemical shift anisotropy (CSA) in BO_3 and BO_4 sites, although precise experimental determinations of CSA and asymmetry parameter η_{CS} of the CSA remain experimentally challenging [11–13].

First-principles calculation of the NMR shielding tensor and quadrupolar coupling parameters C_Q and asymmetry parameter η_Q are extremely useful in understanding the correlation between atomic structure and NMR parameters and can be used to interpret the solid-state NMR spectra to the fullest extent. Such calculations in the past have been largely limited to molecular clusters of various sizes that were used to approximate various local structural environments in periodic solids [14–16]. First-principles calculations of NMR parameters in periodic solids have become feasible with the use of density functional theory (DFT) and the recently developed gauge-including projector augmented wave (GIPAW) method with plane-wave basis sets and pseudopotential approximation [17–20]. NMR shielding tensor parameters are obtained via calculation of the magnetic response of the all-electron wavefunction. We present here the results of such first-principles calculations of ^{11}B NMR parameters for a variety of borate crystals structure characterized by a range of B coordination environments (Table 19.1). These results agree well with the available

TABLE 19.1

Composition and Structural Characteristics of the Borate Crystals Studied in This Work

Chemical Composition	Short-Range Structure	Superstructural Ring Units [References]
1. B_2O_3	Two corner-linked BO_3 sites with all bridging oxygens; B–O–B angles range between ~128° and 133°	None [19]
2. SrB_4O_7	Two highly asymmetric BO_4 units	None [20]
3. $LiBO_2$	One asymmetric BO_3 site with two bridging and one nonbridging oxygen	None [21]
4. $Mg_3B_2O_6$	One symmetric orthoborate BO_3 site with all nonbridging oxygens	None [22]
5. CsB_3O_5	Two BO_3 and one BO_4 site with all bridging oxygens forming a planar B_3O_7 triborate ring; in-plane B–O–B angles are ~120°	Triborate ring [23]
6. α-Li_3BO_3	One symmetric orthoborate BO_3 site with all nonbridging oxygens	None [24]
7. $Cs_2B_{18}O_{28}$	Eight BO_3 and one BO_4 sites with all bridging oxygens. Six BO_3 sites belong to two planar B_3O_6 boroxol rings, whereas the other two BO_3 sites belong to a planar B_3O_7 triborate group; in-plane B–O–B angles are ~120°	Boroxol and triborate rings [25]

experimental data on these materials and provide fundamental underpinning of the empirical correlations between structure and experimental NMR parameters reported in the literature.

19.2   CALCULATION METHODOLOGY

The DFT-based codes CASTEP and CASTEP–NMR (Accelrys) were used for calculations of the ^{11}B NMR parameters for B sites in B_2O_3, $LiBO_2$, $Mg_3B_2O_6$, CsB_3O_5, α-$Li_3B_2O_3$, α-CsB_9O_{14}, and SrB_4O_7 crystal structures [19,21–28]. The short-range structure around B atoms and the nature of the superstructural borate group in these materials are listed in Table 19.1. The unit cell parameters and atom positions for all crystal structures were taken from diffraction-based structural refinement studies published in the literature and were used for calculation of ^{11}B NMR parameters without further geometry optimization [21–27]. The GIPAW algorithm and the generalized gradient approximation (GGA) simplified by Perdew–Burke–Ernzerhof (PBE) functional were used [17–20]. An energy cutoff of 600 eV–650 eV was used for the plane wave basis expansions. The Brillouin zone was sampled using the Monkhorst–Pack scheme and a $4 \times 3 \times 3$ k-point grid [19]. All core–valence interactions are modeled with ultrasoft pseudopotentials. Recent studies have demonstrated the excellent accuracy of the GIPAW method in calculating the NMR parameters for [29]Si, [17]O, [25]Mg, [23]Na, and [51]V nuclides in a variety of crystals and glasses [28–34]. These calculations yield the absolute shielding tensor principal components σ_{xx}, σ_{yy}, and σ_{zz}. The isotropic chemical shift δ_{iso} was obtained from isotropic shielding $\sigma_{iso} = 1/3(\sigma_{xx} + \sigma_{yy} + \sigma_{zz})$ using the relationship $\delta_{iso} = -(\sigma_{iso} - \sigma_{ref})$, where σ_{ref} is the isotropic shielding of a reference material. The calculated [11]B isotropic shielding of the BO_3 site in $LiBO_2$ crystal (78.03 ppm) has been used as a reference in this study and its δ_{iso} has been equated to the experimentally determined value of $\delta_{iso} = 17.08$ ppm [11]. The CSA and asymmetry parameter η_{CS} have been calculated using the relationships $CSA = (\sigma_{zz} - \sigma_{iso})$ and $\eta_{CS} = (\sigma_{yy} - \sigma_{xx})/(\sigma_{zz} - \sigma_{iso})$. The principal components of the electric field gradient tensor V_{ii} are reported as C_Q and η_Q for enabling comparison with the experimentally determined values of these two parameters, where the relationships among V_{ii}, C_Q, and η_Q can be expressed as $C_Q = eQV_{zz}/h$ and $\eta_Q = (V_{xx} - V_{yy})/V_{zz}$. The convention $|V_{zz}| \geq |V_{yy}| \geq |V_{xx}|$ and the literature-reported quadrupole moment Q of 40.59 mB for [11]B were used [35].

19.3   RESULTS AND DISCUSSION

The calculated ^{11}B NMR parameters δ_{iso}, CSA, η_{CS}, C_Q, and η_Q for all crystal structures are listed in Table 19.2. δ_{iso}, C_Q, and η_Q are the three NMR parameters that are most readily measured experimentally with ^{11}B NMR spectroscopy. The calculated values of these three parameters for all crystals are compared with the corresponding available experimental data in Table 19.2 and in Figures 19.1 and 19.2. The calculated values of δ_{iso}, C_Q, and η_Q agree within ±0.7 ppm, ±0.08 MHz, and ±0.05, respectively, in all cases.

It may be noted here that similar good agreement between theory and experiment has also been observed in a recent study based on DFT calculations of C_Q and η_Q values for B sites in a number of borate and borosilicate crystals [35].

TABLE 19.2

Calculated and Experimental ¹¹B NMR Parameters for BO_3 and BO_4 Sites in Different Crystal Structures

Crystal [Reference for NMR Experiment]	Calculated ¹¹B NMR Parameters						Experimental ¹¹B NMR Parameters			
	δ_{iso} (ppm)	CSA (ppm)	η_{CS}	Ω (ppm)/κ	C_Q^a (MHz)	η_Q	δ_{iso} (ppm)	Ω (ppm)/κ^b	C_Q (MHz)	η_Q
					B_2O_3 [11]^c					
BO_3–B1	14.50	11.68	0.95	23.07/0.04	2.661	0.15	14.6 ± 0.1^b	15 ± 2/1.0	2.690 ± 0.005	<0.05
BO_3–B2	14.89	11.14	0.97	22.11/0.02	2.638	0.18				
					SrB_4O_7					
BO_4–B1	0.49	–12.74	0.90	24.84/–0.08	1.149	0.42	n.a.	n.a.	n.a.	n.a.
BO_4–B2	0.14	12.25	0.51	21.50/0.42	0.677	0.70				
					$LiBO_2$ [10,11]					
BO_3–B1	17.08	43.58	0.90	84.98/0.08	2.552	0.54	17.08 ± 0.06	n.a.	2.47 [10] / 2.56 [11]	0.50 [10] / 0.60 [11]
					$Mg_3B_2O_6$ [11]					
BO_3–B1	23.18	9.17	0.05	13.98/0.93	2.863	0.05	22.5 ± 0.1	n.a.	2.94 ± 0.02	<0.05

				CsB₃O₅ [5]					
TR–BO₃–B1	17.69	−18.95	0.91	37.05/−0.07	2.469	0.34	17.8	n.a.	2.55	0.30
TR–BO₃–B2	19.90	−18.16	0.67	33.32/−0.27	2.810	0.27	19.1	n.a.	2.75	0.27
BO₄ site	1.31	7.57	0.62	13.70/0.31	0.177	0.49	0.5	n.a.	0.17	0.50
					α-Li₃BO₃ [10]					
BO₃–B1	21.25	−5.60	0.50	9.80/−0.43	2.687	0.05	n.a.	n.a.	2.64	0.035–0.048
					Cs₂B₁₈O₂₈ [11]ᶜ					
TR–BO₃–B2	16.80	−15.30	0.75	28.69/−0.20	2.539	0.32	16.7 ± 0.2ᶜ	21 ± 2/1.0	2.50 ± 0.05	0.2ᵇ
TR–BO₃–B3	16.76	−14.97	0.72	27.84/−0.23	2.535	0.33				
BR–BO₃–B4	16.52	13.28	0.83	25.43/0.13	2.509	0.54				
BR–BO₃–B5	16.75	14.53	0.81	27.68/0.15	2.533	0.14				
BR–BO₃–B6	17.00	14.21	0.76	26.70/0.19	2.547	0.22				
BR–BO₃–B7	16.61	13.80	0.79	26.15/0.17	2.523	0.55				
BR–BO₃–B8	17.23	14.46	0.76	27.18/0.19	2.589	0.21				
BR–BO₃–B9	17.45	14.68	0.73	27.38/0.21	2.640	0.09				
BO₄ site–B1	1.06	−5.2	0.46	9.00/−0.47	0.164	0.59	0.95	n.a.	0.20 ± 0.05	>0.50

Note: Experimental values when not available are indicated by "n.a." "TR" and "BR" represent triborate and boroxol rings, respectively.

ᵃ Absolute values are reported.

ᵇ Value was fixed in experimental line shape calculation [11].

ᶜ Experimental values are reported as averages of multiple sites due to lack of spectral resolution [11].

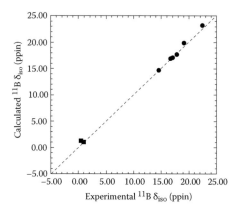

FIGURE 19.1 Comparison between the calculated and experimentally obtained ^{11}B δ_{iso} values for BO$_3$ (circles) and BO$_4$ (squares) sites in borate crystals. In case of B$_2$O$_3$ and Cs$_2$B$_{18}$O$_{28}$, the mean of the calculated δ_{iso} values for all BO$_3$ sites is compared with the experimentally determined average δ_{iso}. Dashed diagonal line through the plot represents the locus of all points with equal values of abscissa and ordinate.

In addition to δ_{iso}, C_Q, and η_Q that are typically measured in NMR experiments on quadrupolar nuclides, these calculations also provide the principal values of the ^{11}B shielding tensor and, hence, the span of the shielding anisotropy Ω defined as $\Omega = \sigma_{zz} - \sigma_{xx} = \delta_{xx} - \delta_{zz}$ [12]. The principal components of the chemical shift tensor are ordered such that $\delta_{xx} \geq \delta_{yy} \geq \delta_{zz}$ and therefore Ω is always positive. The corresponding

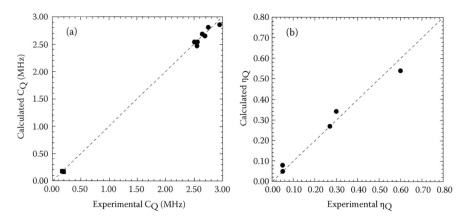

FIGURE 19.2 Comparison between the calculated and experimentally obtained (a) C_Q values for BO$_3$ (circles) and BO$_4$ (squares) sites and (b) η_Q values for BO$_3$ (circles) sites in borate crystals. In case of B$_2$O$_3$ and Cs$_2$B$_{18}$O$_{28}$, the mean of the calculated C_Q values for all BO$_3$ sites is compared with the experimentally determined average values. Calculated and experimental η_Q values for BO$_3$ sites in Cs$_2$B$_{18}$O$_{28}$ and for BO$_4$ sites in all crystals are not compared because experimental values for these sites are not accurately known. Dashed diagonal line through each plot represents the locus of all points with equal values of abscissa and ordinate.

skew of the tensor is defined as $\kappa = 3(\delta_{yy} - \delta_{iso})/\Omega$. Experimental measurement of the parameter Ω for a quadrupolar nuclide such as [11]B is not common, and it has recently become feasible with NMR spectroscopy at high magnetic fields of ~14.1 T and higher [11]. A comparison of the calculated values of Ω with the available experimental data for this parameter for BO_3 sites in B_2O_3 and $Cs_2B_{18}O_{28}$ crystals shows a somewhat poor agreement, with the calculated values being significantly higher than the experimental values (Table 19.2). However, such discrepancy may arise due to the approximations involved in obtaining Ω values from simulations of experimental central and satellite spinning sideband line shapes. For example, in the case of B_2O_3 crystal, the small value of η_Q was taken as evidence for a similarly small value of η_{CS} and the shielding tensor was taken to be axially symmetric for the calculation of Ω [11]. However, the calculated ^{11}B NMR parameters for various BO_3 sites in Table 19.2 indicate the lack of any clear correlation between η_Q and η_{CS}. Therefore, caution should be exercised in making unwarranted assumptions to reduce the number of unknowns in the analyses of powder ^{11}B NMR data for both chemical shift and quadrupolar coupling tensor, since such analyses will always be underconstrained at a single magnetic field. The largest Ω value of ~85 ppm is observed for BO_3 sites with a mixture of bridging and nonbridging oxygen atoms in $LiBO_2$ crystal, and smaller Ω values (~22 ppm–37 ppm) are found to be characteristic of BO_3 sites with all three bridging oxygens in B_2O_3, CsB_3O_5, and $Cs_2B_{18}O_{28}$ (Table 19.2). The lowest Ω values (~10 ppm–14 ppm) are found for BO_3 sites with all three nonbridging oxygens in Li_3BO_3 and $Mg_3B_2O_6$ (Table 19.2). This general trend is consistent with previous experimental results [11]. A similar trend is also observed for the variation of η_Q for the BO_3 sites in these crystal structures (Table 19.2).

The theoretically calculated Ω values for the BO_4 sites in CsB_3O_5 and $Cs_2B_{18}O_{28}$ are rather small (~9 ppm–14 ppm) and are comparable with the Ω values characteristic of BO_3 sites with all three nonbridging oxygens (Table 19.2). Not surprisingly, the C_Q values for these BO_4 sites are also rather small (Table 19.2). This result is consistent with the regular site symmetry of these BO_4 tetrahedra characterized by maximum variation of B–O bond lengths of ~0.03 Å for any tetrahedron [25,27]. The magnitude of the corresponding CSA values for these BO_4 sites is found to range between ~5 ppm and 8 ppm (Table 19.2). A recent high-field (14.1 T) ^{11}B NMR study has reported a similar range of CSA values for BO_4 sites in a number of borate and borosilicate crystals [13]. One interesting exception is the case of SrB_4O_7 crystal where the two BO_4 sites are characterized by relatively large Ω and C_Q values (Table 19.2) that are consistent with the highly distorted nature of these two tetrahedra where the maximum variations of B–O bond lengths are as high as ~0.10 Å to 0.23 Å [22].

The [11]B δ_{iso} of BO_3 sites in borates have been shown in previous experimental studies to systematically increase with progressive replacement of bridging oxygen atoms with nonbridging oxygens as well as with decreasing cation–oxygen bond strength sum for all three O atoms bonded to a B atom [11]. However, a large variation in δ_{iso} ranging from 14.5 ppm to up to 19.9 ppm has been observed in this study even for BO_3 sites with all three bridging oxygens in three different crystal structures, namely, those of B_2O_3, CsB_3O_5, and $Cs_2B_{18}O_{28}$. Our calculations show that the [11]B δ_{iso} of nonring BO_3 sites in B_2O_3 ranges between 14.5 ppm and 14.9 ppm, whereas those of the BO_3 sites in three-membered triborate (B_3O_7) and boroxol (B_3O_6) type rings in CsB_3O_5 and $Cs_2B_{18}O_{28}$ are deshielded and range between 16.5 ppm and 19.9 ppm (Table 19.2). It may be noted that BO_3 sites in triborate and boroxol rings are characterized by average B–O–B angles of ~120° [25,27]. In contrast, the B–O–B angles in nonring BO_3 sites in B_2O_3 are significantly larger (~128°–133°) [21]. A previous DFT study of boroxol-ring and nonring BO_3 sites in molecular clusters had suggested that such difference in B–O–B angles is responsible for the corresponding difference between the ^{11}B NMR parameters of boroxol ring and nonring sites [14]. Interestingly, the δ_{iso} values for eight different BO_3 sites in triborate and boroxol rings in CsB_3O_5 and $Cs_2B_{18}O_{28}$, as determined in this study, vary over a narrow range (~17 ppm–20 ppm) that is distinctly higher than the δ_{iso} for nonring BO_3 sites in crystalline B_2O_3. This result clearly demonstrates that B–O–B angle may indeed be an important controlling factor for ^{11}B δ_{iso} of BO_3 sites with all three bridging oxygen atoms. On the other hand, the ^{11}B δ_{iso} for such sites in ring and nonring configurations do not show any simple systematic correlation with the cation–oxygen bond strength sums, unlike the trend that has been previously observed experimentally for BO_3 sites with nonbridging oxygens in crystalline borates [11]. For example, the cation–oxygen bond strength sums for the two BO_3 sites in CsB_3O_5 with δ_{iso} = 17.7 ppm and 19.9 ppm are 5.95 and 6.18, respectively, displaying an increasing trend of δ_{iso} with increasing cation–oxygen bond strength sum, which is the opposite of the decreasing trend observed in other borates [11]. It is also interesting to note in this regard that the CSA values of the BO_3 sites in the triborate rings are consistently negative, whereas those for the BO_3 sites in boroxol-ring or nonring geometries are consistently positive (Table 19.2).

High-field ^{11}B MAS NMR and ^{11}B dynamic-angle-spinning NMR studies of glassy B_2O_3 have shown the presence of two BO_3 sites with δ_{iso} ~17.8 and 13.3 (±1.0) ppm that have been assigned to boroxol-ring and nonring environments, respectively [14]. This assignment is fully consistent with the observed trend of deshielding of the ring BO_3 sites in CsB_3O_5 and $Cs_2B_{18}O_{28}$ structures with respect to the nonring BO_3 sites in crystalline B_2O_3 (Table 19.2). It may be noted that within experimental error, the ^{11}B δ_{iso} of the deshielded BO_3 site in glassy B_2O_3 agrees remarkably well with the rather tight range of chemical shifts (~16.5 ppm to 17.5 ppm) obtained in this study for the six different BO_3 sites in boroxol rings in crystalline $Cs_2B_{18}O_{28}$ (Table 19.2). Therefore, the boroxol-ring geometry in glassy B_2O_3 would have to be similar to those present in the structure of crystalline $Cs_2B_{18}O_{28}$.

19.4 CONCLUSIONS

The results of the present study demonstrate good agreement between the experimental ^{11}B NMR shielding and electric field gradient parameters and those obtained from first-principles calculations using DFT along with the GIPAW algorithm, for a wide variety of borate crystal structures. The ^{11}B δ_{iso} appears to be controlled by B–O–B angles for BO_3 sites with all three bridging oxygen atoms. This result is in contrast with the case for BO_3 sites with nonbridging oxygens where previous experimental studies have shown that ^{11}B δ_{iso} is controlled primarily by the number of nonbridging oxygen nearest neighbors and cation–oxygen bond strength sums. The characteristic ^{11}B δ_{iso} for BO_3 sites in boroxol ring geometries indicates that such rings in glassy B_2O_3 must be similar in geometry to those in crystalline $Cs_2B_{18}O_{28}$.

REFERENCES

1. Y. Mori, Y. K. Yap, T. Kamimura, M. Yoshimura and T. Sasaki. Recent development of nonlinear optical borate crystals for UV generation. *Opt. Mater.* **19**, 2002, 1.
2. E. Cavalli, D. Jaque, N. I. Leonyuk, A. Speghini and M. Bettinelli. Optical spectra of Tm^{3+}-doped $YAl_3(BO_3)_4$ single crystals. *Phys. Stat. Sol.* **C4**, 2007, 809.
3. M. Cable and J. M. Parker, editors. *High-Performance Glasses*. New York: Chapman & Hall, 1992.
4. P. J. Bray. NMR and NQR studies of boron in vitreous and crystalline borates. *Inorg. Chim. Acta* **289**, 1999, 158.
5. P. M. Aguiar and S. Kroeker. Medium-range order in cesium borate glasses probed by double-resonance NMR. *Solid-State NMR Spectr.* **27**, 2005, 10.
6. R. Martens and W. Muller-Warmuth. Structural groups and their mixing in borosilicate glasses of various compositions—An NMR study. *J. Non-Cryst. Solids* **265**, 2000, 167.
7. G. L. Turner, K. A. Smith, R. J. Kirkpatrick and E. Oldfield. Boron-11 nuclear magnetic resonance spectroscopic study of borate and borosilicate minerals and a borosilicate glass. *J. Magn. Reson.* **67**, 1986, 544.
8. G. Kunath-Fandrei, D. Ehrt and C. Jäger. Progress in structural elucidation of glasses by ^{27}Al and ^{11}B satellite transition NMR spectroscopy. *Z. Naturforsch., A: Phys. Sci.* **50a**, 1995, 413.
9. S. Kroeker, S. A. Feller, M. Affatigato, C. O'Brien, W. Clarida and M. Kodama. Multiple four-coordinated boron sites in cesium borate glasses and their relation to medium-range order. *Phys. Chem. Glasses* **44**, 2003, 54.
10. D. Holland, S. A. Feller, T. F. Kemp, M. E. Smith, A. P. Howes, D. Winslow and M. Kodama. Boron-10 NMR: What extra information can it give about borate glasses? *Phys. Chem. Glasses* **48**, 2007, 1.
11. S. Kroeker and J. F. Stebbins. Three-coordinated Boron-11 chemical shifts in borates. *Inorg. Chem.* **40**, 2001, 6239.
12. D. L. Bryce, R. E. Wasylishen and M. Gee. Characterization of tricoordinate boron chemical shift tensors: Definitive high-field solid-state NMR evidence for anisotropic boron shielding. *J. Phys. Chem. A* **105**, 2001, 3633.
13. M. R. Hansen, T. Vosegaard, H. J. Jakobsen and J. Skibsted. ^{11}B chemical shift anisotropies in borates from ^{11}B MAS, MQMAS, and single-crystal NMR spectroscopy. *J. Phys. Chem. A* **108**, 2004, 586.
14. J. W. Zwanziger. The NMR response of boroxol rings: A density functional theory study. *Solid-State NMR Spectr.* **27**, 2005, 5.
15. J. A. Tossell. Calculation of the structural and spectral properties of boroxol ring and non-ring B sites in B_2O_3 glass. *J. Non-Cryst. Solids* **183**, 1995, 307.
16. J. A. Tossell. Calculation of B and O NMR parameters in molecular models for B_2O_3 and alkali borate glasses. *J. Non-Cryst. Solids* **183**, 1997, 236.
17. F. Mauri, B. G. Pfrommer and S. G. Louie. Ab initio theory of NMR chemical shifts in solids and liquids. *Phys. Rev. Lett.* **77**, 1996, 5300.
18. C. J. Pickard and F. Mauri. All-electron magnetic response with pseudopotentials: NMR chemical shifts. *Phys. Rev. B* **63**, 2001, 245101.
19. M. D. Segall, P. J. D. Lindan, M. J. Probert, C. J., Pickard, P. J. Hasnip and S. J. Clark. First-principles simulation: Ideas, illustrations and the CASTEP code. *J. Phys.: Condens. Matter* **14**, 2002, 2717.

20. J. P. Perdew, K. Burke and M. Ernzerhof. Generalized gradient approximation made simple. *Phys. Rev. Lett.* **77**, 1996, 3865.

21. G. E. Gurr, P. W. Montgomery, C. D. Knutson and B. T. Gorres. The crystal structure of trigonal diboron trioxide. *Acta Crystallogr.* **B26**, 1970, 906.

22. A. Perloff and S. Block. The crystal structure of the strontium and lead tetraborates, $SrO.2B_2O_3$ and $PbO.2B_2O_3$. *Acta Crystallogr.* **20**, 1966, 274.

23. A. Kirfel, G. Will and R. F. Stewart. The chemical bonding in lithium metaborate, $LiBO_2$. Charge densities and electrostatic properties. *Acta Crystallogr.* **B39**, 1983, 175.

24. S. V. Berger. The crystal structure of the isomorphous orthoborates of cobalt and magnesium. *Acta Chem. Scand.* **3**, 1949, 660.

25. J. Krogh-Moe. Refinement of the crystal structure of caesium triborate, $Cs_2O.3B_2O_3$. *Acta Crystallogr.* **B30**, 1974, 1178.

26. F. Stewner. Die Kristallstruktur von α-Li_3BO_3. *Acta Crystallogr.* **B27**, 1971, 904.

27. N. Penin, M. Touboul and G. Nowogrocki. Refinement of α-CsB_9O_{14} crystal structure. *J. Solid-State Chem.* **175**, 2003, 348.

28. S. E. Ashbrook, L. Le Polles, C. J. Pickard, A. J. Berry, S. Wimperis and I. Farnan. First-principles calculations of solid-state ^{17}O and ^{29}Si NMR spectra of Mg_2SiO_4 polymorphs. *Phys. Chem. Chem. Phys.* **9**, 2007, 1587.

29. L. Truflandier, M. Paris, C. Payen and F. Boucher. First-principles calculations within periodic boundary conditions of the NMR shielding tensor for a transition metal nucleus in a solid state system: The example of V-51 in $AlVO_4$. *J. Phys. Chem. B* **110**, 2006, 21403.

30. T. Charpentier, S. Ispas, M. Profeta, F. Mauri and C. J. Pickard. First-principles calculation of O-17, Si-29, and Na-23 NMR spectra of sodium silicate crystals and glasses. *J. Phys. Chem. B* **108**, 2004, 4147.

31. E. Balan, F. Mauri, C. J. Pickard, I. Farnan and G. Calas. The aperiodic states of zircon: An ab initio molecular dynamics study. *Am. Mineral.* **88**, 2003, 1769.

32. I. Farnan, E. Balan, C. J. Pickard and F. Mauri. The effect of radiation damage on local structure in the crystalline fraction of $ZrSiO_4$: Investigating the ^{29}Si NMR response to pressure in zircon and reidite. *Am. Mineral.* **88**, 2003, 1663.

33. S. Rossano, F. Mauri, C. J. Pickard and I. Farnan. First-principles calculation of O-17 and Mg-25 NMR shieldings in MgO at finite temperature: Rovibrational effect in solids. *J. Phys. Chem. B* **109**, 2005, 7245.

34. A. Soleilhavoup, M. R. Hampson, S. J. Clark, J. S. O. Evans and P. Hodgkinson. Using ^{17}O solid-state NMR and first principles calculation to characterise structure and dynamics in inorganic framework material. *Magn. Reson. Chem.* **45**, 2007, S144.

35. M. R. Hansen, G. K. H. Madsen, H. J. Jakobsen and J. Skibsted. Refinement of borate structures from ^{11}B MAS NMR spectroscopy and density functional theory calculations of ^{11}B electric field gradients. *J. Phys. Chem. A* **109**, 2005, 1989.

20 Study of Synthesis Gas Conversion to Methane and Methanol over an Mo_6P_3 Cluster Using Density Functional Theory

Sharif F. Zaman and Kevin J. Smith

CONTENTS

20.1   INTRODUCTION

The conversion of synthesis gas ($CO + H_2$) to alcohols and hydrocarbons using heterogeneous catalysts is well known. The production of CH_3OH using Cu/ZnO catalysts is practiced commercially [1], as is the production of gasoline and diesel fuels via the Fischer–Tropsch synthesis using Fe or Co catalysts [2,3]. The selective conversion of synthesis gas to liquid fuels provides a route to renewable fuels that is almost CO_2 neutral if the synthesis gas is produced from biomass. Our interest is in the selective conversion of synthesis gas to ethanol for use as a fuel or fuel additive. Several catalysts, based on metals such as Cu, Co, Pd, and Fe, have been investigated for the higher alcohol synthesis [4], but few reports on the synthesis of ethanol from synthesis gas are available. Rhodium-based catalysts are able to produce oxygenates from synthesis gas [4,5]. The addition of appropriate promoters such as Mn enhances the rate of formation of these oxygenates, especially in the case of ethanol [5–10]. Mo-based catalysts also have high selectivity toward higher alcohols when Mo is doped with alkali metals [11]. Mo_2O_3 has also been used for the syngas conversion reaction [12], although high ethanol selectivity was not achieved. The highest selectivity for ethanol has been reported on MoS_2 catalysts [13]. Interestingly, MoP supported on metal oxides (Al_2O_3 and SiO_2) has been investigated as an alternative to MoS_2 catalysts for hydrodenitrogenation and hydrodesulfurization reactions [14,15]. Although there are no reports on the use of MoP for synthesis gas conversion to alcohols or hydrocarbons, previous researchers have suggested that metal phosphides may have good activity in other hydrogenation reactions, such as synthesis gas conversion to hydrocarbons and alcohols [16].

The use of computational chemistry in heterogeneous catalyst research and development has increased recently because of improved computational power and accuracy. Density functional theory (DFT) can be used to calculate the formation energy of molecules and solids with high accuracy [17]. Information related to the surface reaction, such as heat of reaction, the reaction energy barrier, and transition state structure, can also be determined. Computational chemistry can be used as a tool for catalyst design by calculating the catalyst's suitability for a particular reaction, without experimentation. Thus, a computational approach toward screening potential catalysts for a particular reaction is available, and this principle has been reported in the literature [18]. Kubo et al. [17] used DFT to identify new catalyst formulations for methanol and Fischer–Tropsch synthesis based on the adsorption and formation energies of surface stable species on potential catalysts. Greeley and Mavrikakis [19] investigated the competitive methanol decomposition pathway on Cu(111) considering all combinations of stable surface species. Alcala et al. [20] used DFT to generate the reaction energy diagram for ethanol decomposition on Pt(111). Similarly, a kinetic model of methanol decomposition on Pt(111) using DFT has been investigated by Gokhale et al. [21] and Kandoi et al. [22], who also reported the potential energy surface (PES) for this reaction.

As a first step in assessing MoP as a catalyst for synthesis gas conversion, especially to ethanol, we report herein on the reaction pathway for CH_3OH and CH_4 synthesis from CO and H_2 over an Mo_6P_3 cluster, determining the PES of the reactions. Due to limited experimental evidence of stable surface species over MoP, we investigated several likely stable surface species in each step of the reaction network and used the results of these calculations to determine the PES.

20.2 METHODS

20.2.1 CALCULATION PROCEDURE

The DMol^3 module of Materials Studio® (version 4.0) from Accelrys was used to complete the DFT calculations [23]. Accordingly, the electronic wave functions are expanded in numerical atomic basis sets defined on an atomic-centered spherical-polar mesh. The double numerical plus d-function (DND) all-electron basis set was used for all the calculations. The DND basis set includes one numerical function for each occupied atomic orbital and a second set of functions for valence atomic orbitals, plus a polarization d-function on all atoms. The Becke exchange [24] plus Perdew–Wang approximation [25] nonlocal functional (GGA–PW91) was used in all the calculations. Each basis function was restricted to a cutoff radius of 4.5 Å, allowing for efficient calculations without loss of accuracy. The Kohn–Sham equations [26] were solved by a self-consistent field procedure. The techniques of direct inversion in an iterative subspace [27] with a size value of six and thermal smearing [28] of 0.005 Ha were applied to accelerate convergence. The optimization convergence thresholds for energy change, maximum force, and maximum displacement between the optimization cycles were 0.00002 Ha, 0.004 Ha/Å, and 0.005 Å, respectively. The k-point set of (1 × 1 × 1) was used for all calculations. The activation energy between two surface species was identified by complete linear synchronous transit and quadratic synchronous transit search methods [29,30], followed by transition state confirmation through the nudge elastic band method [31]. Spin polarization and symmetry were imposed in all the calculations.

20.2.2 MODELING APPROACH

The Mo_6P_3 cluster model and the reactant and product species were created using the Materials Studio Visualizer. The Cartesian positions of the atoms of the Mo_6P_3 cluster were fixed in a vacuum after performing geometry optimization. The reactants and products were placed on the cluster in several different configurations, based on probable surface structures reported in the literature for CH_4 and CH_3OH synthesis. Geometric optimization of each structure was then done with the atoms of the Mo_6P_3 cluster fixed and no constraints placed on the reactants and products. The DFT simulation generates a field around the atoms placed in the vacuum to perform the calculations. For the DMol^3 electrostatic potential calculation, the xyz dimensions of the field were 10 × 12 × 12 Å. For the DMol^3 lowest unoccupied molecular orbital (LUMO) calculation, the xyz dimensions of the field were 9.8 × 12.14 × 11.8 Å. Reaction pathway modeling was approached by calculating the adsorption energies of all probable surface species. The adsorption energy was calculated by subtracting the energies of the gas phase species and the cluster from the energy of the adsorbed species according to the equation $E_{ad} = E_{(adsorbate/cluster)} - (E_{adsorbate} + E_{cluster})$. With this definition, a negative E_{ad} corresponds to a stable surface species.

The activation energy was calculated by using the transition state search (TS search) tool in DMol^3, applied to the reactant, a stable surface species plus an adsorbed H atom (H_{ad}) on the Mo_6P_3 cluster, and the product.

20.3 RESULTS AND DISCUSSION

20.3.1 AN Mo_6P_3 CLUSTER MODEL OF MoP

MoP has a hexagonal crystal structure, belonging to the $P_{\bar{6}m2}$ space group with lattice parameters $a = 3.22$ Å and $c = 3.19$ Å [16,33]. In the present study, an MoP crystal was built using the above information from which the (100) face was cleaved and a four-atom layer was taken as the cluster model, after geometric optimization. Note that the cluster building unit resembles the (100) crystal face of MoP, as shown in Figure 20.1. The distances and angles between atoms for the cluster are tabulated in Table 20.1, and these values compared favorably (within ±95%) to those of the MoP (100) slab. Mulliken population analysis showed a positive charge density (0.048e) on the Mo atoms and a negative charge density (–0.128e) on the P atoms of the cluster. Liu and Rodriguez [33] reported Mulliken charge densities of the MoP (001) crystal plane as 0.045e for Mo and –0.077e for P. Although the electron charge on Mo is similar for the cluster and the (001) plane of MoP, the P atoms have higher

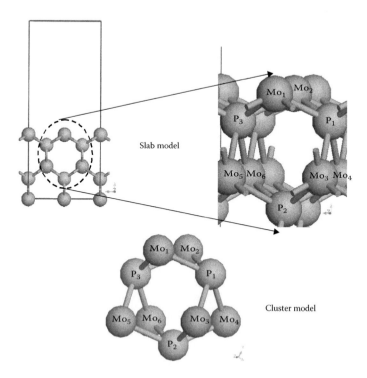

FIGURE 20.1 **(See color insert.)** Comparison between MoP slab (100) face and the Mo_6P_3 cluster model of the present study.

TABLE 20.1

Comparison between Mo_6P_3 Cluster and MoP (001) Slab Dimensions after Geometric Optimization

	Slab	Cluster[a]
Distances	Å	Å
$Mo_{(1)}–Mo_{(2)}$	3.19	3.19
$Mo_{(1)}–P_{(1)}$	2.45	2.45
$Mo_{(3)}–P_{(1)}$	2.45	2.45
$Mo_{(3)}–P_{(2)}$	2.51	2.45
$Mo_{(1)}–P_{(2)}$	4.18	4.05
Angles	**deg**	**deg**
$\theta_{Mo(1)–P(1)–Mo(3)}$	82.18	82.04
$\theta_{Mo(3)–P(2)–Mo(5)}$	79.74	82.18
$\theta_{Mo(1)–P(3)–Mo(2)}$	81.39	81.26
$\theta_{Mo(3)–P(2)–Mo(4)}$	78.85	81.26

[a] See Figure 20.1 for atom locations.

electronegativity in the cluster compared to the MoP (001) plane. The difference is due to the metal-rich stoichiometry of the Mo_6P_3 cluster.

20.3.2 CO ADSORPTION ON THE Mo_6P_3 CLUSTER

The adsorption energy of CO on the Mo_6P_3 cluster was calculated as –50.73 kcal/mol, in very good agreement with values of –50.5 kcal/mol [32] and –45.66 kcal/mol [33] reported for CO adsorption on the (001) plane of MoP. The CO adsorption energies on Cu(111), Pd(111), Pt(111), and Ni(111), as reported in the literature, are summarized in Table 20.2. These data show that CO is adsorbed more strongly on MoP than on any of these metals. For CO, the highest energy occupied molecular orbital (HOMO) is 5σ, a lone pair orbital, localized on the C atom. The LUMO is the $2\pi^*$ orbital, a C–O π antibonding orbital also localized on the C atom. Hence, CO adsorbs on the Mo atom through the C atom. The LUMO energy of CO adsorbed

TABLE 20.2

Adsorption Energy of CO and CH_3OH on Transition Metals

Metal	CO Adsorption Energy (kcal/mol)	CH_3OH Adsorption Energy (kcal/mol)	Reference
Cu(111)	–16.14	–4.38	[19,37]
Pt(111)	–41.97	–7.61	[22]
Pd(111)	–33.90	–6.46	[34]
Ni(111)	–35.98	–0.46	[36]

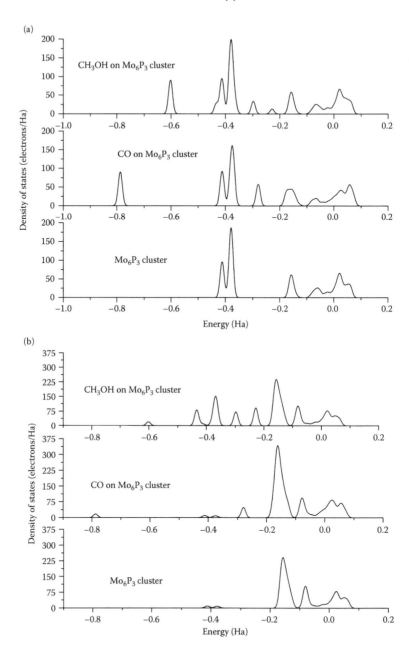

FIGURE 20.2 (a) Density of states (s-orbital) of Mo_6P_3 cluster and CO and CH_3OH adsorbed on Mo_6P_3 cluster. (b) Density of states (p-orbital) of Mo_6P_3 cluster and CO and CH_3OH adsorbed on Mo_6P_3 cluster.

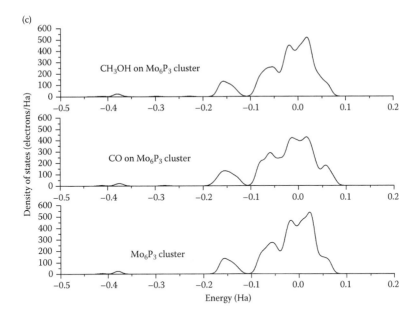

FIGURE 20.2   (Continued) (c) Density of states (d-orbital) of Mo$_6$P$_3$ cluster and CO and CH$_3$OH adsorbed on Mo$_6$P$_3$ cluster.

on Mo$_6$P$_3$ is –75.18 kcal/mol, the HOMO energy is –76.56 kcal/mol, and the Fermi energy level is –76.33 kcal/mol. The HOMO energy is lower than the Fermi energy level, and the LUMO is above the Fermi energy level, typical for surface chemisorbed species; hence, CO is strongly adsorbed on the Mo$_6$P$_3$ cluster. The density of states (DOS) of the Mo$_6$P$_3$ cluster compared to the DOS of the CO–Mo$_6$P$_3$ system shows that all the s (Figure 20.2a), p (Figure 20.2b), and d (Figure 20.2c) orbitals are altered. The d-orbital energy distribution, being the most affected, implies that the d-orbital of Mo is the main contributor to the adsorption process.

20.3.3   Determining the Potential Energy Surface for CH$_4$ Formation

The search for the PES of CH$_4$ formation from H$_2$ + CO on the Mo$_6$P$_3$ cluster was accomplished by evaluating the adsorption energy of several possible surface intermediates and calculating the activation energy between two successive species. The reaction pathway for CH$_4$ (and CH$_3$OH) formation is depicted in Figure 20.3, where the surface reaction propagates by addition of H$_{ad}$ to each stable adsorbed surface species. We have considered all combinations of H$_{ad}$ attachment with the C and O atoms of CO. Bond angle, bond length, and adsorption energies of adsorbed surface species and transition state structures, heats of reaction, and heats of adsorption are reported in Tables 20.3 through 20.5. The structure of the reactants, products, and transition states are shown in Figures 20.4 through 20.6.

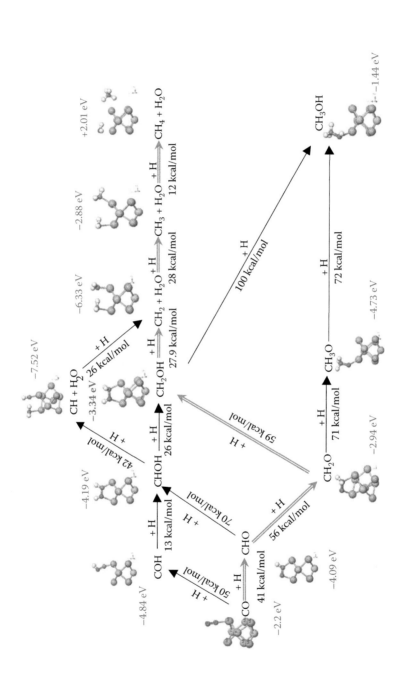

FIGURE 20.3 Reaction network and activation energies (kcal/mol) and adsorption energies (eV) of stable surface species on Mo_6P_3 cluster for syngas conversion to methanol and methane over the Mo_6P_3 cluster. Gray (large) spheres = Mo_6P_3 cluster atoms; dark gray (small) sphere = oxygen; gray (small) sphere = carbon; white (small) sphere = hydrogen.

TABLE 20.3
Properties of Surface-Adsorbed Species on Mo_6P_3 Cluster from DFT Calculations: Angle to the Surface, Distance between Atoms, Total Energy, and Adsorption Energy of Stable Surface Species Relevant to Methane and Methanol Formation from Syngas

Species	Figure	$\theta_{C-Mo-Mo}$ (deg)	$\theta_{O-Mo-Mo}$ (deg)	d_{C-Mo} (Å)	d_{O-Mo} (Å)	d_{C-O} (Å)	$E_{element}$ (au)	E_{ad} (kcal/mol)
CO_{ad}	–	78.28	–	1.97	–	1.18	–113.345	–50.73
CHO_{ad}	4-I.c	57.68	66.22	1.99	2.10	1.31	–113.857	–94.32
COH_{ad}	4-II.c	77.27	–	1.82	–	1.34	–113.782	–111.62
$CHOH_{ad}$	4-III.c	68.15	63.22	1.95	2.33	1.44	–114.403	–96.63
CH_2O_{ad}	4-IV.c	59.70	70.43	2.22	1.96	1.41	–114.504	–67.79
CH_2OH_{ad}	4-VI.c	67.50	67.95	2.22	1.96	1.41	–115.052	–77.02
CH_3O_{ad}	6-XIV.c	–	119.32	–	1.92	1.41	–115.047	–109.08
CH_3OH_{ad}	6-XIII.c	–	110.03	–	2.32	1.45	–115.711	–33.21
$CH_{ad} + H_2O_{ad}$	5-VIII.c	97.67	78.86	1.82	2.29	3.02	–38.457	–173.42
$CH_{2.ad} + H_2O_{ad}$	5-IX.c + 5-X.c	97.60	79.28	1.96	2.30	3.05	–39.130	–145.97
$CH_{3.ad} + H_2O_{ad}$	5-XI.c	123.65	85.80	2.16	2.36	4.26	–39.814	–66.42
$CH_{3.ad} + H_2O_{free}$	–	122.60	Free	2.16	2.37	4.21	–39.814	–66.42
$CH_4 + H_2O_{free}$	5-XII.c	Free	Free	2.65	2.39	–	–40.492	46.35

Note: Energy of free H_2O = –76.422778 au. Adsorption energy of hydrogen atom = –0.58213 au [au = atomic unit].

20.3.3.1 Formyl (CHO) and Hydroxymethylidyne (COH)

The addition of H_{ad} to the C of CO_{ad} on Mo yields CHO_{ad} species with the C and O attached to two nearby Mo atoms in a bridged structure (Figure 20.4-I). The adsorption energy of this species was calculated as –94.32 kcal/mol (Table 20.3). The addition of H_{ad} to the C atom of adsorbed CO species decreases the Mo–C bond strength, as indicated by an increase in bond length (1.99 Å) with respect to CO_{ad} on-top adsorption (1.97 Å). The Mo–O bond length is 2.10 Å. The hydroxymethylidyne (COH_{ad}) species is formed by addition of H_{ad} to the O atom of adsorbed CO_{ad} (Figure 20.4-II). The C atom is bound to a Mo atom (on-top adsorption) and the Mo–C bond length decreases to 1.82 Å, indicating a more tightly bound surface species with higher adsorption energy (–111.62 kcal/mol) compared to CHO_{ad}. A higher adsorption energy for COH_{ad}, compared to CHO_{ad}, has also been observed on Cu(111) [19], Pt(111) [22], and Pd(111) [34] surfaces. The C–O bond length for CHO_{ad} is 1.31 Å vs. 1.34 Å for the COH_{ad} species, whereas for CO_{ad}, it is 1.18 Å (Table 20.3). Increasing the C-O bond length increases the C and O reactivity, as electrons are being accumulated on the atoms rather than being shared, and H_{ad} added to CO_{ad} also weakens the C–O covalent bond strength. The activation energy associated with CHO_{ad} formation from $CO_{ad} + H_{ad}$ is 41.37 kcal/mol, whereas for COH_{ad}, a value of 50.00 kcal/

TABLE 20.4

Properties of Surface-Adsorbed Species on Mo_6P_3 Cluster from DFT Calculations: Angle to the Surface, Distance between Atoms, Total Energy, and Adsorption Energy of Stable Surface Species Relevant to Methane and Methanol Formation from Syngas

Species	Figure	$\theta_{C-Mo-Mo}$ (deg)	$\theta_{O-Mo-Mo}$ (deg)	d_{C-Mo} (Å)	d_{O-Mo} (Å)	d_{C-O} (Å)	$E_{element}$ (au)	E_{ad} (kcal/mol)
$CO_{ad} + H_{ad}$	4-I.a	78.23	–	1.98	–	1.18	–113.345	–62.03
$CO_{ad} + H_{ad}$	4-II.a	78.89	–	1.99	–	1.18	–113.345	–61.11
$CHO_{ad} + H_{ad}$	4-IV.a	52.98	66.41	1.99	2.10	1.30	–113.857	–106.31
$CHO_{ad} + H_{ad}$	4-V.a	57.47	66.31	1.99	2.10	1.30	–113.857	–106.31
$COH_{ad} + H_{ad}$	4-III.a	78.40	–	1.82	–	1.33	–113.782	–123.61
$CHOH_{ad} + H_{ad}$	4-VI.a	67.89	63.57	1.97	2.23	1.44	–114.403	–117.84
$CHOH_{ad} + H_{ad}$	5-VIII.a	68.53	63.26	1.94	2.29	1.47	–114.403	–85.09
$CH_2OH_{ad} + H_{ad}$	5-IX.a	69.17	66.46	2.20	2.28	1.49	–115.052	–54.88
$CH_2OH_{ad} + H_{ad}$	6-XIII.a	67.70	67.03	2.21	2.23	1.48	–115.052	–86.94
$CH_2O_{ad} + H_{ad}$	4-VII.a	58.21	71.71	2.23	1.96	1.41	–114.504	–81.86
$CH_3O_{ad} + H_{ad}$	6-XV.a	–	128.83	–	1.92	1.41	–115.047	–120.61
$CH_{ad} + H_2O_{ad} + H_{ad}$	5-X.a	96.02	78.83	1.81	2.27	2.98	–38.457	–192.56
$CH_{2.ad} + H_2O_{ad} + H_{ad}$	5-XI.a	96.08	78.89	1.97	2.29	2.97	–39.130	–103.31
$CH_{3.ad} + H_2O_{ad} + H_{ad}$	5-XII.a	123.18	–	2.16	2.36	4.23	–39.814	–81.64

mol was obtained (Table 20.5). Formation of the CHO_{ad} is thermodynamically more favorable than the formation of COH_{ad}, and both reactions are endothermic, with a heat of formation of 10.13 kcal/mol for CHO_{ad} species and 49.50 kcal/mol for COH_{ad} species. Nunan et al. [35] reported CHO_{ad} as a precursor for alcohol production on copper-based catalysts. The results presented herein suggest that CHO_{ad} is also the energetically favored precursor for CH_4 and CH_3OH formation on the Mo_6P_3 cluster.

20.3.3.2 Formaldehyde (CH_2O) and Hydroxymethylene (CHOH)

Addition of H_{ad} to the C atom of the formyl species forms CH_2O_{ad} (Figure 20.4-IV), with an adsorption energy of –67.80 kcal/mol. The Mo–C bond length is 2.22 Å and the Mo–O bond length is 1.96 Å. Compared to the CHO_{ad} species, the Mo–C bond

TABLE 20.5

Properties of Transition State for Reactions Shown: Structure, Angle with the Surface, Distance between Atoms, Energy of Reaction, and Activation Energy for Methane and Methanol Formation from Syngas

Reactions	$\theta_{C-Mo-Mo}$ (deg)	$\theta_{O-Mo-Mo}$ (deg)	d_{C-Mo} (Å)	d_{O-Mo} (Å)	d_{C-O} (Å)	ΔE_r (kcal/mol)	ΔE (kcal/mol)
Common: (see Figure 20.4)							
$CO_{ad} + H_{ad} \rightarrow CHO_{ad}$	70.99	–	1.98	–	1.18	10.13	41.37
$CO_{ad} + H_{ad} \rightarrow COH_{ad}$	82.83	–	1.92	–	1.24	49.50	50.00
$CHO_{ad} + H_{ad} \rightarrow CHOH_{ad}$	65.08	62.87	1.98	2.32	1.32	32.95	69.53
$COH_{ad} + H_{ad} \rightarrow CHOH_{ad}$	40.40	–	2.07	–	1.36	2.59	13.22
$CHO_{ad} + H_{ad} \rightarrow CH_2O_{ad}$	56.09	67.76	2.01	2.11	1.30	–3.25	56.24
$CHOH_{ad} + H_{ad} \rightarrow CH_2OH_{ad}$	68.23	63.29	1.96	2.23	1.47	0.49	25.67
$CH_2O_{ad} + H_{ad} \rightarrow CH_2OH_{ad}$	70.21	62.48	2.22	2.15	1.45	34.56	59.12
Methane Synthesis: (see Figure 20.5)							
$CHOH_{ad} + H_{ad} \rightarrow CH_{ad} + H_2O_{ad}$	96.09	82.61	1.80	1.98	3.16	–5.87	42.16
$CH_{ad} + H_{ad} + H_2O_{ad} \rightarrow CH_{2.ad} + H_2O_{ad}$	93.83	79.96	1.83	2.23	2.94	–5.45	60.94
$CH_2OH_{ad} + H_{ad} \rightarrow CH_{2.ad} + H_2O_{ad}$	94.31	76.84	1.80	1.99	2.88	–6.19	27.88
$CH_{2.ad} + H_{ad} + H_2O_{ad} \rightarrow CH_{3.ad} + H_2O_{ad}$	110.36	82.63	1.97	2.35	3.62	–6.01	18.67
$CH_{3.ad} + H_{ad} + H_2O_{ad} \rightarrow CH_{4.ad} + H_2O_{ad}$	118.95	–	2.11	2.36	4.39	8.70	12.16
Methanol Synthesis: (see Figure 20.6)							
$CH_2O_{ad} + H_{ad} \rightarrow CH_3O_{ad}$	63.04	83.21	2.38	2.13	1.33	–6.22	71.13
$CH_3O_{ad} + H_{ad} \rightarrow CH_3OH_{ad}$	–	112.98	4.74	2.03	1.42	36.65	71.54
$CH_2OH_{ad} + H_{ad} \rightarrow CH_3OH_{ad}$	62.62	73.94	2.30	2.28	1.52	8.45	100.91

length is increased, whereas the Mo–O bond length is decreased, CH_2O_{ad} is tightly bound through the Mo–O, and electrons are withdrawn from the substrate by the O atom. The C–O bond length (1.41 Å) increases compared to CHO_{ad} (Table 20.3). The molecular orbital of the CH_2O_{ad} species weakens the C–O bond strength. The π electron interaction between CH_2 (π bonding and antibonding orbitals) and O (π-type loan pair electron) gives rise to a new molecular orbital. Because the energies of these interacting orbitals are similar, the new orbital is C–O antibonding and C–H bonding. If an H atom adds to the O atom of adsorbed CHO_{ad} species, $CHOH_{ad}$ is formed (Figure 20.4-V) with an adsorption energy of –96.63 kcal/mol. The Mo–C bond length (1.95 Å) decreases and the Mo–O bond length (2.33 Å) increases compared

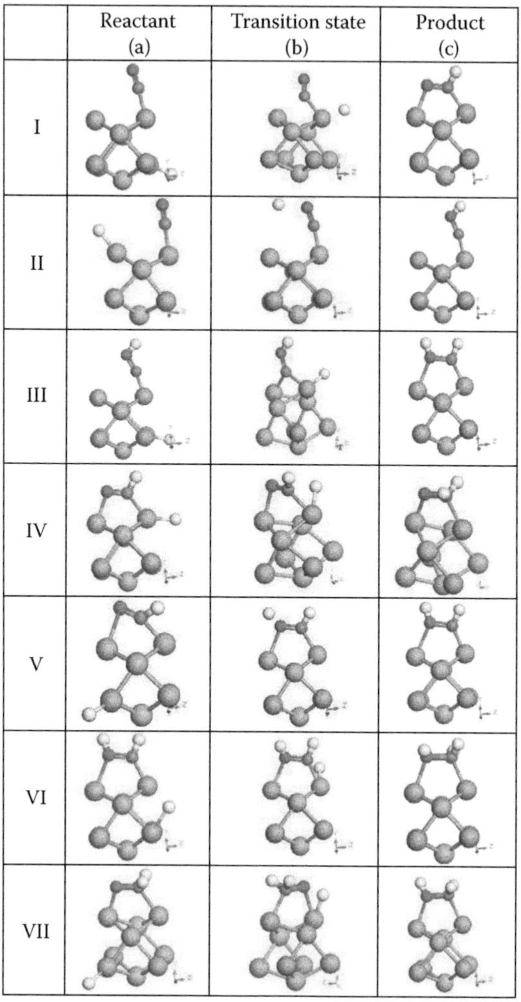

	Reactant (a)	Transition state (b)	Product (c)
I			
II			
III			
IV			
V			
VI			
VII			

FIGURE 20.4 **(See color insert.)** Methane and methanol formation reaction steps on Mo_6P_3 cluster. Gray (large) spheres = Mo_6P_3 cluster atoms; dark gray (small) sphere = oxygen; gray (small) sphere = carbon; white (small) sphere = hydrogen.

with the adsorbed CHO_{ad} species. The C–O bond length for CH_2O_{ad} species (1.41 Å) is higher than that of CHO_{ad} but lower than that of $CHOH_{ad}$ (Table 20.3).

The activation energy for CH_2O_{ad} formation is 56.24 kcal/mol compared with 69.53 kcal/mol for $CHOH_{ad}$ (Table 20.5). Formation of CH_2O_{ad} is thermodynamically more favorable; formation of CH_2O_{ad} is exothermic ($\Delta E_r = -3.25$ kcal/mol), whereas formation of $CHOH_{ad}$ is endothermic ($\Delta E_r = 2.58$ kcal/mol). $CHOH_{ad}$ can also be formed by the addition of H_{ad} to COH_{ad} (Figure 20.4-III) with a lower activation energy (13.22 kcal/mol) and formation energy of 2.59 kcal/mol (Table 20.5) compared to H_{ad} addition to CHO_{ad}. This route is important for the decomposition of CH_4 and CH_3OH to CO and H_2.

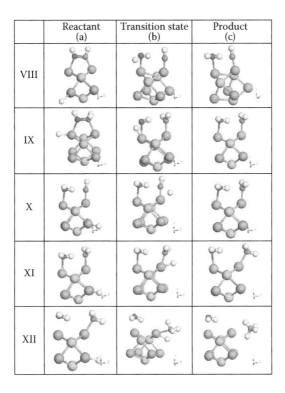

	Reactant (a)	Transition state (b)	Product (c)
VIII			
IX			
X			
XI			
XII			

FIGURE 20.5 Methane formation reaction steps on Mo_6P_3 cluster. Gray (large) spheres = Mo_6P_3 cluster atoms; dark gray (small) sphere = oxygen; gray (small) sphere = carbon; white (small) sphere = hydrogen.

	Reactant (a)	Transition state (b)	Product (c)
XIII			
XIV			
XV			

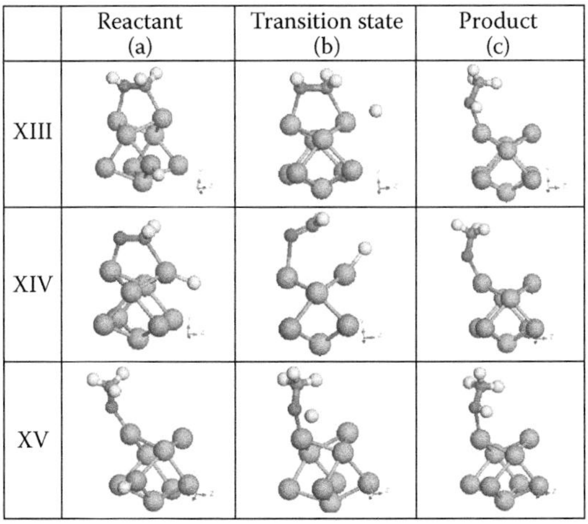

FIGURE 20.6 (**See color insert.**) Methanol formation reaction steps on Mo_6P_3 cluster. Gray (large) spheres = Mo_6P_3 cluster atoms; dark gray (small) sphere = oxygen; gray (small) sphere = carbon; white (small) sphere = hydrogen.

20.3.3.3 Hydroxymethyl (CH₂OH)

CH_2OH_{ad} species can evolve either by addition of H_{ad} to the C atom of $CHOH_{ad}$ species or by the addition of H_{ad} to the O atom of adsorbed CH_2O_{ad} (Figure 20.4-VI and 20.4-VII). The calculated adsorption energy of CH_2OH_{ad} on the Mo_6P_3 cluster was −77.02 kcal/mol. The bond lengths are Mo–C 2.22, Mo–O 1.96, and C–O 1.41 Å. These lengths are close to those calculated for the CH_2O_{ad} species, whereas compared to $CHOH_{ad}$, the Mo–C bond length is increased, and the Mo–O and the C–O bond lengths are decreased. The activation and reaction energies for $CH_2O_{ad} + H_{ad} \rightarrow CH_2OH_{ad}$ are 59.12 kcal/mol and 34.56 kcal/mol, respectively, and for the $CHOH_{ad} + H_{ad} \rightarrow CH_2OH_{ad}$, they are 25.67 kcal/mol and 0.49 kcal/mol, respectively.

20.3.3.4 C–O Bond Scission and the Formation of CH₄

Three intermediates are relevant for C–O bond scission: COH_{ad}, $CHOH_{ad}$, and CH_2OH_{ad}. Bond scission from COH_{ad} is described in a later section that discusses carbon formation on the catalyst surface. Adding H_{ad} to $CHOH_{ad}$ yields CH_{ad} (carbene) species and H_2O_{ad}, whereas adding H_{ad} to CH_2OH_{ad} forms $CH_{2.ad}$ (methylene) and H_2O_{ad}. These steps are shown in Figure 20.5-VIII and 20.5-IX, respectively. The adsorption energies of $CH_{ad} + H_2O_{ad}$ and $CH_{2.ad} + H_2O_{ad}$ are −173.42 kcal/mol and −145.97 kcal/mol, respectively. The Mo–C bond length is 1.82 Å for CH_{ad} species and 1.96 Å for $CH_{2.ad}$ species. CH_{ad} species are more strongly adsorbed on the surface than $CH_{2.ad}$ species. With two H atoms attached to the O atom, the octet condition for O, the most stable condition, is satisfied. The H_2O_{ad} molecule subsequently desorbs from the catalyst surface. The activation energy for C–O bond scission via the $CHOH_{ad}$ route is 42.16 kcal/mol, whereas, via the CH_2OH_{ad} route, it is 27.88 kcal/mol (Table 20.5). Bond breakage results in heat generation with $CHOH_{ad} + H_{ad} \rightarrow CH_{ad} + H_2O_{ad}$ yielding $\Delta E_r = -5.87$ kcal/mol, and $CH_2OH_{ad} + H_{ad} \rightarrow CH_{2.ad} + H_2O_{ad}$ yielding −6.19 kcal/mol.

Addition of H_{ad} to CH_{ad} species yields $CH_{2.ad}$ species (Figure 20.5-X) with an adsorption energy −106.08 kcal/mol. The H_2O_{ad} molecule produced by the C–O bond scission is adsorbed on a Mo atom, and the total adsorption energy of $CH_{2.ad} + H_2O_{ad}$ is −145.98 kcal/mol with an adsorption energy of H_2O_{ad} on the cluster of −32.52 kcal/mol. Separate calculations for the CH_{ad}, $CH_{2.ad}$, and $CH_{3.ad}$ species on the cluster yielded adsorption energies of −116.23, −106.08, and −80.94 kcal/mol, respectively. The total adsorption energy for $CH_{2.ad} + H_2O_{ad}$ is marginally lower than the sum of the energies of the separately adsorbed species in the system, likely due to interaction effects between the adsorbed species. The activation energy for this reaction step is 60.94 kcal/mol with an exothermic reaction energy $\Delta E_r = -5.45$ kcal/mol.

Addition of H_{ad} to $CH_{2.ad}$ yields adsorbed $CH_{3.ad}$ species. The Mo–C bond length associated with the $CH_{3.ad}$ intermediate increased to 2.16 Å, and the adsorption energy decreased to −66.42 kcal/mol compared to the adsorbed $CH_{2.ad}$ species. The activation energy for the process is 18.67 kcal/mol with an exothermic heat of reaction $\Delta E_r = -6.01$ kcal/mol. The reaction step is depicted in Figure 20.5-XI.

Adding another H_{ad} to the C atom of $CH_{3.ad}$ species forms CH_4. Both the CH_4 molecule and the H_2O_{ad} molecule are desorbed from the surface. CH_4 has a positive adsorption energy (46.35 kcal/mol), indicating that CH_4 is not adsorbed on the surface at the simulation temperature (273 K). Consequently, once CH_4 is formed, it will not engage in further reaction and will emerge as a product. The activation energy

FIGURE 20.7   Carbon formation on the Mo_6P_3 cluster.

of this step is 12.16 kcal/mol. The reaction is endothermic and the reaction energy is 8.7 kcal/mol. Figure 20.5-XII shows this reaction step.

20.3.3.5   Carbon Formation on the Surface

Addition of H_{ad} to the O atom of CO, rather than to the C atom, yields hydroxymethylidyne (COH_{ad}) species. COH_{ad} has the C atom attached to Mo (CO on-top adsorption), whereas the O is not bonded to the catalyst surface. The adsorption energy for COH_{ad} is -111.62 kcal/mol, making it more stable than CHO_{ad} species on the Mo_6P_3 cluster. COH_{ad} proceeds in the reaction by adding another H atom to C, forming hydroxymethylene species. The COH_{ad} route is not energetically favored ($\Delta E = 50.00$ kcal/mol) compared with the CHO_{ad} route. If H_{ad} adds to the O atom of COH_{ad}, the C–O bond breaks and H_2O_{ad} and surface-adsorbed carbon are produced. Carbon is strongly bound to the surface between two molybdenum atoms and a phosphorus atom. This carbon atom is very difficult to remove from the surface and will eventually deactivate the catalyst (Figure 20.7).

Based on the above analysis of the reaction intermediates, the PES for CH_4 formation is constructed and depicted in Figure 20.8, in which the thermochemical data

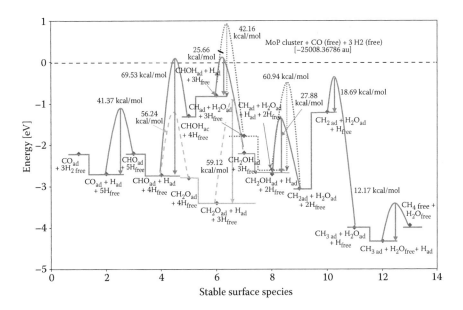

FIGURE 20.8   Kinetic pathway of methane formation over Mo_6P_3 cluster.

and activation energies of the elementary reaction steps are also shown. Accordingly, the favorable PES pathway for the formation of CH_4 may be summarized as (H_{ad} addition of each step is assumed but not shown) $CO_{ad} \rightarrow CHO_{ad} \rightarrow CH_2O_{ad} \rightarrow CH_2OH_{ad} \rightarrow [CH_{2.ad} + H_2O_{ad}] \rightarrow [CH_{3.ad} + H_2O_{ad}] \rightarrow [CH_4 + H_2O]$.

20.3.4 Determining the Potential Energy Surface for CH_3OH Formation

Methanol formation occurs via two precursors, CH_2OH_{ad} and CH_2O_{ad}, which are common to CH_4 formation as well. Methanol can be generated from the bridge-bonded CH_2OH_{ad} species by adding H_{ad} to the C atom (Figure 20.6-XIII). The Mo–C bond breaks, the C–O bond length is 1.45 Å, and the Mo–O bond length is 2.32 Å. The CH_3OH_{ad} adsorption energy is –33.21 kcal/mol which is relatively high compared with adsorption energies on other metals as shown in Table 20.2. CH_3OH_{ad} is attached to a Mo atom through the O atom. The DOS for the $CH_3OH–Mo_6P_3$ system (Figure 20.2) shows that the d-orbital (which belongs to Mo) distribution is not altered to a great extent, suggesting that the d-orbital does not have a strong overlap with the O atom. Similar behavior has been observed with other transition metals [38,39]. The main interaction occurs between the p-orbital and the s-orbital. The p-orbital interaction comes mostly from the P atom of the Mo_6P_3 cluster and the oxygen atom. Hence, the adsorption energy of CH_3OH_{ad} on the Mo_6P_3 cluster is higher than other transition metals, that is, Pt(111), Pd(111), Cu(111), and Ni(111) (Table 20.2) that do not have p-orbitals available. Most of the transition metals show low adsorption energy of CH_3OH due to a small p-orbital contribution to the bond with the O atom. The HOMO energy of $CH_3OH–Mo_6P_3$ system is –2.48 kcal/mol and the LUMO is –2.246 kcal/mol, whereas the Fermi energy is –2.25 kcal/mol. The high adsorption energy of CH_3OH_{ad} on the Mo_6P_3 cluster suggests that it may be available for further reaction to form ethanol and other higher carbon-number products. The activation and formation energies for CH_3OH_{ad} via the CH_2OH_{ad} route are 100.91 kcal/mol and 8.45 kcal/mol, respectively.

20.3.4.1 Methoxy (CH_3O) and Methanol (CH_3OH)

Addition of H_{ad} to bridge-bonded CH_2O_{ad} yields CH_3O_{ad} species (Figure 20.6-XIV), with the carbon atom detached from the metal surface. The C–O bond length is 1.41 Å and the Mo–O bond length is 1.92 Å. The CH_3O_{ad} adsorption energy is –109.08 kcal/mol and the activation energy of this step is 71.13 kcal/mol. Addition of another H_{ad} to the O atom of CH_3O_{ad} yields CH_3OH_{ad} (Figure 20.6-XV) with an activation energy of 71.54 kcal/mol and energy of reaction of 36.65 kcal/mol.

The PES pathway for CH_3OH_{ad} formation (Figure 20.9) follows the route (H_{ad} addition of each step is assumed but not shown) $CO_{ad} \rightarrow CHO_{ad} \rightarrow CH_2O_{ad} \rightarrow CH_2OH_{ad} \rightarrow CH_3OH_{ad}$. This route has the lowest energy surface species up to CH_2OH_{ad}. However, the final step has the largest energy barrier, 100.86 kcal/mol, compared with all other reaction steps. Hence, we conclude that adsorbed CO_{ad} and H_{ad} will form hydroxymethyl, but the C–O bond will then break, as the energy barrier is too high to form methanol, resulting in the formation of $CH_{2.ad}$ and water species. This pathway explains the higher selectivity to CH_4 than CH_3OH when synthesis gas is reacted over MoP catalysts [40]. Blocking the CH_4 production pathway would enhance alcohol production, and this may be possible by hindering the approach of

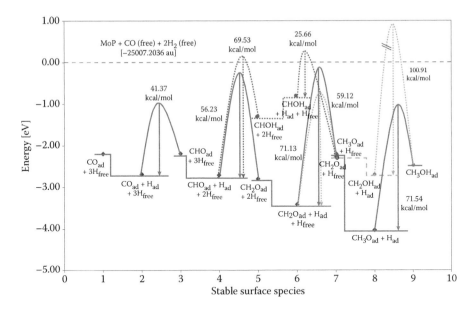

FIGURE 20.9   Kinetic pathway of methanol formation over Mo_6P_3 cluster.

a H atom toward the O atom to prevent cleavage of the C–O bond and inhibit water formation. Alkali metals such as K should promote this kind of reaction because K provides additional electrons to the nearby O atom and hence can hinder the formation of H_2O and enhance alcohol production.

Note that the reverse order of the reactions, that is, the decomposition of methanol and methane over MoP catalysts, can also be examined from the data presented herein. Formation of H_2 and CO from CH_3OH follows the decomposition pathway $[CH_3OH_{ad} \rightarrow CH_3O_{ad} + H_{ad} \rightarrow CH_2O_{ad} + H_{ad} \rightarrow CHO_{ad} + H_{ad} \rightarrow CO_{ad} + H_{ad}]$ and from $CH_4 + H_2O$, the pathway is $[CH_4 + H_2O \rightarrow CH_{3.ad} + H_2O_{ad} \rightarrow CH_{2.ad} + H_2O_{ad} \rightarrow CH_2OH_{ad} \rightarrow CHOH_{ad} \rightarrow COH_{ad} \rightarrow CO_{ad}]$. The same CH_3OH decomposition pathway is observed on Pt(111) [22] and Ni(111) [36].

20.3.5   HYDROGEN DISSOCIATION ENERGY

Both reaction pathways to CH_4 and $CH_3OH_{(ad)}$ formation show that most of the intermediate reaction steps are endothermic. However, both CH_4 and $CH_3OH_{(ad)}$ formation from $CO + H_2$ are exothermic. Heat evolution occurs from adsorption and bond dissociation, whereas bond formation and desorption are endothermic. Bond dissociation can occur between the C–O bond (CH_4 formation) and the H–H bond. The exothermic nature of the reactions is mainly attributed to the dissociation of the hydrogen molecule, which was not included in the energy calculations reported herein. The simulation was accomplished by taking an H atom adsorbed on a Mo atom with a stable surface carbon-bearing species. The hydrogen molecule adsorbs on Mo, dissociates into atoms, and adsorbs on two different Mo atoms. The energy released by H–H bond dissociation is –20.56 kcal/mol and the activation energy is 89.34 kcal/mol.

20.4 CONCLUSIONS

A DFT study of CH_4 and CH_3OH formation over an Mo_6P_3 cluster model is described. The Mo_6P_3 cluster was representative of the (100) face of MoP and had similar adsorption energies to MoP. Hydroxymethyl (CH_2OH) is a common intermediate for both CH_4 and CH_3OH formation. However, the energy barrier for CH_3OH formation from CH_2OH was significantly higher than for the formation of methylene and water that leads to CH_4. Thus, the simulation predicts the formation of CH_4 rather than CH_3OH over MoP.

ACKNOWLEDGMENTS

Financial support from the Natural Sciences and Engineering Research Council of Canada is gratefully acknowledged.

REFERENCES

1. A. Gotti and R. Prins. Basic metal oxides as cocatalysts for Cu/SiO_2 catalysts in the conversion of synthesis gas to methanol. *J. Catal.* **178**, 1998, 511–519.
2. E. Iglesia. Design synthesis, and use of cobalt-based Fischer–Tropsch synthesis catalysis. *Appl. Catal., A* **161**, 1997, 59–78.
3. R. B. Anderson. *The Fischer–Tropsch Synthesis.* New York: Academic Press, 1984.
4. J. P. Hindermann, G. J. Hutchings and A. Kiennemann. Mechanistic aspects of the formation of hydrocarbons and alcohols from carbon monoxide hydrogenation. *Catal. Rev.—Sci. Eng.* **35**, 1993, 1–127.
5. P. C. Ellgen, W. J. Bartley, M. M. Bhasin and T. P. Wilson. Rhodium-based catalysts for the conversion of synthesis gas to two-carbon chemicals. *Adv. Chem.* **78**, 1979, 147–157.
6. S. C. Chuang, J. G. Goodwin Jr. and I. Wender. The effect of alkali promotion on CO hydrogenation over Rh/TiO_2. *J. Catal.* **95**, 1985, 435–446.
7. A. Kiennemann, R. Breault, J. P. Hindermann and M. Laurin. Ethanol promotion by the addition of cerium to rhodium–silica catalysts. *J. Chem. Soc., Faraday Trans. I* **83**, 1987, 2119–2128.
8. S. J. Rieck and A. T. Bell. Studies of the interactions of H_2 and CO with Pd/SiO_2 promoted with La_2O_3, CeO_2, Pr_6O_{11}, Nd_2O_3, and Sm_2O_3. *J. Catal.* **99**, 1986, 278–292.
9. M. Ichikawa. Catalysis by supported metal crystallites from carbonyl clusters. II. Catalytic ethanol synthesis from carbon monoxide and hydrogen under atmospheric pressure over supported rhodium crystallites prepared from Rh carbonyl clusters deposited on titanium dioxide, zirconium oxide, and lanthanum oxide. *Bull. Chem. Soc. Jpn.* **51**, 1978, 2273–2277.
10. B. J. Kip, P. A. T. Smeets, J. Grondelle and R. Prins. Hydrogenation of carbon monoxide over vanadium oxide promoted rhodium catalysts. *Appl. Catal.* **33**, 1987, 181–208.
11. Y. Zhang, Y. Sun and B. Zhong. Synthesis of higher alcohols from syngas over ultrafine Mo–Co–K catalysts. *Catal. Lett.* **76**, 2001, 249–253.
12. G. Bian, L. Fan, Y. Fu and K. Fujimoto. High temperature calcined $K–MoO_3/\gamma$-Al_2O_3 Catalysts for mixed alcohol synthesis from syngas: Effect of Mo loading. *Appl. Catal., A* **170**, 1998, 255–268.
13. J. Iranmahboob, H. Toghiani, D. O. Hill and F. Nadim. The influence of clay on $K_2CO_3/$Co–MoS_2 catalyst in the production of higher alcohol fuel. *Fuel Proc. Tech.* **79**, 2002, 71–75.

14. V. Zuzaniuk and R. Prins. Synthesis and characterization of silica supported transition-metal phosphides as HDN catalysts. *J. Catal.* **219**, 2003, 85–96.
15. A. C. Paul and S. T. Oyama. Alumina-supported molybdenum phosphide hydroprocessing catalysts. *J. Catal.* **218**, 2003, 78–87.
16. S. T. Oyama. Novel catalysts for advanced hydroprocessing: Transition metal phosphides. *J. Catal.* **216**, 2003, 343–352.
17. M. Kubo, T. Kubota, C. Jung, M. Ando, S. Sakahara, K. Yajima, K. Seki, et al. Design of new catalysts for ecological high-quality transportation fuels by combinatorial computational chemistry and tight-binding quantum chemical molecular dynamics approaches. *Catal. Today* **89**, 2004, 479–493.
18. C. J. H. Jacobsen, S. Dahl, B. S. Clausen, S. Bhan, A. Logadottir and J. K. Norskov. Catalyst design by interpolation in the periodic table: Bimetallic ammonia synthesis catalysts. *J. Am. Chem. Soc.* **123**, 2001, 8404–8405.
19. J. Greeley and M. Mavrikakis. Methanol decomposition on C(111): A DFT study. *J. Catal.* **208**, 2002, 291–300.
20. R. Alcala, M. Mavrikakis and J. A. Dumesic. DFT studies for cleavage of C–C and C–O bonds in surface species derived from ethanol on Pt(111). *J. Catal.* **218**, 2003, 178–190.
21. A. A. Gokhale, S. Kandoi, J. P. Greeley, M. Mavrikakis and J. A. Dumesic. Molecular-level descriptions of surface chemistry in kinetic models using density functional theory. *Chem. Eng. Sci.* **59**, 2004, 4679–4691.
22. S. Kandoi, J. Greeley, A. M. Sanchez-Castillo, S. T. Evans, A. A. Gokhale, J. A. Dumesic and M. Mavrikakis. Prediction of experimental methanol decomposition rates on platinum from first principles. *Top. Catal.* **37**, 2006, 17–28.
23. B. Delley. From molecules to solids with the DMol^3 approach. *J. Chem. Phys.* **113**, 2000, 7756–7764.
24. A. D. Becke. A multicenter numerical integration scheme for polyatomic molecules. *J. Chem. Phys.* **8**, 1988, 2547–25453.
25. J. P. Perdew and Y. Wang. Accurate and simple analytic representation of the electron–gas correlation energy. *Phys. Rev. B* **45**, 1992, 13244–13249.
26. W. Kohn and L. J. Sham. Self-consistent equations including exchange and correlation effects. *Phys. Rev. A* **140**, 1965, 1133–1138.
27. P. Pulay. Improved SCF convergence acceleration. *J. Comp. Chem.* **3**, 1983, 4556–4560.
28. B. Delley. *Modern Density Functional Theory: A Tool for Chemistry, Theoretical and Computational Chemistry*, Volume 2, edited by J. M. Seminorio and P. Politzer. Amsterdam: Elsevier Science, 1995.
29. S. Bell and J. S. Crighton. Locating transition states. *J. Chem. Phys.* **80**, 1984, 2464–2475.
30. S. Fischer and M. Karplus. Conjugate peak refinement: An algorithm for finding reaction paths and accurate transition states in systems with many degrees of freedom. *Chem. Phys. Lett.* **194**, 1992, 252–261.
31. G. Henkelman and H. Jonsson. Improved tangent estimate in the nudged elastic band method for finding minimum energy paths and saddle points. *J. Chem. Phys.* **113**, 2000, 9978–9985.
32. Z. Feng, C. Liang, W. Wu, Z. Wu, A. R. Santen and C. Li. Carbon monoxide adsorption on molybdenum phosphides: Fourier transform infrared spectroscopic and density functional theory studies. *J. Phys. Chem. B* **107**, 2003, 13698–13702.
33. P. Liu and J. A. Rodriguez. Catalytic properties of molybdenum carbide, nitride and phosphide: A theoretical study. *Catal. Lett.* **91**, 2003, 247–252.
34. M. Neurock. First-principles analysis of the hydrogenation of carbon monoxide over palladium. *Top. Catal.* **9**, 1999, 135–152.
35. J. G. Nunan, C. E. Bogdan, K. Klier, K. J. Smith, C. Young and R. G. Herman. Higher alcohol and oxygenate synthesis over cesium-doped Cu/ZnO catalysts. *J. Catal.* **116**, 1989, 195–221.

36. I. N. Remediakis, F. Abild-Pedersen and J. K. Norskov. DFT study of formaldehyde and methanol synthesis from CO and H_2 on Ni(111). *J. Phys. Chem. B* **108**, 2004, 14535–14540.

37. J. Greeley, A. A. Gokhale, J. Kreuser, J. A. Dumesic, H. Topsoe, N. Y. Topsoe and M. Mavrikakis. CO vibrational frequencies on methanol synthesis catalysts: A DFT study. *J. Catal.* **213**, 2003, 63–72.

38. R. Hoffmann. A chemical and theoretical way to look at bonding on surfaces. *Rev. Mod. Phys.* **60**, 1988, 601–628.

39. D. Zeroka and R. Hoffmann. Adsorption of methoxy on Cu(100). *Langmuir* **2**, 1986, 553–558.

40. S. F. Zaman and K. J. Smith. A study of K promoted MoP-SiO$_2$ catalyst for syngas conversion. *Appl. Catal.* **378**, 2010, 59–68.

21 Glass Simulations in the Nuclear Industry

Shyam Vyas, Scott L. Owens, and Mark Bankhead

CONTENTS

21.1 INTRODUCTION

In recent years, there has been a renewed interest in the use of nuclear power to meet the world's energy needs; this stems from concerns over climate change, energy security, and a forecasted increase in the global demand for electricity [1]. The generation of electricity from nuclear power therefore offers a relatively cheap, reliable, and stable method to meet these energy needs, with minimal carbon emissions.

The primary concern for the broader public regarding nuclear power has been the disposal of radioactive waste materials. Such materials remain radioactive for thousands of years and are often chemically toxic. Most of the highly active waste generated comes from the reprocessing of spent nuclear fuel rods, which is done to reduce the volume of waste generated and to potentially recycle the spent fuel. This is due to the fact that 97% of the fuel is unreacted at the point where most reactor types are unable to continue burning the fuel efficiently. In the current generation of civilian reprocessing plants (in the U.K., France, and Japan), fuel rods are dissolved in nitric acid, and the uranium and plutonium are separated from the fission products using a solvent extraction process (the PUREX process). The uranium and plutonium are initially partitioned into the organic phase, whereas the fission products remain in the aqueous phase—termed the aqueous raffinate or liquor [2]. The volume of liquor generated per tonne of fuel processed is quite small but accounts for over 90% of the radioactivity of all the waste generated by the civil nuclear industry.

Vitrification is used worldwide as a means of immobilizing the highly active liquor (HAL) into a nonleachable and physically stable monolith for long-term repository or surface interim storage [3]. In some instances, vitrification is also used to immobilize less active streams due to its ability to deal with a broad composition range compared to other waste forms. The process used as a typical example here is derived from the UK's HAL immobilization process; however, many of the features described for the process are common to all radioactive vitrification processes used worldwide. During the process operated at the Sellafield Waste Vitrification Plant, the highly active nitrate liquors are dried and then calcined within a kiln to give an oxide powder. This is then heated with an alkali borosilicate glass powder in a melter vessel (described later). The glass melt is poured into a container where rapid cooling leads to a mechanically stable and chemically durable product. While borosilicate glass is primarily used in the U.K., other related glass compositions have been shown to offer advantages for certain waste types in other countries.

The melting and pouring process appears superficially simple, in that the glass powder is added to an Inconel melter and heated to melt it. Then further glass powder and waste oxide powder are added. Depending on the type of melter used, mixing is achieved by (or a combination of) buoyancy-driven convection, mechanical mixing, or air sparge.

There are a number of specific problems in the operation of this process, including melter corrosion, stresses in the melter imparted by heating and cooling, and the settling of the less soluble fission products—resulting in shorter melter lifetimes and lower product quality. Operating experience shows that plant performance and lifetime is strongly dependent on the rheology of glass melts as a function of feed composition. In particular, little is understood with regard to the effect of solid particles within the melts, or their phase equilibria, which may result in the formation of precipitates or "heels." A predictive knowledge of the glass material properties, particularly the melt rheology, thermal conductivity, and density yielded by a certain glass composition, is highly desirable to understand the optimum conditions for plant operation.

Overlaid upon this is the drive to improve the efficiency of reprocessing operations to deal with legacy wastes arising from the current generation of nuclear power stations and new build. Planning is underway to modify existing plants such as the Thermal Oxide Reprocessing Plant in the U.K. to deal with future wastes arising from the postoperational clean out of legacy facilities. Therefore, future vitrification processes will have to incorporate a range of feed compositions due to the chemical variability of the wastes. Consequently, a great deal of research is being done in understanding the scientific and engineering characteristics of vitrification glasses and their melts.

Materials modeling can provide a technological solution to the operational challenges associated with waste vitrification. Experience within the nuclear industry has shown that this cannot be practically solved by experimentation alone. Thus, the nuclear industry has devoted significant resources to modeling and simulation. These simulations reduce the number of comparatively expensive experimental programs required. They can also lead to a stronger technical underpinning of the processes and properties, and how these change with composition. Experiments alone would be unable to do this in a reasonable time frame or at an acceptable cost. Thus, simulations are an integral part of most technical programs within the industry. As such, there are numerous scientists within the industry doing a range of different types of simulations, including computational chemistry and materials science, engineering simulations, nuclear physics, environmental modeling, and statistical analysis. As a result of this, the industry has developed the capability to create very complex models of systems that can combine and integrate many of these subdisciplines.

An example of such a multidisciplinary and multiscale approach is the simulation of glass production. Techniques range from thermal studies of heat and mass redistribution within a glass melter [4] to electronic structure calculations of small silicate clusters [5,6]. Much of the modeling development effort, however, is on understanding the rheological properties and stability of these materials. The current chapter presents three previously published studies [7–9] using atomistic and particle-based methods to study the fundamental properties of glass materials. Section 21.2 covers the application of atomistic simulations and mesoscale simulations to study the rheology and thermal properties of glass melts [8,9] in support of engineering simulations. Section 21.3 is a molecular modeling study of glass surfaces to understand the stability of glass formulations [7].

The purpose of these examples is twofold: (1) to provide the reader an overview of the modeling capabilities in the broader nuclear industry and specifically at the U.K.'s National Nuclear Laboratory (NNL) and (2) to provide protocols and recipes that can be used with "out-of-the-box" computational chemistry tools to aid the work of industrial material modelers when studying glasses that have their own unique issues.

21.1.1 SIMULATIONS IN SUPPORT OF PLANT OPERATIONS

It is important to recognize at this point the overall modeling framework in which these developments are applied. An understanding of atomic- and particle-scale processes is most useful when it can be directly linked to problems and processes observed on plant and can be used to make improvements. In the case of highly

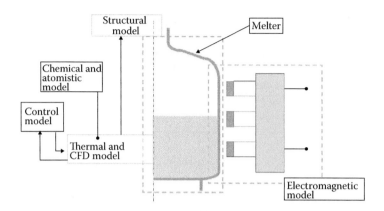

FIGURE 21.1   Links between the modeling approaches.

active waste vitrification, even small improvements in durability of plant or process throughput can lead to savings of millions of dollars over the lifetime of the plant.

A good example of such an application is in studying the lifetime of a vitrification melter. The lifetime of these melters is determined by corrosion—which is known to be significantly enhanced by thermal stress in the melter material. The thermal stresses arise as a result of temperature gradients through the wall material, and this is crucially influenced by movement of the glass melt within the melter.

The heart of any melter simulation will be an engineering model (Figure 21.1). This focuses on the fluid dynamics behavior of a vitrification melt. Simulations of the thermal and mass transport within the melter present a quantitative picture of the dynamic temperature profile of the glass. The boundary conditions arising in these computational fluid dynamics (CFD) simulations act as an input to finite element (FE) simulations of material stresses developed within the melter material—which in turn can be used to advise on how to input heat to the melt and melter to reduce the thermal gradients that cause stress.

Figure 21.2 shows the temperature, heat generated, and thermal stresses, in a previous design of melter that led to significant degradation in melter lifetime [8]. It

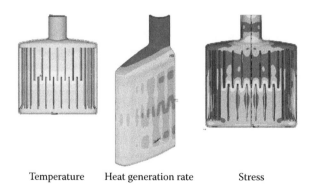

Temperature   Heat generation rate   Stress

FIGURE 21.2   FE models showing relationship between stress and thermal properties.

is also interesting to note that the combination of CFD and properties derived from atomistic methods helped to dispel the premise that heat transfer within these melters was conduction-driven—and led to the removal of the fins seen in the diagram that were the major site for stress-led corrosion.

To develop credible engineering models of vitrification melt movement and heat transfer, the thermal and rheological properties of the melt with composition must be quantitatively understood. Much of the following discussion centers around the need to provide good properties and understanding for such models.

21.1.2 ATOMISTIC AND OTHER PARTICLE-BASED SIMULATIONS OF GLASS MELTS

The previous section has described how engineering simulations are valuable in improving plant operations. To produce accurate results, several composition-based properties are required to model the melt—its density, specific heat capacity, and dissolution thermodynamics—but the most important properties are the viscosity and thermal conductivity.

However, there are two problems in obtaining the required values for these properties to be used in CFD simulations. First, the properties (particularly viscosity) depend on material structure at a number of different length scales (from atomic structure to micron-sized particles), and these are highly dependent on the local melt composition. Measuring the viscosity is impractical, as the wide composition range (containing much of the periodic table) and prohibitively large cost of many components (e.g., Ru, Rh, Pt, and Pd) would require an extremely expensive experimental program. In addition, the nonlinear dependence of viscosity on composition would not guarantee to give better than an indicative viscosity for a complex mixture at a given temperature. The process is operated at a temperature that is low enough to reduce the volatility of the components in the glass melt; however, this means that the consequences of a high viscosity component could be catastrophic to the process. For example, a 5% addition of aluminum oxide to the melt would result in a viscosity increase of three orders of magnitude—effectively removing any fluid mixing.

Thus, while simulating glass melt properties is difficult (due to the complex composition and multiscale structure of the melt), the payback is large: reducing melter downtime and lengthening its operational lifetime. Molecular mechanics and particle-based techniques can be tied into fluid dynamics and stress calculations of the melt and the melter, so using the strengths of each technique. This results in a unified and consistent simulation approach to the description of the process. This is in opposition to many modeling developments, where researchers attempt to solve multiscale problems with one approach—we attempt to use a number of techniques, playing to their strengths, and integrate the results and approaches as much as possible.

21.2 PROPERTY PREDICTION OF GLASS MELTS USING ATOMISTIC AND PARTICLE SIMULATIONS

In the present section, we describe how atomistic simulations can be used to calculate a number of important properties that are used to support CFD and other engineering scale simulations [8]. All of the properties being simulated are using

TABLE 21.1
Glass Composition Used in the Study

Glass Component	mol%
SiO_2	60–65
B_2O_3	15–20
Li_2O	10
Na_2O	10

techniques that have not previously been applied to silicate glasses, so a test system was chosen to allow the modeling techniques to be calibrated against known experimental values. The composition is shown in Table 21.1, which was chosen for its similarity to the base glass composition used to form the melt matrix. It was in some cases difficult to get literature values for exactly the same composition quoted above; however, the values quoted are for those that are structurally as close as possible.

21.2.1 BUILDING AMORPHOUS BULK MATERIALS

Glass is an amorphous meta-stable solid and is formed by removing heat rapidly from a melt. This "freezes" the position of the atoms in a liquid-like arrangement, as they have had insufficient time to form any long-range order. On an atomistic level, it consists of a framework (typically silicate) in which the non-framework ions (e.g. sodium and calcium) are dispersed. Thus, the first phase of any atomistic property calculation of bulk glasses is to accurately represent the atomic structure of the bulk glass. Typically, such structures are built by using a simulated annealing scheme; this allows the ions to thoroughly explore conformational space at high temperatures, and as the temperature is reduced, they come to a local equilibrium. The procedure consists of two stages: the creation of a periodic cell containing the glass ions and annealing the system. Initially, one must first select the fragments to be used as building blocks to create an amorphous cell of the material, for example, in a pure silicate glass, the fragments would be SiO_2 units. Occasionally, Si_2O_4 fragments can also be included to produce a reasonable starting distribution of both 2- and 4-coordinate silicon atoms that can bring the system to equilibrium in a shorter time frame. The fragments are then packed into a periodic cell at an appropriate density using a Monte Carlo algorithm [10]. A cell containing between 700 and 1000 atoms is built using the appropriate fragment concentration, within a three-dimensional (3D) periodic cell at a density of about 2.0 g/cm^{-3}. Using these parameters, the periodic cell would be approximately 25 × 25 × 25 Å in size, which is a sufficient size to ensure that atoms within the cell will not interact with their periodic images in adjacent cells during the annealing process. Such an approach has become standard when studying amorphous polymers.

Once the cell is built, a quenched molecular dynamics approach is used to optimize the positions of the atoms. This method involves running NVT molecular dynamics (constant number, volume, and temperature) at 7000 K for 20 ps, followed by another 20 ps of NPT (constant number, pressure, and temperature) molecular dynamics. The purpose of this initial two-stage process is to randomize the structure at an atomic scale followed by relaxing the unit cell structure during the NPT phase. The system is brought to equilibrium by gradually reducing the temperature and re-equilibrating for 20 ps, after each temperature reduction, until it reaches 300 K. This allows the ions to thoroughly explore conformational space at the higher temperatures, and as their energy is gradually reduced, they come to low-energy metastable positions in the glass. In the current chapter, the calculations were performed using the Discover molecular mechanics code [11]. This atomic representation of the bulk acts as a starting point for the property calculations (e.g., the viscosity).

It is important to note that this procedure is considerably faster than would be possible in making a real glass, and so the fictive temperature of a simulated glass will always be higher than for a real glass of the same composition. The consequence of this is that a simulated glass will have a structure that is topologically similar to a liquid of a considerably higher temperature than a real glass.

21.2.2 VISCOSITY

Viscosity is one of the principal physicochemical characteristics of melts. As such, it is one of the key properties for understanding the behavior of many chemical engineering and geological processes. Applications where viscosity is an important parameter include geology [12] and the vitrification of highly active nuclear waste [13]. Industry requires mechanistic models to predict viscosities, both from an operational sense and as a means of accelerating the design of new processes. Mechanistic models of glass viscosity models have been reviewed extensively in the literature [14]. The simplest empirical models for silicate glass viscosities are based on a first-order Arrhenius equation with constant activation energy of viscosity [15]. These equations offer a reasonable representation of viscosity trends in the liquid phase but poorly describe the behavior of viscosity at intermediate temperatures between the strain point and the melting point. Improvements to this basic model include the Vogel, Tamman, and Fulcher (VTF) equation (Equation 21.1); this method includes a modifier term to the basic exponential equation to better describe the viscosities over these intermediate temperatures [16]. Several theories have been developed for describing the dependence of the terms in these empirical equations on the underlying phase composition and molecular structure of a glass. These include cooperative motion, percolation, kinetic constraint, and ordering [17,18].

The VTF relationship is described as

$$\log \eta = A + \left(\frac{B_{act}}{19.15\left(T - T_g\right)} \right) \tag{21.1}$$

where η is the intrinsic viscosity and T is the temperature. T_g is the glass transition temperature, and A and B_{act} are fitted constants. The physical basis is an Arrhenius scaling of the property with temperature, and it seems to be a reasonable approximation for any simple glass. The problem is that one must have sufficient data for the composition of interest to fit such an equation.

TABLE 21.2
Comparison of Viscosity Values

	log(η/Pa·s)	Reference
Literature (at 1000°C)[a]	2.17	[19]
Internal NNL (at 1050°C)	1.08	
Lakatos analytical model	1.62	[21]
Priven analytical model	1.22	[20]

[a] For mol%: SiO_2—64.22, B_2O_3—10.98, Na_2O—9.91, Li_2O—7.3; however, this glass contained a measurable amount of Al_2O_3—which stiffens glass considerably at any temperature.

There are two analytical models that are applied generally to this type of glass composition: the methods of Priven and Lakatos et al. [20,21]. These give viscosities for the composition and temperature required (1050°C); the results of these approaches can be seen in Table 21.2. For comparison, two experimentally measured viscosity values are also included.

As can be seen, the literature values and the derived analytical values vary widely. As a note of caution about analytical models of viscosity, generally, a well-known phenomenon for alkali glasses is the "mixed alkali effect." This is where glasses containing a mixture of two alkali oxides, in equimolar quantities, show markedly lower viscosities than would be expected from extrapolating from the single alkali or the mixed alkali away from equimolar quantities. This has been explained in terms of alkali partitioning in the glass by Greaves [22,23] and Vessal et al. [24,25]. The composition chosen takes advantage of this effect, so it should be noted that our simulations of this composition are aiming to show lower values of viscosity than analytical models should give.

21.2.2.1   Confined Shear Calculations

Molecular modeling has been used to calculate viscosity using a variety of different well-developed approaches, including velocity autocorrelation [26], kinetic energy exchange [27], and simulating Poiseuille or Couette flow (also called the confined shear method) [28–34].

The method discussed here is the confined shear method. In this method, a large thermally equilibrated periodic model of the required glass composition was constructed using the methods described previously. This model was then divided into three regions along the *C* crystal axis (Figure 21.3), consisting of a central region and two walls.

The bulk of the material is in the central region—the region that undergoes simulated shear. There are two thinner regions at the top and bottom of the model (shown in black) representing the confining walls (or the plates causing the shear). The walls were constructed with the same composition as the glass, though they could have been constructed of any material. The atoms in the wall material were tethered in place and were not allowed to translate significantly from their original position—they were allowed to only vibrate about it. During the simulations, the walls are

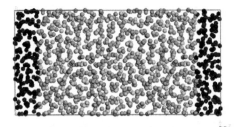

FIGURE 21.3   (See color insert.) Model of glass for confined shear. Oxygen atoms—red, silicon—yellow, boron—brown, the alkali ions—purple (lithium smaller than sodium).

moved in opposite directions to each other, thus subjecting the atoms in the central region to a shear force. The two walls were then sheared at realistic shear rates (of the order of 0.5 ms^{-1}–5 ms^{-1}).

The confined shear simulation was undertaken using the Materials Studio® Discover molecular mechanics package [11]—using a modification of the cvff_aug force field [35–37]. This is a nonbonded potential set—allowing free movement of ions. The atom types used were originally intended for use in describing clay structures; however, they also seem to predict glass formation well, possibly as the clay structures are more "open" than other silicate and aluminosilicate structures.

Following a sheared molecular dynamics simulation at the required temperature (including an initial period of equilibration), a graph of applied shear stress (i.e. the wall shear) vs. glass strain (i.e. the translation of the atoms in the bulk liquid) was constructed. From this, the viscosity was calculated. Figure 21.4 shows a snapshot from later in the molecular dynamics simulation. The image on the left shows a periodic cell containing the alkali borosilicate composition specified in Table 21.1. The image on the right shows the sheared system (with the original periodic boundary left for reference). In addition to the calculation yielding the property value, a number of features of the material under shear become evident. One obvious observation is that the alkali ions have displaced much faster than the other atoms and so are having a large effect in decreasing the viscosity (as the viscosity is inversely

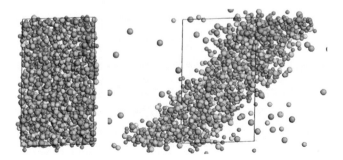

FIGURE 21.4   (See color insert.) Shearing of borosilicate glass using confined shear method. Oxygen atoms—red, silicon—yellow, boron—brown, the alkali ions—purple (lithium smaller than sodium).

proportional to the average mobility of the atoms). This is a well-known observation, but it is interesting to see the simulation reproduce this effect.

The viscosity calculated by the confined shear simulation is 1.11 Pa·s at 1050°C. This agrees very well with the value measured previously for this composition within the (NNL). It also seems to have accounted for the "mixed alkali effect," which could not be modeled using the simpler analytical methods. Comparisons at other temperatures and with other silicate, borosilicate, and aluminosilicate melts showed similar agreement.

21.2.3   THERMAL PROPERTIES

The thermal properties of a glass melt are very difficult to simulate or measure experimentally. This is because the overall heat transfer behavior is a combination of all three possible mechanisms: conduction, convection, and radiation (and in a glass melt, all can be significant). Conduction and convection are fairly straightforward to model at all scales; however, the radiative heat transfer mechanism is very much harder to describe. This is exacerbated (or perhaps simplified—see later discussion) in the case of waste loaded melts, as the transparency of the glass melt reduces and the absorption characteristics change.

No peer-reviewed literature values have been quoted for thermal conductivity for glasses with compositions that are close to the required composition and none for elevated temperature. This creates a problem, as there are no reliable reference values to compare the simulation answers to.

A simple analytical model based on molar concentration has been developed by Choudhary [38]. This gives a calculated thermal conductivity at 300 K of 0.91 W m^{-1} K^{-1}. This value does not contain any temperature scaling, which limits its usefulness in calculating the value for the melt.

Searching product specifications on the internet (from glass manufacturers and suppliers websites), a number of values for thermal conductivity for alkali borosilicate glasses have been quoted. It seems that below the glass transition temperature, values for thermal conductivity are quoted around 1.0 ± 0.25 W m^{-1} K^{-1} at around room temperature, increasing to a value of about 2.5 ± 0.25 W m^{-1} K^{-1} at about 1400 K. These values seem reasonably insensitive to composition and so may be used as markers for our glass thermal conductivity.

Above the glass transition temperature, a much wider variation in conductivity with composition is expected—due to different compositions having different transition temperatures and due to the domination of fluid thermal conductivity. For the fluid dynamics model, it is important to decouple all modes of heat transfer (to ensure that convection heat transfer is not included in the simulation twice).

The estimation of thermal properties from atomic level calculations is difficult because there is no suitable way of modeling the radiative component for heat transfer. However, as the glass is loaded with waste, this component becomes less important—because the mean path length of radiation from emission to absorption becomes shorter as the transparency of the glass decreases. Hence, in glass–waste oxide mixtures, atomistic simulations of thermal diffusivity will be more applicable as conduction will be the dominant mechanism.

Atomistic calculations can be used to simulate the local (Å scale) conduction components of heat transfer, and these are both captured within the thermal diffusivity of the species within a glass melt.

Thermal conductivity can be related to the thermal diffusivity of a material through the following relation:

$$K = D_T \rho C_P \tag{21.2}$$

where K is the thermal conductivity (diffusive component), D_T is the thermal diffusivity, ρ is the density, and C_P is the specific heat capacity at constant pressure. All of these values are known or can be calculated reasonably simply for a given glass composition, except for the thermal diffusivity. The materials properties for the compositions given for our test glass can be obtained easily from similar systems in the literature and are given by

Density—2.31×10^{-3} kg m^{-3}
C_P—1914 J kg^{-1} K^{-1}

The thermal diffusivity can be simulated at an atomic level using the method described below. This involves taking a cell containing the glass composition and dividing it into two. One-half of the cell is equilibrated at one temperature, and the other half is equilibrated at another temperature using constant volume adiabatic dynamics (shown in Figure 21.5). Then the two halves of the cell are allowed to come into thermal contact and to reach thermal equilibrium. The thermal diffusivity can be calculated from the rate at which the temperature of the system comes to equilibrium.

The simulated value for thermal diffusivity for this glass composition at 1050°C was found to be 0.615 m^2 s^{-1}. By substituting this in Equation 21.2 and using the literature values for density and heat capacity, the derived thermal conductivity value is 2.72 W m^{-1} K^{-1}.

This value is a little higher than seen experimentally—however, the value simulated for thermal diffusivity is averaged over the time period of the simulation—and the most accurate answer is obtained when there are higher thermal gradients (at the beginning of the simulation). Thus, it may be better in the future to either use larger temperature differences or possibly use larger simulation boxes.

FIGURE 21.5   Two linked simulation boxes, equilibrated at different temperatures.

The radiative heat transfer mechanism has been discussed in a number of papers (for a good summary, see Kiyohashi et al. [39]). As a general statement, the radiative component within a glass (i.e., from one point in a glass to another) is thought to be very difficult to estimate—but is not considered to be of significance. The major radiative component is usually considered to be from the walls of the melter vessel. It will be shown that this is not usually a significant component (for most melter metals) until about 1000 K. Again, the review [39] shows a graph containing the calculated correction to the thermal conductivity due to radiative heat transfer. The radiative heat transfer component can be calculated for a given experimental setup (the component is obviously sensitive to the detector arrangement and the wall distance, among other things). The radiative heat transfer contribution for the glass composition mentioned in that paper is approximately 5% of the thermal conductivity at the melt temperature.

21.2.4 PARTICLE-BASED SIMULATIONS FOR VISCOSITY CALCULATIONS

So far, we have discussed property calculations using molecular dynamics. The principal limitation of molecular dynamics as a tool to model the viscosities of complex silicate glasses is computational. Modern parallel computers can routinely solve simulations containing several million atoms over short timescales (less than a microsecond) and at a reasonable computational cost. However, for molecular dynamics to be used to explore very complex silicate melts on longer timescales (e.g., engineering or geological applications), or to study important features such as solid-phase inclusions within a melt, other methods must be used. Given the importance of modeling the viscosity of glasses, another approach is the use of particle-based methods, which, instead of simulating individual atoms or a pure continuum, use a collection of particles to represent the system.

Recent advances in simulation techniques have led to the development of meso-scale modeling approaches including smooth particle applied mechanics [40] and dissipative particle dynamics (DPD) [41]. These related techniques involve the simulation of a set of particles carrying the properties of the underlying material using efficient computational algorithms. They can be used to simulate length and timescales on a par with FE methods. However, they retain features common to molecular dynamics, particularly with regard to the relationship between the materials properties in terms of interparticle forces that can be easily expressed in terms of individual particle chemistries. DPD has gathered considerable interest as a tool to model fluid flow and has been extensively used in the studies of organic materials [42]. As will be discussed, the technique could potentially represent many of the features present in theories of silicate glass viscosity. In this section, we explore the application of DPD to model a simple silicate glass to assess its relative strengths, weaknesses, and applicability.

21.2.4.1 Background to the DPD Approach

The form of DPD applied in this study is where the equations of motion for an ensemble of beads correspond to Langevin-Brownian dynamics with pair-wise central forces [43,44]. DPD conserves momentum which is essential for recovering the correct hydrodynamic behavior at sufficiently large length and timescales. In contrast with molecular dynamics, a DPD model describes a reduced number of degrees of

freedom by considering a simple description of matter, the most significant difference being the absence of an explicit description of the molecular structure in a DPD
simulation. In this coarse-grained description of matter, the chemistry imparted by
the underlying molecular structure has been included implicitly through the internal
dissipative, stochastic, and conservative forces acting on the bead. Each bead then
represents, on a molecular scale, a cluster of fluid atoms. The beads can represent different chemical species or phases, and a distribution of different beads can represent
a complex mixture. For DPD beads positioned at $\dot{r}_i$, with velocities $\dot{v}_i$ and momenta
$\dot{p}_i$, the equations of motion can be written for any pair of beads, i and j, where

$$\hat{r}_{ij} = \left| \dot{r}_i \cdot \dot{r}_j \right| \text{ and } \hat{v}_{ij} = \dot{v}_i - \dot{v}_j \tag{21.3}$$

$$\dot{p}_i = \sum_{j \neq i} \left(\mathbf{F}_{ij}^C + \mathbf{F}_{ij}^D + \mathbf{F}_{ij}^R \right) \tag{21.4}$$

$$\mathbf{F}_{ij}^C = a_{ij} \omega^C \hat{r}_{ij} \tag{21.5}$$

$$\mathbf{F}_{ij}^R = \sigma_{ij} \, \omega^R \, \xi \, \hat{r}_{ij} \tag{21.6}$$

$$\mathbf{F}_{ij}^D = -\zeta_{ij} \, \omega^D \left(\mathbf{v}_{ij} \cdot \hat{r}_{ij} \right) \hat{r}_{ij} \tag{21.7}$$

The conservative force (Equation 21.5) between the beads accounts for chemical
interactions in the DPD model system and is described by a repulsive soft potential
in terms of an interaction parameter a_{ij} and a weight function ω^C. The weight function is a simple linear ramp, giving rise to very soft repulsive forces between the
particles. Varying the strength of a_{ij} changes the response of the fluid to an applied
stress and spatial arrangement of different bead types.

The stochastic force (Equation 21.6) represents the movement arising from
molecular collisions caused by random thermal motion and, as a result, is related to
the temperature in the system. The dissipative force (Equation 21.7) represents the
viscous drag introduced by the underlying molecular structure on the bead. Together,
the stochastic and dissipative force terms introduce Brownian motion. In order for
DPD to have the statistical mechanics of the canonical ensemble, the weight constants in the dissipative and conservative force terms are related by $\omega^D = (\omega^R)^2$ and
the constants by $\sigma_{ij}^2 = 2\zeta_{ij}k_{\mathrm{B}}T$. This relationship acts as a system thermostat controlling the kinetic energy due to random thermal motion. Additional thermostats may
be required to control additional kinetic energy due to the application of external
forces (nonequilibrium conditions). The force constant ζ_{ij} controls the strength of the
dissipative and random forces in the simulation and hence the underlying transport
properties of the DPD model fluid by controlling its time response to an applied
strain rate or pressure.

The thermodynamic expression for the Helmholtz free energy of mixing of a
DPD binary fluid is given by Equation 21.8. The first term on the right-hand side of

this expression represents the entropy of mixing for an ideal gas. The second part of the expression represents the enthalpy of mixing. This expression forms the basis of empirical parameterization methods for DPD simulation of organic polymers from Flory–Huggins theory [45] or Regular Solution Theory (RST) [46]

$$\frac{A^M}{n_T} = RT \sum_i x_i \ln \phi_i - r_c^4 \nu N_A^2 \alpha \phi_i \phi_j \left[\frac{a_{ii}}{v_i^2} - 2\frac{a_{ij}}{v_i v_j} + \frac{a_{ii}}{v_{jj}^2} \right] \tag{21.8}$$

The radius of a DPD particle r_c is defined in terms of the molar volume v_i and the reduced number density $\bar{\rho}_i$ (Equation 21.9). The units of a DPD simulation are re-expressed in a reduced form for numerical considerations. Reduced units are defined with r_c set to 1 and the energy $\varepsilon = k_B T = 1$. For example, the expression for the reduced number density is given by

$$\bar{\rho}_i = \frac{r_c^3}{v_i} \tag{21.9}$$

This in turn leads to the specification of reduced units for the force constants that control the behavior of a distribution of DPD particles. Reduced units are noted with a bar-overstrike. This defines both the length scale and the degree of discretization of the DPD material in the model.

By varying the model parameters $\bar{a}$, $\bar{\xi}$, and $\bar{\rho}$, one can describe an arbitrary equation of state allowing DPD to be applied to the modeling of any material. By applying a set of thermodynamic laws to these equations, the method can be applied to solve quantitative problems. In determining the value of the strength of the conservative forces ($\bar{a}$) and number densities ($\bar{\rho}$), we have followed the method first described by Travis et al. [46], derived from regular solution theory, adapted for glasses, and described fully in Ref. [9].

The selection of the value of the dissipative force constant $\bar{\varsigma}_{ij}$ has been discussed by various authors, for example, Evans [47]. There is currently no generally satisfactory mechanistic approach to derive the value of this parameter from the absolute transport properties of the material. The preferred method is to effectively match the experimentally measured transport properties of the fluid against a specific value of $\bar{\varsigma}_{ij}$. Hence, the approach to be taken in this study will be to carry out a series of numerical simulations where, for a constant set of conservative forces and number densities, the value of $\bar{\varsigma}_{ij}$ for the fluid will be varied so that a calibration curve can be calculated. Using this curve, a value of $\bar{\varsigma}_{ij}$ can then be selected that allows the transport properties of the fluid to be set, allowing comparison with experiment and taking into account of numerical accuracy and computational considerations.

21.2.4.2 Calculating Viscosity by Simulating Poiseuille Flow with Dissipative Particle Dynamics

A silica melt was chosen as the benchmark system, as it is the principle base component of many inorganic glasses; producing a validated DPD model of this glass is

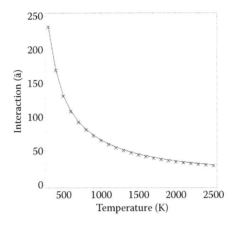

FIGURE 21.6 Plot of the conservative interaction parameters for SiO_2.

essential before more complex glasses can be modeled. Additionally, the viscosity of a silicate glass shows Arrhenius behavior, and as a result, a thermodynamic model based on the simple method described can be expected to reproduce this behavior with a reasonable level of accuracy. The model parameters $\bar{a}$, $\bar{\xi}$, and $\bar{\rho}$ were determined from the methods described in the references. The $\bar{a}$ and $\bar{\rho}$ parameters for silica are plotted as functions of temperature in Figures 21.6 and 21.7. A value of $\bar{\xi} = 36.0$ was chosen based on a match of the transport properties of the DPD model to molten silica and taking into account computational issues [9].

As with an atomistic study, the approach taken to calculate viscosity with DPD is to simulate Poiseuille flow between two infinite parallel plates, to measure the velocity of the fluid within the plates, and then to apply the Navier–Stokes method to calculate the fluid's viscosity from the velocity. DPD simulations were solved using an Open MPI compiled version of the Sandia National Laboratory LAMMPS code

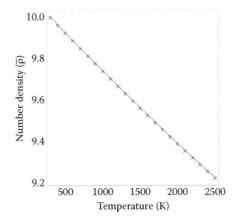

FIGURE 21.7 Variation of the DPD number density with temperature (for SiO_2 beads).

[48]. Simulations were run using a 240-CPU cluster, typically over 24–32 cores per run. A model was created by dividing an oblique simulation cell into three parallel domains to create a starting geometry. Two regions, each 1/5 of the cell z-axis wide, represented the walls. The lengths of the cell sides were in the ratio $2 \times 2 \times 1$. A method was implemented within a computer program whereby a physical domain is divided into a number of voxels of definable dimensions. Each voxel is filled with random uniform distributions of DPD beads to define a reasonable starting condition retaining the layered geometry. The number of DPD beads to introduce into each voxel is the number or mole fraction of each component. Values of the bead number density, $\rho r_c^3 > 7$, represent a good balance between the error introduced by using the α approximation of 0.101 against computational efficiency. The walls are represented as a set of frozen DPD particles for which the initial force and velocities are set to zero and which are excluded from the integration of the equations of motion. A force was applied to the z direction to induce motion into the fluid. The magnitude of this force can be varied to simulate different flow conditions.

The fluid phase was simulated by creating an initial velocity distribution in the set of DPD particles representing the fluid, applying an NVE ensemble, and solving the model measuring dynamic and structural properties over an appropriate timescale. The time step used during the gathering of statistics was 0.0005 DPD units, with a time step of 0.01 DPD units used during the equilibration period. The short time step is required during data gathering runs due to the limitation of the DPD thermostat to control the temperature in wall-bounded systems (e.g., see Ref. [49]). Sensitivity runs with the time step confirmed the convergence of the system pressure and average temperature to ±1 and ±0.1, respectively, with these parameters.

Each DPD simulation was carried out according to the following basic recipe. The starting conditions were determined as described. The regular solution [9] approach was used to determine the input parameters for the simulation $\bar{a}_{ii}$ and $\bar{\rho}_i$ and shown in Figures 21.6 and 21.7. The wall boundary conditions were defined by setting the wall beads as representing a subcooled liquid that is immiscible with the model silica fluid. This was achieved by defining the walls as representing silica at 300 K, with bead number density set to be eight. The fluid interactions were set relative to this value to allow direct computation of trends from the numerical results in reduced units. The cross term $\bar{a}_{ij}$ was calculated via Equation 21.8, with respect to the properties of the given fluid. Similar boundary conditions have been tested by other authors with respect to generic DPD fluids (e.g., see Refs. [49] and [50]). It is noted that a no-slip boundary condition was not specifically imposed as has been introduced by other authors [51]. A validation of the wall boundary conditions was carried out to ensure the local density (and hence the thermodynamic properties of the fluid) and temperature profile across the channel (an indication of local temperature control).

Simulations consisted of an initial equilibration period of 4000 DPD time steps where the fluid phase is unconstrained and where the wall beads were constrained in the x–y plane. Following the initial equilibration period, a large shearing gravitational force, $\bar{g}_z = 0.1$, was applied to the system for 20,000 time steps; the system was again relaxed for 4000 time steps; finally, a gravitational force was applied to introduce flow into the fluid. The velocity of the fluid phase beads was measured over 400,000 time steps. The velocity within the confined fluid was measured by a spatial averaging method. The fluid was divided into a series of bins running in the X–Y plane parallel to the walls. The velocity and number of particles within each bin was cumulatively averaged over the simulation to avoid the development of a systematic error in the velocity [52]. The cumulative average of the velocity was allowed to develop over 10,000 time steps until a constant value was obtained. The standard deviation in the measured velocities was recorded to provide a direct measure of the error. The resulting velocities were fitted to the theoretical Navier–Stokes quadratic equation for flow in a confined channel. The maximum streaming velocity from the peak of the parabolic velocity profile $\langle \bar{v}_z \rangle$ was used to calculate the viscosity $\bar{\eta}$ using Equation 21.10 [53–55], where D is the width of the fluid phase between the walls, $\bar{\rho}$ is the number density of the fluid phase, and the applied force is $\bar{g}_z$. The shear stress is related to $\bar{g}_z$ by Equation 21.11.

$$\langle \bar{v}_z \rangle = \frac{\bar{\rho}\bar{g}_z D^2}{12\bar{\eta}}$$ (21.10)

$$\bar{\tau}_{xz} = \bar{\rho}\left(x - \tfrac{1}{2}D\right)\bar{g}_z$$ (21.11)

In the following sections, a series of results benchmarking the silica DPD fluid properties will be presented. First, the value of the dissipative force term $\bar{\varsigma}_{ij}$ was determined by generating an appropriate calibration curve. Second, the fluid flow regime was determined by calculating the viscosity as a function of applied force, $\bar{g}_z$. Finally, the trend in viscosity as a function of temperature was calculated and compared directly against experimental trends.

21.2.4.3 Simulations

21.2.4.3.1 Stress/Strain Behavior of the SiO₂ DPD Fluid Model

21.2.4.3.1 Stress/Strain Behavior of the SiO_2 DPD Fluid Model

A series of Poiseuille flow simulations were run to benchmark the applied DPD force $\bar{g}_z$ against the properties of the fluid. The system chosen was silica in a fluidic state, at 1800 K, and at atmospheric pressure. The simulations were run using the standard DPD method, varying the applied force from 0.05 DPD to 0.02 DPD units. A plot of the shear stress $\bar{f}_x$ calculated using Equation 21.10 against the strain rate is shown in Figure 21.8. The linear response to the strain rate shows that the fluid exhibits Newtonian behavior over the range of applied force considered. Further checks of the silicate fluid models at 800 K and 2200 K confirmed that a value of $\bar{g}_z$ of 0.015 represented flow in a Newtonian regime across the temperature range studied. At higher values of the applied force, a nonlinear response may be observed; however, this has not been explored further.

The velocity profiles were measured and fitted to the Poiseuille equation and the viscosity was calculated. The velocity profiles, $\bar{v}_z$ vs. $\bar{z}$, are shown for an applied force $\bar{g}_z = 1.0$, $\bar{g}_z = 0.075$, $\bar{g}_z = 0.04$, and $\bar{g}_z = 0.01$ in Figure 21.9. For the values of $\bar{g}_z$ explored, the velocity profile shows excellent agreement with the Navier–Stokes parabolic profile. At low values of $\bar{g}_z$, a feature possibly characteristic of wall slip

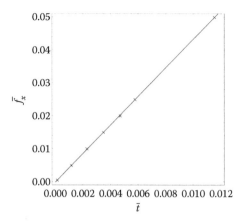

FIGURE 21.8 Shear stress vs. strain rate from model for SiO_2 at 1800 K for an applied force of 0.05 DPD–0.02 DPD units.

is observed. Wall slip has been studied in relation to DPD fluids [56,57], similar to density and temperature fluctuations, and is found to be a function of the strength of the conservative interactions and of the particle number densities.

21.2.4.3.2 Validation of the SiO_2 DPD Fluid Model against Experimental Data
A series of model runs were carried out to fit the DPD calculated silica viscosity against experimental data. The viscosities of silica from 800 K to 2200 K were calculated using the Poiseuille flow method at 200 K intervals. An additional simulation was carried out at the glass transition temperature of 1500 K. The force applied to the fluid was 0.015 DPD units. The dissipative force constant was set to be $\bar{\varsigma}_{ij} = 36.0$.

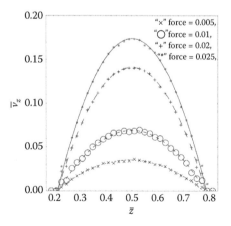

FIGURE 21.9 Velocity profiles perpendicular to the system walls for increasing shear force showing DPD model averages and NS solutions.

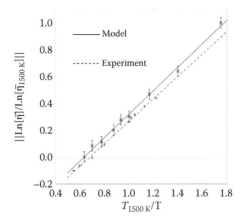

FIGURE 21.10 Normalized plot of SiO_2 viscosity calculated from the DPD model and experimental data from reference (see text).

The DPD viscosity was initially converted to an Arrhenius form, $\ln(\overline{\eta}_{T_g})/\ln(\overline{\eta})$ against T_g/T, where T_g represents the experimental value of the glass transition temperature of 1500 K. A Min–Max normalization was then applied to the viscosity data. A numerical error in the viscosities computed by DPD of 4% was estimated from the magnitude of the deviation in the mean velocity as a function of simulation time step. A first-order Arrhenius fit was then calculated to the trend. A similar normalization and Arrhenius fit was then applied to SiO_2 viscosity data obtained from published sources [58,59]. The resulting data can be compared with the DPD model calculations, as shown in Figure 21.10. The DPD model and the experimental data both show excellent agreement with a first-order Arrhenius model. The difference in the slopes and intercepts of the two trends was found to be less than 3% in both cases.

21.2.5 DISCUSSION AND LIMITATIONS OF MELT PROPERTY CALCULATIONS

As mentioned previously, the driver for undertaking this type of modeling is to optimize the mixing within the melter, to supply temperature boundary conditions to the melter walls for stress simulations, and to explore the relative importance of the different heat transfer mechanisms.

Several discussion points have been raised during the course of this work that relate to the applicability of the results to the real process and particularly to the scale-up through to the engineering modeling methods.

21.2.5.1 Discussion and Limitations of the Atomistic Methods

21.2.5.1.1 Complex Phase Behavior

One issue that is pertinent to the addition of waste oxide to the glass melt is the possibility of unknown phases being formed. In the current glass melt simulations, we have a single, well-mixed phase (as all of the components are soluble in each other).

This is not true once the complex waste oxide has been added to the melt. These have consequences for both the property prediction and for the way they are represented in a fluid dynamics simulation (multiphase fluid models and the interaction between these phases are beyond the scope of this paper). The problem is quite fundamental in that any phase can be included—as long as it is recognized that it is there. Currently, there is no robust way to predict from *ab initio* calculations that a phase will exist under a given set of conditions—one must assume that the phase is possible and include it within any calculations.*

21.2.5.1.2   Parameterization of New Chemical Species in Molecular Simulations

The major issue that arises with the inclusion of waste oxide is that some of the chemical species required in the simulations will not be parameterized for use. Some of these contributions may be included by substitution of chemically similar species, and some may be possible to ignore; however, some will be significant enough that they will need to be included explicitly and, therefore, parameterization will be required.

For molecular mechanics simulations, there are a number of fairly important species for which parameters are not available. To add to this problem, in some cases, the actual chemical nature of these species in glass may be extremely sensitive to the local environment and their bonding complex, making force field parameters difficult to derive.

21.2.5.1.3   Viscosity

There are a number of ways, both directly and indirectly, to predict the viscosity of a homogeneous melt (in addition to the atomic level simulations undertaken here, viscosity can be calculated thermodynamically). However, many of the mixtures encountered in real vitrification melters contain significant amounts of particulate material (and the amount evolves with temperature, time, and composition).

There are two consequences of this. First, the mechanical properties may exhibit highly nonlinear behavior in these inhomogeneous melts. For example, a highly viscous fluid will exhibit considerable "necking" as it forms a droplet (the rate of droplet formation can be directly related to the viscosity). The same fluid containing a nonsticking particulate will neck considerably less before the drop is detached. Second, the particulate may be at a size level that falls into the "mesoscopic" size regime (~µm). This region is not accessible to either CFD or atomistic methods. This then leads to inclusion of other techniques into the loop that can better describe this region, for example, DPD. This is not a problem *per se* and is preferable to operating one of the other techniques outside its "comfort zone." The multiscale approach discussed in this chapter provides a possible solution to this problem.

---

* It is possible to calculate the thermodynamic properties of a given phase from first principles. It is also possible to use Monte Carlo methods to predict the possible phases that might form for a given elemental composition. However, it is not possible to determine whether the kinetics of formation of any given phase will make it likely to form under a given set of conditions.

21.2.5.1.4   Thermal Properties

These properties undoubtedly present the greatest problems for calculation. This is because there is still much uncertainty about the underlying physics of the property and how it can be represented for a glass. One possible advantage presented by real glass/waste oxide mixtures is that the radiative component will become less significant—as most melts are completely opaque and highly absorbing. This then leaves the diffusive component to calculate (as the convection component is contained within the fluid dynamics model).

It is unlikely that all inhomogeneous compositions will have similar thermal conductivities the way that simple glasses do. Thus, the full atomic simulation may be the only way at present for obtaining this value. However, like viscosity, it may be possible to obtain a relation between the thermal diffusivity and some other atomic property of the melt that will make the simulation simpler.

21.2.5.2   Discussion and Limitations of the Particle Methods for Melts

The fluidic behavior observed in the DPD approach taken here shows broad agreement with previously published studies of qualitative and quantitative simulation with DPD. The experimentally determined Newtonian rheology of a silica melt is reasonably well reproduced by the DPD model. This builds a degree of confidence that the approach could be extended to model a broader range of liquid silicate glasses. The study of non-Newtonian fluid behavior, due to different glass chemistries or different applied shear stresses, is certainly within the potential capabilities of DPD. Given the approximations inherent in the thermodynamic model used to obtain the DPD model parameters, one cannot eliminate the possibility that systematic errors in the methods may fortuitously cancel to result in such a good agreement between model and experiment. These errors could easily be explored through a more detailed investigation of the thermodynamic model and by exploration of other molten materials using the same approach. It would also be useful to apply the method to a well-characterized ideal system, for example, a Lennard–Jones solid, where the cohesive forces and viscosity can be independently computed using a more established method such as nonequilibrium molecular dynamics (NEMD). DPD has already been partially validated in this manner [60].

Regular solution theory, in the form used to calculate the conservative parameters, contains a number of assumptions, which would affect any predictions of systems that show significant deviations from regular solution behavior. As has been discussed, Hildebrand's basic theories have been extended over many years to correct most of the deficiencies inherent in the RST approach. Universal methods, such as that proposed by Hansen [61], and covered extensively by Barton [62], could be adapted to enable the modeling of multicomponent silicate glasses. Using multicomponent solubility parameter models is possible via a simple substitution into Equation 21.8. Corrections for a volume change of mixing directly affect the calculation of $\bar{\rho}_i$ in the DPD method and would also affect the strength of the cohesive forces within the fluid. This information can only be obtained from accurate measurements of the partial molar volumes of individual components in the melt. These principles have already been applied to the thermodynamic modeling of silicate melts [63].

The DPD approach discussed in this chapter may be compared against established empirical approaches for modeling viscosity. The principal advantage over empirical viscosity models is that the approach could be relatively easily adapted to study viscous flow in different flow regimes. The modeling of viscous flow in different flow regimes is possible using conventional continuum approaches such as CFD; however, this approach is dependent on an *a priori* knowledge of the fluidic behavior of a glass with a specific chemistry.

Compared with the numerous empirical/semiempirical models, and molecular dynamics simulation, we find that DPD falls between these two approaches in terms of methodology and applicability. Compared to an empirical approach, the methods used in this study offer a simplistic, and in all likelihood less accurate, description of the underlying thermodynamics of the mixture than some empirical viscosity models. However, the DPD method is more flexible in that the thermodynamics can be readily extended and has more practical applications than the simple calculation of viscosities. Measuring the viscosities of glass melts across the broad phase composition range covered by borosilicate melts incorporating wastes is a challenging problem for molecular dynamics due to the computational efficiency of the approach. In contrast, the DPD approach is generally more computationally efficient due to the simpler force terms and the coarse graining, which simplifies the representation of the structural dependence of the chemistry. These factors combine so that the length and timescales that can be explored are orders of magnitude larger than what can be practically modeled using molecular dynamics. Scoping studies carried out in support of the research presented here have supported the view that adding components to the melt that can affect the viscosity shows at least some qualitative agreement with experiment. Addition of "harder" DPD beads, for example, to represent Al_2O_3 increases the viscosity of the melt in a similar way to that observed for lowering the temperature of the fluid.

The major limitation in the DPD approach, compared against molecular dynamics, is the inability to quantitatively predict viscosities. This limitation stems from the current inability to determine a suitable value for $\bar{\varsigma}_{ij}$ from first principles, which would allow the viscosity of the fluid to be set appropriately. However, the normalization approach as used here could be used in a practical sense to extrapolate between a limited experimental data set for a complex glass melt. In this way, the basic methodology as set out here could easily be used for exploring variations in viscosity with temperature and composition for a wide range of chemical engineering and geological applications. The DPD method has been extended to potentially solve these problems [64]. Finally, the issue of energy conservation has been addressed [65,66] to allow the method to be extended to model heat transfer.

In terms of computational efficiency, each of the typical DPD simulations presented in this paper was solved in approximately 2 h on 32 cores of a Linux cluster. Reasonable statistics can be obtained to as few as 40,000 particles, reducing the simulation times by a factor of 10. A performance comparison was carried out against an equivalent NEMD simulation of approximately 4000 SiO_2 molecules using a Buckingham potential for 30,000 time steps using the LAMMPS code. The timings produced in solving this simple test case show that for the same number of particles, an equivalent DPD simulation would be approximately 70 times faster. An individual DPD bead can represent many molecules, so in practice, far longer length and timescales can be simulated with DPD than molecular dynamics.

21.2.6 CONCLUSIONS

To conclude, the present section has highlighted the approaches and limitation to calculating the properties of glass melts. Atomistic and particle simulation can predict the general relationships between composition and properties (e.g., viscosity). Thus, the fluid dynamics models generated will have an implicit description of the underlying chemistry. This enables plant improvements to be made based on a sound chemical and physical understanding.

The results presented clearly show that DPD shows some promise as a tool for predicting viscosity trends as a function of temperature. By varying the conservative parameters as a function of chemistry and temperature, the flow properties of the resulting fluid vary to a similar degree to what is observed in real fluids. Pressure dependence of viscosity is also readily accessible, provided that the equation of state is known with reasonable accuracy. Variation of viscosity with chemical composition would enable DPD to be a far more useful predictive tool.

21.3 ATOMISTIC SIMULATION STUDIES OF GLASS SURFACES

The vast majority of chemical reactions occur at an interface or surface. Thus, the structure of a surface can be very important in understanding how reactions will occur. For a vitrified waste glass, surface simulation can become very important when trying to understand the stability and leaching characteristics of the glass. It can aid in the understanding of how materials may diffuse from the bulk glass to the surface, and how these may diffuse into the environment. In addition, the reaction of glass surfaces with water and other molecules will be important in determining how they degrade over time. The chemical degradation of the surface also has a controlling effect on the mechanical stability of the glass surface.

Hence, it is imperative that we understand the surface atomistic structure of the glass. Molecular modeling has been widely used in developing an understanding of the atomic structure of glasses and their surfaces and has proved a powerful tool for numerous systems [67–73]. In the example presented in this section, molecular dynamics simulations are used to build surfaces of sodium silicate glass and

TABLE 21.3
AEBS Compositions Expressed as Mole Percent

AEBS Components	Concentration (Cation %)
SiO_2	55.9
Al_2O_3	18.7
B_2O_3	14.8
CaO	5.3
BaO	5.2

TABLE 21.4
Sodium Silicate Glass Compositions Expressed as Mole Percent

Sodium Silicate Components	Concentration (Cation %)
SiO_2	66.6
Na_2O	33.3

an alkaline earth borosilicate (AEBS) glass to improve our understanding of these substrates. The composition of the glasses are listed in Tables 21.3 and 21.4. The surfaces we create do not include the effect of hydroxylation as we would first like to understand the behavior of a fresh fracture surface.

Previous simulations in this area have been primarily done by expert computational chemistry groups and have typically required significant amounts of time (both in terms of computational time and "research time") to do the studies [74–77]. However, with the advent of cheaper and better computer hardware and the availability of desktop simulations packages [78], this area is starting to become more accessible for the nonspecialist. Although this solves many of these issues, it does not answer the questions on how to do these types of simulations in a straightforward manner. In this section, we present what we believe to be a holistic methodology to study such materials with commercially available computational tools, on relatively modest hardware.

21.3.1 MODELING APPROACH FOR SURFACES

The requirements to build glass surfaces are very similar to those discussed for the building of bulk glasses. A simulated annealing approach needs to be used to optimize the surface atom positions. However, there are some minor differences that are described in the following section.

The modeling of surfaces requires the use of a technique called the "vacuum slab model." This involves placing a "slab" of material into a 3D cell and then extending one side of the cell (typically the C-axis) such that slabs in neighboring cells will not interact with each other (usually a distance of 50 Å–60 Å). In addition, the number of atoms in the cell is increased by extending the A and B directions to create 30 × 40 Å cell. Thus, each surface will consist of approximately 2000–3000 atoms.

The simulated annealing protocol adopted to build the glass surfaces is similar to those used for building the bulk glass. However, subjecting them to such high temperatures would result in the atoms leaving the surface slab and entering the vacuum. Thus, the samples are initially annealed at 600 K (as opposed to 7000 K) and cooled to 300 K in 100 K steps using 20-ps equilibration times. Such a process allows a relaxation of the ions at the surface to energetically more favorable positions. Considering that the materials are amorphous, the surfaces should be built several times to produce a statistically averaged system. Hence, the surface-simulated annealing calculation is repeated 10 times, which is a reasonable number of surfaces from which to draw conclusions. Such a modeling procedure can be experimentally likened to the behavior of a fresh fracture surface of glass, where the ions will try and move to their energy minima when the surface is created. As with the previous work on the bulk, a modified version of the cvff_aug force field parameter set is used that contains additional parameters for boron. Although this force field contains no valence terms and is purely based on the nonbonded forces, it has successfully been applied to a range of systems [35–37]. For the Columbic forces, an Ewald [79] summation is used, whereas a direct cutoff of 12 Å is applied for the Van der Waals forces. It would be worth noting that other force fields have been used to model glass materials, such as Buckingham-based potentials; the reader is pointed to several reviews on the topic [80,81].

To ensure that any results for ion migration or rearrangement were not the result of surface "vaporization," the surfaces were annealed at successively higher temperatures to find the temperature at which this occurred. This was seen to happen at 4000 K, implying that the annealed surfaces we have created are stable and not an artifact of the simulation method.

21.3.2 POST-RUN ANALYSIS

Once the calculations were complete, several types of analysis of the glass structures were carried out to develop an understanding of the behavior of the glass surfaces at an atomistic level. Each individual property calculated gives some information and can be pieced together to give an overall picture. The analysis is described below.

Structural visualization—The ability to visualize the surface slab is a very powerful tool, as it allows the examination of the surface structure: which atoms are at the surface, how they exist at the surface, and where they are on the surface. In addition, such analysis can guide the decision as to what further analysis is required.

Diffusion coefficient—During a molecular dynamics run, it is possible to save a trajectory file of the atom positions and velocities as a function of time. Thus, it is possible to map out the movement of ions over the period of the calculation. This trajectory file can be analyzed to give the mean square displacement (MSD) of the ions of interest over a specific period of time. The gradient of the MSD vs. time approximates the Einstein self-diffusion coefficient over sufficiently long simulation timescales. This property is a good measure of the mobility of individual ionic species in the glass. Knowledge of this in concert with other properties can help predict future structural changes such as surface segregation and the relative rates at which they may happen. In the present scenario, the MSD is calculated by taking averages over all of the ions in the system using a time frame of 20 ps (when the ions are at 300 K). This is a relatively short time frame for an accurate prediction of the diffusion coefficient; however, the system size makes longer calculations prohibitive. This being said, the average is over a large number of runs, and as the primary interest is in the relative mobility of the different ionic species in the system, this value will still provide useful information. Thus, it would be better to treat the results as qualitative rather than quantitative. More detailed description of diffusion simulations in glasses can be found in the work of Cormack et al. [82] and Litton [76] who carry out more extensive studies in this area. In addition, such calculations only give the self-diffusion coefficient of the ions; they do not take into account other types of diffusive behavior occurring over longer time and length scales such as glass flow (which was discussed earlier in the chapter).

Concentration profile—The concentration profile, as the name implies, is a profile of composition or concentration of a species as a function of points along the axes of the periodic repeat unit. Hence, for a surface, a profile of the concentration along the axis normal to the surface slab will allow the study of changes in the composition before and after relaxation, for example, to study surface segregation.

Coordination numbers—The coordination of ions in the system is a good indication of how "polymerized" the glass system is and of what units they are composed. Glass is a metastable material, which consists of framework ions, in which other nonframework ions are distributed. The framework ions will lower their energies by trying to assume coordinations similar to those found in crystalline systems; for example, silicon will try to be four-coordinate and aluminum ions will try to be either a four- or a six-coordinate structure. Thus, the numbers of framework ions that have "nonstandard" coordination are good indicators of the stability of the glass. In addition, where such ions are found can be of importance. For example, a large number of three coordinate silicon ions at the surface would imply that such a glass would react more readily with molecules than a surface with a large number of four-coordinate silicon atoms.

21.3.3   COMPARISON OF AEBS AND SODIUM SILICATE GLASS

Structure—Simulated annealing results of sodium silicate glass show that silicon and oxygen ions form linked tetrahedra; these act as the primary network formers in the glass. The sodium atoms form clusters throughout the glass, both in the bulk and at the surface. These sodium-rich regions are surrounded by singly coordinated oxygen ions, so-called nonbridging oxygen ions (NBO). Visually, it can be seen that the sodium ions show some degree of surface segregation. Figures 21.11 and 21.12 show an example of a glass surface, from two views, side on and top down, where the network structure of the glass and the sodium-rich regions can clearly be seen. This is consistent with the previous modeling results from Cormack and Huang [77] who used shell model simulations to study bulk sodium silicate glasses, as well as the work of Garofalini and Levine [83,84] who used molecular dynamics to study the surfaces of lithium-, sodium-, and potassium-based glass surfaces. In addition, extended x-ray absorption fine structure studies of caesium- and potassium-based glasses are consistent with this interpretation [85,86].

The resulting structures from the simulated annealing of AEBS glass show a different structure compared with sodium silicate glass, with many more of the ions involved in the formation of the framework, including aluminum, boron, silicon, and oxygen. The alkaline earth ions (calcium and barium) show a uniform distribution through the material and do not seem to cluster like the sodium ions in sodium silicate glass. Figures 21.13 and 21.14 show an example of the AEBS glass surface, with views parallel and perpendicular to the A- and B-axis planes. The figures show a highly networked structure akin to a highly cross-linked polymer resin. The surface shows a predominance of the network forming ions, with no sign of the alkaline earth ion segregation to the surface. However, it must be kept in mind that the

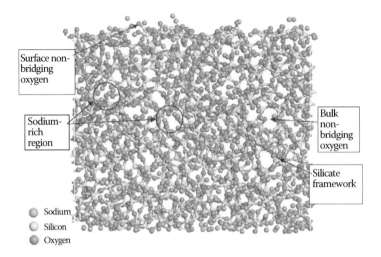

Surface non-bridging oxygen

Sodium-rich region

Bulk non-bridging oxygen

Silicate framework

Sodium
Silicon
Oxygen

FIGURE 21.11 View of the soda lime glass surface parallel to the *C*-axis.

concentration of these ions is approximately a third of that in the sodium silicate glass. Hence, clustering and segregation may occur at higher concentrations. The aluminum and silicon cations are known to form four-coordinate tetrahedra with oxygen ions. Boron can form both three- and four-coordinate structures. Presumably, ions that form four-coordinate structures relatively easily are suitable coupling points for the silane reagents. Another point of note is that the surface is much smoother and shows a greater similarity to the structure of bulk AEBS.

Diffusion—The results for the diffusion coefficients for the ions in sodium silicate glass are shown in Figure 21.15. The general trend is very clear in that sodium ions are roughly 10 times more mobile than the silicon or oxygen ions. The oxygen ions are a little more mobile than the silicon ions, but both are essentially static on

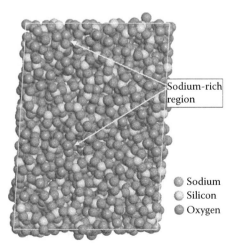

Sodium-rich region

Sodium
Silicon
Oxygen

FIGURE 21.12 View of the soda lime surface slab normal to the *A*- and *B*-axis planes.

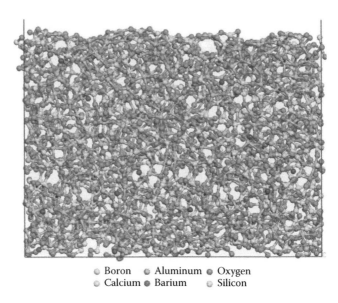

○ Boron	◎ Aluminum	● Oxygen
○ Calcium	● Barium	○ Silicon

FIGURE 21.13   View of the AEBS glass surface parallel to the *C*-axis.

the time frame of the sodium ions. The values are probably underestimates, and no doubt, the sodium ion is much more mobile than these values.

The diffusion coefficients for the AEBS glass are plotted in Figure 21.16; to aid comparison, the same scale is used for Figure 21.15. The results seem less conclusive than the results for sodium silicate in that no one species is consistently more mobile than another. However, it does show that, in general, the ionic species in AEBS are less mobile than in sodium silicate, with the exception of Run 5, which we believe to be a "rogue" point. Moreover, it shows that the diffusion coefficients of

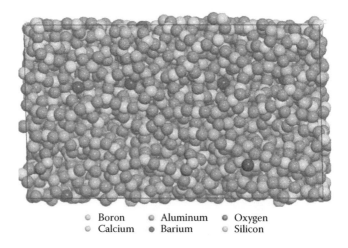

○ Boron	◎ Aluminum	● Oxygen
○ Calcium	● Barium	○ Silicon

FIGURE 21.14   View of the AEBS surface normal to the *A*- and *B*-axis planes. Note the homogeneity of the surface.

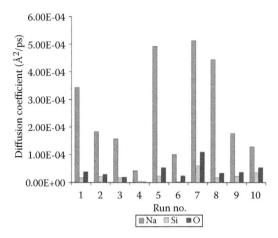

FIGURE 21.15 Diffusion coefficients for the ions in sodium silicate glass as a function of run number.

the framework species are closer to each other. The alkali ions, calcium and barium, are more mobile. However, from the graph, it would be hard to give a quantitative number on how much more mobile these alkali ions are. The most likely reason for the poor quality of these results is that as the ions are highly networked, they are practically immobile on the timescale of our annealing simulations. Thus, to get better quality data, we would have to run the calculations for a much longer time, possibly hundreds of picoseconds to get any reliable data. No doubt, the result of longer calculation would give much better quantitative values. However, the overall conclusion would be the same: the framework ions will have very small diffusion coefficients, and the calcium and barium ions will be more mobile, but not as mobile as the alkali ions in sodium silicate.

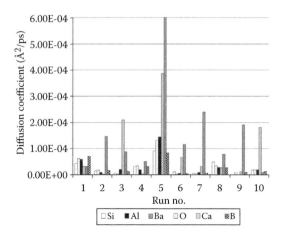

FIGURE 21.16 Diffusion coefficients of the ions in the AEBS glass as a function of run number.

Concentration profile—The profiles of the changes in concentration for the sodium silicate can be seen in Figures 21.17 and 21.18, which show an example from one of the runs. It is immediately clear that the annealing process encourages the segregation of sodium ions to the surface of the slab. This helps substantiate the conjectures made during the visual analysis of the slabs and analysis of the MSDs.

Given that there are more species in the AEBS glass, the concentration profiles are more complicated and features are more difficult to discern. The concentration profile before annealing shows that there are several peaks of high alkaline earth concentration and an essentially continuous distribution of the network-forming ions through the material (an example is shown in Figure 21.19). The annealing process essentially makes the glass slightly more homogeneous in composition (Figure 21.20), but overall, it remains relatively unchanged. The network-forming ions show very little change in their concentration during the annealing process. This is not surprising given that they are essentially static on the annealing timescale, implying very little segregation of any ions to the surface. As the individual alkaline earth concentration is much lower than in the sodium silicate glass, their peaks represent one or two ions as opposed to large concentrations of ions. Hence, one must be careful when drawing general conclusions. However, the annealing process reduces the two high concentration calcium sites, creating more sites with lower concentrations. Barium behaves in a similar way and the ions become more dispersed throughout the lattice. Instead of isolated sharp peaks in the concentration profile, we see more peaks with lower concentrations.

Coordination—The coordination numbers of the bulk sodium silicate glass are listed in Table 21.5. It is clear from these numbers that the glass is made up of a silicate framework in which the sodium ions are dispersed. The majority of the silicon atoms form four-coordinate structures. Two-thirds of the oxygen ions form bridging atoms between the silicate tetrahedra, which can be seen by their twofold

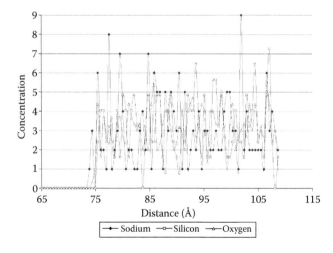

FIGURE 21.17 Concentration profile of soda lime glass parallel to the *C*-axis after annealing.

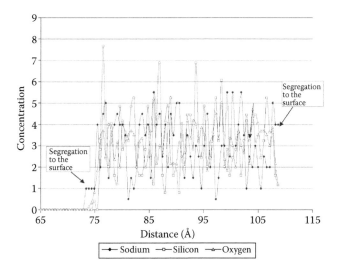

FIGURE 21.18 Concentration profile of soda lime glass parallel to the *C*-axis before annealing.

coordination. The rest are singly coordinated and can be classified as NBO, which coordinate with the alkali ions. In fact, from a purely numerical point of view, there are 140 NBO atoms and 140 sodium ions, implying that all the NBO atoms are associated with a sodium ion.

The average ionic coordination for the surface slabs is listed in Table 21.6. Generally, the coordination for the ions does not change significantly. This is not surprising if one considers the size of the surface slab. It is about 35 Å in size, and

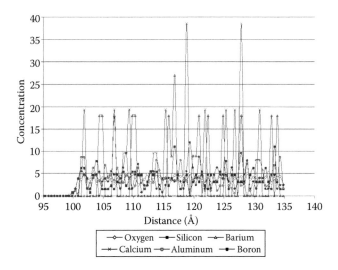

FIGURE 21.19 Concentration profile for the AEBS glass parallel to the *C*-axis before annealing.

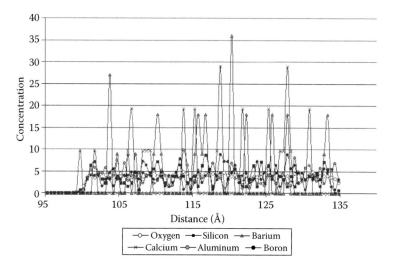

FIGURE 21.20 Concentration profile for the AEBS glass parallel to the *C*-axis after annealing.

hence, the number of ions at the actual surface will be smaller than the number of ions in the bulk. However, the results do prove useful in that one can pick up subtle changes in the coordination as a result of there being a surface. There is about a 3% increase in both the numbers of three- and five-coordinate silicon atoms. Likewise, for oxygen, there is a small increase in the amount of zero coordinate oxygen (1%), and from the decrease in singly coordinated oxygen, these zero coordinate ions must have been NBO atoms that moved further into the sodium-rich regions. The three-coordinate silicon ions no doubt act as coupling points for silane molecules.

For the AEBS glass, the coordination of the bulk silicon ions is very similar to that of sodium silicate glass with over 99% being in four-coordinate positions (Table 21.7). Similarly, the majority of oxygen ions adopt a two-coordinate bridging structure. However, unlike sodium silicate, there are virtually no single coordinate ions, with approximately 30% of ions being three-coordinate, which is commensurate with observation of an extensive network structure within the glass. Boron and aluminum show high coordination numbers because they are also part of the network

TABLE 21.5

Ionic Coordination (in %) for Bulk Sodium Silicate Glass

Coordination	0	1	2	3	4	5	6
Si	0	0	0	0	99.43	0.57	0
O	0	33.10	66.90	0	0	0	0
Na	100	0	0	0	0	0	0

TABLE 21.6
Average Ionic Coordination (in %) for Surfaces of Sodium Silicate Glass

Coordination	0	1	2	3	4	5	6
Si	0	0	0	2.6	94.21	3.17	0.19
O	0.94	32.82	67.04	0.14	0	0	0
Na	100	0	0	0	0	0	0

structure. The aluminum ions show a predominance of four-coordinate structure, which is analogous to aluminum coordination in aluminosilicate materials. It is unusual that there is very little octahedral coordination (the other stable coordination state for aluminum), which can be explained in terms of the significant amount of five-coordinate aluminum. These five-coordinate species are possibly frustrated coordination states: ions that were going from a tetrahedral to octahedral coordination but were trapped in the five-coordinate state during the annealing process. Hence, as also indicated in the diffusion studies, the highly networked structure of AEBS prevents the aluminum ions from migrating to their global minima. This ratio may change if the annealing was run for longer. This being said, NMR studies of AEBS glasses [87] do show significant amount of five-coordinate aluminum. Boron has a roughly 2:1 split of three and four-coordinate ions, both of which are also observed in the NMR studies.

The average results for the surface slabs are shown in Table 21.8. The silicon ions, like those in sodium silicate, show increased three and five-coordinate ions with similar amounts becoming three- and five-coordinate. The aluminum ions show about a 3.5% increase in the number of three-coordinate ions, which must be a result of undercoordinated tetrahedral sites at the surface, with a resulting decrease in the amount of four-coordinate aluminum. Analogously, the boron ions also show an approximately 3.5% increase in the three-coordinate concentration, which corresponds to a similar decrease

TABLE 21.7
Ionic Coordination (in %) of Bulk AEBS Glass

Coordination	0	1	2	3	4	5	6
Si	0	0	0	0	99.29	0.71	0
Al	0	0	0	0.53	71.28	25	3.19
O	0	0.49	71.05	28.40	0.37	0	0
Na	0	0	0	0	0	0	0
Ca	100	0	0	0	0	0	0
Ba	100	0	0	0	0	0	0
B	0	0	0	68.24	31.76	0	0

TABLE 21.8

Average Ionic Coordination (in %) of AEBS Glass Surfaces

Coordination	0	1	2	3	4	5	6
Si	0	0	0	1.36	95.81	2.83	0
Al	0	0	0	3.96	68.33	25.46	2.27
O	0	0.91	70.13	28.68	0.28	0	0
Ca	100	0	0	0	0	0	0
Ba	100	0	0	0	0	0	0
B	0	0	0.3	71.94	27.91	0	0

in the four-coordinate concentration. This would imply that these ions are at the surface of the slab.

21.3.4 SUMMARY AND DISCUSSION

The results from the previous sections show that within sodium silicate glass, the silicon and oxygen ions form a framework in which the sodium ions are dispersed. These form sodium-rich regions that are surrounded by nonbridging oxygen atoms. These characteristics are seen both in the bulk and on the surface of the glass. In fact, these clusters can interlink to form channel structures and can extend from the surface into bulk. This is also seen in potassium- and rubidium-based glasses [25,88]. This can cause problems for the encapsulation of nuclear wastes because this may be a potential leaching pathway for the fission products. Analysis of the concentration profiles shows that sodium migrates to the surface of the slab during the annealing process. This is further reinforced by the diffusion coefficient calculations which show that sodium is the most mobile of the ions. Thus, over time, the sodium concentration at the surface will increase and concentrate in specific areas. These sodium ions would readily interact with water or other molecules to degrade the glass.

The AEBS glass shows a highly networked structure with boron, silicon, aluminum, and oxygen forming the framework, in which the barium and calcium ions are dispersed. The surface structure is very homogeneous and similar in structure to that of the bulk. The concentration profiles show no major change during the annealing process, implying surface segregation of the ions over a much longer timescale, if at all. The diffusion studies show that the framework ions are essentially static on the timescale of the simulation and the alkaline earth ions do not migrate extensively. This is not to say that the ions do not diffuse; it is just that if they do, they do so on a longer timescale and certainly not at the same rate as the sodium silicate glass. The coordination studies show that the cations at the surface are undercoordinated and are probably good points for the binding of small molecules.

Thus, the stability and leach resistance of borosilicate glasses can be rationalized in part as follows.

1. The surface composition of AEBS does not change significantly over time compared to that of sodium silicate where the sodium rapidly diffuses to the surface. These can readily interact with water or other molecules, thus degrading the glass.

2. The surface of AEBS is homogeneous and shows no clustering of any of the species. On the other hand, the sodium silicate surface is inhomogeneous in that it has regions of high sodium concentration. Thus, the AEBS glass is more resistant to degradation.

21.4 SUMMARY

This chapter has illustrated some examples of how atomistic and mesoscale tools have been used to study the behavior of glasses used in the nuclear industry. These examples use a range of techniques and study bulk, surface, and melt properties of the material.

ACKNOWLEDGMENTS

The authors would like to thank Bruce Hanson (National Nuclear Laboratory) for the provision of time and funding to write this chapter. They would also like to thank Dr. Marc Meunier for his persistence and patience.

REFERENCES

1. International Energy Agency. *World Energy Outlook 2009*. OECD, Paris, Nov. 2009.
2. P. D. Wilson, editor. *The Nuclear Fuel Cycle from Ore to Waste*. Oxford: Oxford University Press, 1996.
3. I. W. Donald. *Waste Immobilization in Glass and Ceramic Based Hosts: Radioactive, Toxic and Hazardous Wastes*. New York: Wiley-Blackwell, 2010.
4. S. Matsuno, Y. Iso, H. Uchida, I. Oono, T. Fukui and T. Ooba. CFD modeling coupled with electric field analysis for Joule-heated glass melters. *J. Power Energy Syst.* **2** (1), 2008, 447.
5. J. Du, L. Rene Corrales, K. Tsemekhman and E. J. Bylaska. Electron, hole and exciton self-trapping in germanium doped silica glass from DFT calculations with self-interaction correction. *Nucl. Instrum. Methods Phys. Res. B* **255**, 2007, 188.
6. E. Apr, E. J. Bylaska, D. J. Dean, A. Fortunelli, F. Ga, P. S. Krstic, J. C. Wells and T. L. Windus. NWChem for materials science. *Comput. Mater. Sci.* **28**, 2003, 209.
7. S. Vyas, J. E. Dickinson and E. Armstrong-Poston. Towards an understanding of the behavior of silanes on glass: An atomistic simulation study of glass surfaces. *Mol. Simul.* **32** (2), 2006, 135.
8. S. L. Owens, A. Morgan and R. Field. Integrated simulation of waste vitrification plant—From atomistic derivation of melt properties to a prediction of melter lifetime and melt mixing. In Proceedings of 7th World Congress of Chemical Engineering, 2005.
9. M. Bankhead. Modelling the flow of silica glass melts with dissipative particle dynamics, submitted to PCCP.
10. Amorphous Cell Module, part of Materials Studio, Accelrys, San Diego, CA, 2002.
11. Discover, part of Cerius2 4.2 and Materials Studio environments, Accelrys, San Diego, CA, 2002.

12. D. B. Dingwell. Volcanic dilemma: Flow or blow? *Science* **273** (5278), 1996, 1054.
13. J. B. Morris, K. A. Boult, J. T. Dalton, M. H. Delve, R. Gayler, L. Herring, A. Hough and J. A. C. Marples. Durability of vitrified highly active waste from nuclear reprocessing. *Nature* **273**, 1978, 215.
14. J. F. Stebbins, P. F. McMillan and D. B. Dingwell, editors. *Mineral. Soc. Am. Rev. Mineral.* **32**, 1995, 505.
15. Y. Bottinga, P. Richet and A. Sipp. Viscosity regimes of homogeneous silicate melts. *Am. Mineral.* **80**, 1995, 305.
16. R. Knoche, D. B. Dingwell, F. A. Seifert and S. L. Webb. Non-linear properties of super-cooled liquids in the system Na_2O-SiO_2. *Chem. Geol.* **116**, 1994, 1.
17. M. I. Ojovan. Viscosity and glass transition in amorphous oxides. *Adv. Condens. Matter Phys.* **2008** Article ID 817829, 23 pages, 2008. doi:10.1155/2008/817829.
18. A. Fluegel. Glass viscosity calculation based on a global statistical modeling approach. *Glass Technol.: Eur. J. Glass Sci. Technol., Part A* **48** (1), 2007, 13.
19. W. C. Brady, R. L. Tiede, F. M. Veazie and W. W. Wolf. US Pat. 3607322-filed 22 May 1969-issued 21 Sep 1971.
20. A. I. Priven. A new equation for describing the temperature dependence of the viscosity of glass-forming melts. *Glass Phys. Chem.* **25** (6), 1999, 441.
21. T. Lakatos, L. G. Johansson and B. Simmingskold. Inverkan av Li2O och B2O3 soda-kalk-silikatglas på viskositeten. *Glastek. Tidskr.* **30** (1), 1975, 7.
22. G. N. Greaves. EXAFS and the structure of glass. *J. Non-Cryst. Solids* **71**, 1985, 203.
23. G. N. Greaves. *Glass Science and Technology*, **4B**, 1, edited by D. R. Uhlman and N. Kreidt. London: Academic Press, 1990.
24. B. Vessal, M. Amini, D. Fincham and C. R. A. Catlow. Water-like melting behaviour of SiO2 investigated by the molecular dynamics simulation technique. *Philos. Mag. B* **60** (6), 1989, 753.
25. B. Vessal, G. N. Greaves, P. T. Marten, A. V. Chadwick, R. Mole and S. Houde-Walter. Cation microsegregation and ionic mobility in mixed alkali glasses. *Nature* **356**, 1992, 504.
26. W. G. Hoover, D. J. Evans, R. B. Hickman, A. J. C. Ladd, W. T. Ashurst and B. Moran. Lennard–Jones triple-point bulk and shear viscosities. Green–Kubo theory, Hamiltonian mechanics, and nonequilibrium molecular dynamics. *Phys. Rev. A* **22** (4), 1980, 1690–1697.
27. F. M. Plathe. Reversing the perturbation in nonequilibrium molecular dynamics: An easy way to calculate the shear viscosity of fluids. *Phys. Rev. E* **59** (5), 1999, 4894–4898.
28. I. Bitsanis, J. J. Magda, M. Tirrell and H. T. Davis. Molecular dynamics of flow in micropores. *J. Chem. Phys.* **87**, 1987, 1733.
29. P. A. Thompson and M. O. Robbins. Shear flow near solids: Epitaxial order and flow boundary conditions. *Phys. Rev. A* **41**, 1990, 6830.
30. E. Manias, G. Hadziioannou, I. Bitsanis and G. ten Brinke. Stick and slip behaviour of con-fined oligomer melts under shear. A molecular-dynamics study. *Europhys. Lett.* **24**, 1993, 99.
31. S. Y. Liem, D. Brown and J. H. R. Clarke. Investigation of the homogeneous-shear non-equilibrium-molecular-dynamics method. *Phys. Rev. A* **45**, 1992, 3706.
32. R. Khare, J. J. de Pablo and A. Yethiraj. Rheology of confined polymer melts. *Macromolecules* **29**, 1996, 7910.
33. S. A. Gupta, H. D. Cochran and P. T. Cummings. Shear behavior of squalane and tetra-cosane under extreme confinement. II. Confined film structure. *J. Chem. Phys.* **107** (23), 1997, 10316.
34. K. Travis, B. D. Todd and D. J. Evans. Poiseuille flow of molecular fluids. *Physica A: Stat. Theor. Phys.* **240** (1–2), 1997, 315–327.
35. J. R. Hill, A. R. Minihan, E. Wimmer and C. J. Adams. Framework dynamics including computer simulations of the water adsorption isotherm of zeolite Na–MAP. *Phys. Chem. Chem. Phys.* **2**, 2000, 4255.

36. J. R. Hill, C. M. Freeman and L. Subramanian. Use of force-fields in materials modeling. In *Rev. Comp. Chem.*, **16**, 141, edited by K. B. Lipkowitz and D. B. Boyd. New York: Wiley-VCH, John Wiley and Sons, 2000.

37. J. R. Hill and J. Plank. Retardation of setting of plaster of Paris by organic acids: Understanding the mechanism through molecular modeling. *J. Comp. Chem.* **25** (12), 2004, 1438.

38. M. K. Choudhary. Mathematical modeling of flow and heat-transfer phenomena in glass furnace channels and forehearths. *J. Am. Ceram. Soc.* **74** (12), 1991, 3091.

39. H. Kiyohashi, N. Hayakawa, S. Aratani and H. Masuda. Thermal conductivity of heat-absorbed soda-lime-silicate glasses at high temperatures. *15th ECTP Proceedings*, 2002, 1395–1404.

40. W. G. Hoover. *Smooth Particle Applied Mechanics—The State of the Art*, volume 25, edited by R. S. MacKay. Singapore: World Scientific Publishing, Nov. 2006.

41. P. J. Hoogerbrugge and J. M. V. A. Koelman. Simulating microscopic hydrodynamic phenomena with dissipative particle dynamics. *Europhys. Lett.* **19** (3), 1992, 155–160.

42. R. D. Groot, T. J. Madden and D. J. Tildesley. On the role of hydrodynamic interactions in block copolymer microphase separation. *J. Chem. Phys.* **110** (19), 1999, 9739.

43. R. D. Groot and P. B. Warren. Dissipative particle dynamics: Bridging the gap between atomistic and mesoscopic simulation. *J. Chem. Phys.* **107** (11), 1997, 4423–4435.

44. P. Español and P. Warren. Statistical mechanics of dissipative particle dynamics. *Europhys. Lett.* **30** (4), 1995, 191–196.

45. A. Maiti and S. McGrother. Bead–bead interaction parameters in dissipative particle dynamics: Relation to bead-size, solubility parameter, and surface tension. *J. Chem. Phys.* **120** (3), 2004, 1594.

46. K. P. Travis, M. Bankhead, K. Good and S. L. Owens. New parametrization method for dissipative particle dynamics. *J. Chem. Phys.* **127** (1), 2007, 014109.

47. G. T. Evans. Dissipative particle dynamics: Transport coefficients. *J. Chem. Phys.* **110** (3), 1999, 1338–1342.

48. S. Plimpton. Fast parallel algorithms for short-range molecular dynamics. *J. Comput. Phys.* **117** (1), 1995, 1–19.

49. D. Visser, H. Hoefsloot and P. Iedema. Modelling multi-viscosity systems with dissipative particle dynamics. *J. Comput. Phys.* **214** (2), 2006, 491–504.

50. M. Revenga, I. Zúñiga, P. Español and I. Pagonabarraga. Boundary conditions in dissipative particle dynamics. *Int. J. Mod. Phys. C* **9** (8), 1998, 1319–1328.

51. D. Duong-Hong, N. Phan-Thien and X. Fan. An implementation of no-slip boundary conditions in DPD. *Comput. Mech.* **35** (1), 2004, 24–29.

52. M. W. Tysanner and A. L. Garcia. Measurement bias of fluid velocity in molecular simulations. *J. Comput. Phys.* **196** (1), 2004, 173–183.

53. J. A. Backer, C. P. Lowe, H. C. J. Hoefsloot and P. D. Iedema. Poiseuille flow to measure the viscosity of particle model fluids. *J. Chem. Phys.* **122** (15), 2005, 154503.

54. K. P. Travis, B. D. Todd and D. J. Evans. Poiseuille flow of molecular fluids. *Physica A* **204**, 1997, 315.

55. J. W. van de Meent, A. Morozov, E. Somfai, E. Sultan and W. van Saarloos. Coherent structures in dissipative particle dynamics simulations of the transition to turbulence in compressible shear flows. *Phys. Rev. E* **78** (1), 2008, 015701.

56. I. V. Pivkin and G. E. Karniadakis. Controlling density fluctuations in wall-bounded dissipative particle dynamics systems. *Phys. Rev. Lett.* **96** (20), 2006, 206001.

57. B. D. Todd and D. J. Evans. Temperature profile for Poiseuille flow. *Phys. Rev. E* **55** (3), 1997, 2800–2807.

58. S. V. Nemilov. *Thermodynamic and Kinetic Aspects of the Vitreous State*. Laser and Optical Science and Technology Series. Boca Raton, FL: CRC Press, 1995.

59. M. J.-F. Guinel and M. G. Norton. Blowing of silica microforms on silicon carbide. *J. Non-Cryst. Solids* **351** (3), 2005, 251–257.

60. H. Noguchi and G. Gompper. Transport coefficients of dissipative particle dynamics with finite time step. *Europhys. Lett.* **79** (3), 2007, 36002.

61. C. M. Hansen. The three dimensional solubility parameter—Key to paint component affinities. I.—Solvents, plasticizers, polymers, and resins. *J. Paint Technol.* **39** (505), 1967, 104–117.

62. A. F. M. Barton. *Handbook of Solubility Parameters and Other Cohesion Parameters*, 2nd edition. Boca Raton, FL: CRC Press, 1991.

63. M. S. Ghiorso and I. S. E. Carmichael. A regular solution model for met-aluminous silicate liquids: Applications to geothermometry, immiscibility, and the source regions of basic magmas. *Contrib. Mineral. Petrol.* **71** (4), 1980, 323–342.

64. C. P. Lowe. An alternative approach to dissipative particle dynamics. *Europhys. Lett.* **47** (2), 1999, 145.

65. P. Español. Dissipative particle dynamics with energy conservation. *Europhys. Lett.* **40** (6), 1997, 631.

66. J. B. Avalos and A. D. Mackie. Dissipative particle dynamics with energy conservation. *Europhys. Lett.* **40** (2), 1997, 141.

67. S. H. Garofalini. Molecular dynamics simulation of the frequency spectrum of amorphous silica. *J. Chem. Phys.* **76**, 1982, 3189.

68. S. H. Garofalini. A molecular dynamics simulation of the vitreous silica surface. *J. Chem. Phys.* **78**, 1983, 2069.

69. B. Vessal, M. Amini and C. R. A. Catlow. Computer simulation of the structure of silica glass. *J. Non-Cryst. Solids* **159**, 1993, 184.

70. B. Vessal. Simulation studies of silicates and phosphates. *J. Non-Cryst. Solids* **177**, 1994, 103.

71. J. M. Delaye and D. Ghaleb. Molecular dynamics simulation of low energy atomic displacements cascades in a simplified nuclear glass. *J. Nucl. Mater.* **244**, 1997, 22.

72. J.-J. Liang, R. T. Cygan and T. M. Alam. Molecular dynamics simulation of the structure and properties of lithium phosphate glasses. *J. Non-Cryst. Solids* **263, 264**, 2000, 167.

73. S. Chaussedent, V. Teboul and A. Monteil. Molecular dynamics simulations of rare-earth-doped glasses. *Curr. Opin. Solid State Mater. Sci.* **7**, 2003, 111.

74. A. Takada, C. R. A. Catlow and G. D. Price. Computer modeling of B_2O_3: Part I. New interatomic potentials, crystalline phases and predicted polymorphs. *J. Phys.: Condens. Matter.* **7**, 1995, 8659.

75. A. Takada, C. R. A. Catlow and G. D. Price. Computer modeling of B_2O_3: Part II. Molecular dynamics simulations of vitreous structures. *J. Phys.: Condens. Matter.* **7**, 1995, 8693.

76. D. A. Litton and S. H. Garofalini. Vitreous silica bulk and surface self diffusion analysis by molecular dynamics. *J. Non-Cryst. Solids* **217**, 1997, 250.

77. A. N. Cormack and C. Huang. The structure of sodium silicate glass. *J. Chem. Phys.* **93** (11), 1990, 8180.

78. Materials Studio Modeling Software, Accelrys, San Diego, CA, 2002.

79. P. P. Ewald. Evaluation of optical and electrostatic lattice potentials. *Ann. Phys.* **64**, 1921, 253.

80. S. Vyas. Computational chemistry studies of silicate materials: A guide for the non-expert. In *Silica and Silicates in Modern Catalysis*, edited by I. Halasz. Kerala, India: Transworld Research Network, 2010.

81. C. R. A. Catlow. Solids: Computer modelling. In *Encyclopaedia of Inorganic Chemistry*, edited by R. B. King, J. K. Burdett, R. H. Crabtree, C. M. Lukehart, R. A. Scott and R. L. Wells. New York: John Wiley and Sons, 1994.

82. A. N. Cormack, J. Du and T. R. Zeitler. Alkali ion migration mechanisms in silicate glasses probed by molecular dynamics simulations. *Phys. Chem. Chem. Phys.* **4**, 2002, 3193.

83. S. H. Garofalini and S. M. Levine. Differences in surface behavior of alkali ions in Li$_2$O.3SiO$_2$ and Na$_2$O.3SiO$_2$ glasses. *J. Am. Ceram. Soc.* **68**, 1985, 376.

84. S. H. Garofalini. Behavior of atoms at the surface of a K$_2$O.3SiO$_2$ glass—A molecular dynamics simulation. *J. Am. Ceram. Soc.* **67**, 1984, 133.

85. G. N. Greaves. EXAFS, glass structure and diffusion. *Philos. Mag. B* **60**, 1989, 793.

86. G. N. Greaves. Structural studies of the mixed alkali effect in disilicate glasses. *Solid State Ionics* **105**, 1998, 204.

87. B. C. Bunker, R. J. Kirkpatrick, R. K. Brow, G. L. Turner and C. Nelson. Local structure of alkaline-earth boroaluminate crystals and glasses: II. ^{11}B and ^{27}Al MAS NMR spectroscopy of alkaline earth boroaluminate glasses. *J. Am. Ceram. Soc.* **74** (6), 1991, 1430.

88. S. H. Garofalini and D. Zirl. Onset of alkali adsorption on the v-SiO2 surface. *J. Vac. Sci. Technol.* **6**, 1988, 975.

22 Molecular Simulations of In-Plane Stiffness and Shear Modulus of Double-Walled Carbon Nanotubes

Abraham Q. Wang

CONTENTS

22.1 INTRODUCTION

Carbon nanotubes (CNTs) are macromolecules of carbon in a periodic hexagonal arrangement with a cylindrical shell shape [1]. They can be viewed as one (or more) graphite sheet(s) rolled into a seamless tube. A pair of indices (n,m), called the chirality, is used to represent the way a graphite sheet is wrapped. When $m = 0$, the nanotubes are called "zigzag," and when $n = $ m, they are called "armchair." It is widely acknowledged that a reasonable and accurate estimate of their material properties, such as Young's modulus, shear strength, Poisson's ratio, and bending rigidity, is critical for the potential applications of the material [2]. Experimental investigations of the material properties of CNTs have been explored intensively. Krishnan et al. [3] estimated the Young's modulus of single-walled carbon nanotubes (SWNTs) to be 0.9 TPa–1.7 TPa by observing their freestanding room-temperature vibrations in a transmission electron microscope. Salvetat et al. [4] used an atomic force microscope and a special substrate to estimate the elastic and shear moduli of a SWNT to be of the order of 1 TPa and 1 GPa, respectively. Besides the research findings on SWNTs, Poncharal et al. [5] observed the static deformation of a multiwalled CNT (MWNT) and indicated that the Young's modulus of the materials is about 1 TPa using a transmission electron microscope. Wong et al. [2] experimentally determined the Young's modulus of individual, structurally isolated silicon carbide nanorods and MWNTs that were pinned at one end to molybdenum disulfide surfaces and found the value to be 0.7 TPa–1.9 TPa. In addition to these experimental endeavors, the mechanical properties of CNTs in closed forms have also been explored. A stick–spiral model [6] was developed to investigate the mechanical behavior of SWNTs, especially the estimate of Young's modulus and shear modulus based on a molecular mechanics concept. Other close-form expressions for mechanical properties of achiral CNTs were attempted via a concept of representative volume element of the chemical structure of a graphite sheet [7–8]. Length-dependent in-plane stiffness and shear modulus of chiral and achiral SWNTs subjected to axial compression and torsion have recently been discovered [9]. The strain energy of the tubes measured from the molecular mechanics and the corresponding calculated second derivative of the energy were used for the estimation of the properties of CNTs based on an elastic rod theory in relating the CNT material properties directly to the molecular mechanics calculations.

This paper reports a modeling method for calculating the in-plane stiffness and shear modulus of double-walled CNTs (DWNTs) via molecular simulations. The strain energy and the corresponding second derivative with respect to displacement subjected to CNTs are used to evaluate the material properties via an elastic rod theory. Length dependence of the mechanical properties is explored for DWNTs. In addition, van der Waals effect on difference of the material properties between DWNTs and SWNTs is investigated.

22.2   ELASTIC ROD THEORY ON MATERIAL PROPERTIES OF DWNTs

Elastic rod theory has been used to link the strain energy stored in SWNTs under axial compression to their material properties [9] in the molecular simulations and will be applied in the current manuscript for the derivation of material properties of DWNTs. From the mechanics of materials, an elastic bar subjected to an axial uniform compression or tension can be simply modeled or represented by a fictitious spring element with the stiffness given as [10]

$$k = AE/L, \qquad (22.1)$$

where A is the area of the cross section, E is the Young's modulus, and L is the length of the elastic rod. For a DWNT, $A = \pi(D_i + D_o)t$ is set, where t is the wall thickness of the DWNT; D_i and D_o are the medium diameters of the cross section of the inner and outer CNTs, respectively. Therefore, the relationship between the in-plane stiffness of the DWNT, Et, and the stiffness of the fictitious spring is obtained by

$$Et = \frac{kL}{\pi\left(D_i + D_o\right)}. \qquad (22.2)$$

On the other hand, it is known that the spring stiffness is expressed as the second derivative of the strain energy stored in the spring, or, equivalently, the strain energy stored in the CNT [9], U, with respect to the corresponding compression, that is, U''. Hence, the in-plane stiffness of the DWNT can be directly obtained as

$$Et = \frac{U''L}{\pi\left(D_i + D_o\right)}. \tag{22.3}$$

The shear stiffness of DWNTs can be investigated through a similar procedure. From the mechanics of the material, an elastic rod under torsion can be directly modeled by a rotary spring with the stiffness given by

$$k_r = GI_p/L, \tag{22.4}$$

where I_p is the polar moment of the inertia of the circular cross section and G is the shear modulus of CNTs. Similarly, spring stiffness is equivalent to the second derivative of the strain energy restored in the spring, or, equivalently, the strain energy stored in CNTs, with respect to the rotation angle applied to the CNTs, that is, U_r''. It has been acknowledged that the thickness of CNTs is normally viewed to be very thin compared to their diameters. Yakobson et al. [11] concluded that the effective thickness of CNTs should even be taken as $h = 0.066$ nm if the classical shell bending theory is applied to the materials. As the polar moment of inertia for a thin circular rod is approximately given by $I_p = \pi D^3 t/4$ [10], the shear stiffness of DWNTs can be easily obtained by

$$Gt = \frac{4U_r''L}{\pi\left(D_i^3 + D_o^3\right)}. \tag{22.5}$$

In view of the current debate on the thickness of CNTs, the investigation of the in-plane stiffness and shear modulus, Et and Gt, rather than the module E and G, would avoid arguments on the values of the effective thickness of CNTs. The application of the second derivative of strain energy for the estimate of material properties of CNTs was also reported [12]. Once the relationship between the material properties of DWNTs and the corresponding second derivative of strain energy is built, molecular simulations will be conducted to collect strain energy of DWNTs subjected to compression and torsion separately via the Materials Studio® developed by Accelrys.

22.3   MOLECULAR SIMULATIONS VIA MATERIALS STUDIO

Materials Studio is a comprehensive suite of modeling and simulation solutions developed by Accelrys for studying chemicals and materials, including crystal structure and crystallization processes, polymer properties, catalysis, and structure–activity relationships. It offers advanced visualization tools and access to the complete range of computational materials science methods [13] and will be applied in the molecular simulations of the manuscript. The interatomic interactions in Materials Studio are described by the COMPASS force field (condensed-phased optimized molecular potential for atomistic simulation studies) [14]. This is the first ab initio force field that was parameterized and validated using condensed-phase properties, and it has been proved to be applicable in describing the mechanical properties of CNTs [15].

To build a DWNT, we first select "multiwall nanotube" under "build nanostructure" with the Materials Studio version 4.2 purchased by the research group. Next, we choose "Individual" from the nanotube definition, and pick zigzag (8,0) @ (17,0) and armchair (8,8) @ (13,13) DWNTs with various lengths. The medium diameters of the two walls of the zigzag DWNTs are 0.626 nm and 1.331 nm, respectively, with the length of repeated units of 0.426 nm. Six zigzag DWNTS, with the lengths of 2.092, 4.185, 6.069, 7.952, 10.044, and 11.717 nm, are simulated via Materials Studio. For armchair DWNTs, the medium diameters of the two walls are 1.085 nm and 1.763 nm, respectively, with the length of repeated units of 0.246 nm. Five armchair DWNTS, with the lengths of 2.657, 4.351, 6.286, 8.697, and 11.353 nm, are simulated. The molecular simulations are carried out at a temperature of 1 K to avoid the thermal effect with an adiabatic process. In the simulations, the two ends of the DWNTs are clamped, as has been done in previous studies [11,15]. The molecular mechanics of the DWNTs subjected to compression and torsion can be identified through a minimizer processor. The minimizer processor enables the atoms in CNTs to rotate and move relatively to each other following a "Smart Minimizer" algorithm, which starts with the steepest descent method, followed by the conjugate gradient method, and ends with a Newton method, to minimize the strain energy so that an equilibrium state can be identified. The simulations are run in parallel of four processors on a Sun workstation. The CPU used for the initial minimization process of the (8,8) @ (13,13) DWNT with the length of 11.353 nm and 4116 atoms is 1237.55 s. The strain energy is collected at every axial compression displacement with the incremental displacement, 0.01 nm, applied at the two clamped ends of CNTs, whereas strain energy is collected at every torsion displacement with the incremental torsion angle, 1.0°, applied at the two ends of CNTs. Once the strain energy at every step is available, the second derivative of the strain energy with respect to applied displacement at the two ends of the DWNTs can easily be obtained through a simple finite difference method. Each item of second derivative data is obtained by every three items of strain energy data in the sequence of the increased enforced deformation. The in-plane stiffness and shear modulus of the DWNTs can then be directly determined from Equations 22.3 and 22.5 nm, accordingly.

The in-plane stiffness vs. length of the zigzag (8,0) @ (17,0) DWNTs is plotted in Figure 22.1 shown by the curve marked by triangle symbols. It is seen that the stiffness increases from an initial value of $Et = 344.92$ J/m^2 to an asymptotic value of $Et = 375.43$ J/m^2 for DWNTs from the shorter size, 2.092 nm, to the larger size, 11.717 nm. Figure 22.2 shows the in-plane stiffness vs. length of the armchair (8,8) @ (13,13) DWNTs by the curve marked by triangle symbols. Similarly, an increasing variation of the stiffness is observed from an initial value of $Et = 347.38$ J/m^2 to an asymptotic value of $Et = 376.72$ J/m^2 for DWNTs from the shorter size, 2.657 nm, to the larger size, 11.353 nm. In both scenarios, obvious scale effect on the in-plane stiffness is secured for tubes shorter than 12 nm. It is also interesting to find that the asymptotic value of the in-plane stiffness of armchair DWNTs is bigger than that of zigzag DWNTs. Similar observation was also reported [6,9] for simulations of SWNTs. Our molecular simulations reveal that the length-dependent in-plane

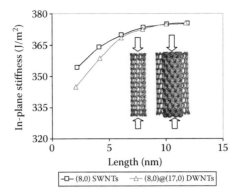

FIGURE 22.1   (See color insert.) Comparison of in-plane stiffness between zigzag DWNTs and SWNTs.

stiffness of DWNTs is in the range of 344 J/m^2–377 J/m^2. Yakobson et al. [11] proposed an estimate of the stiffness to be about 360 J/m^2, based on the data provided by Robertson et al. [16]. Gupta et al. [17] found the in-plane stiffness to be about 420 J/m^2, fairly close to our prediction. In addition, other estimates of Young's modulus based on experimental results [2–5] also confirm the range of the material property once the thickness of CNTs, $t = 0.34$ nm is used in calculations.

Figures 22.1 and 22.2 also demonstrate the results of the stiffness of zigzag (8,0) and armchair (8,8) SWNTs [9], which is shown by the curves marked by square symbols, for comparison purposes. It is naturally presumed that the in-plane stiffness of (8,0) @ (17,17) DWNTs could be same with that of (8,0) SWNTs with the same length if the two walls in the DWNTs are viewed as individual rods arranged in a parallel way. However, it is clearly seen from the two figures that the stiffness of DWNTs is obviously smaller than that of SWNTs at shorter sizes. Such difference is only diminished for longer CNTs. An interpretation of the observation is attempted by virtue of van der Waals effect occurring between the two walls of

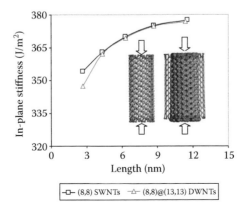

FIGURE 22.2   (See color insert.) Comparison of in-plane stiffness between armchair DWNTs and SWNTs.

DWNTs. In DWNTs, the major interaction of atoms within individual walls is the valence bond; however, the interaction of the atoms between two different walls is the nonbond van der Waals effect. Such van der Waals effect weakens the stiffness of DWNT and MWNTs. The role of the effect is briefly illustrated as follows from an axial compression process of a DWNT. When the DWNT is under compression, the two walls are apparently widened because the cross sections of the inner and outer walls of the DWNT expand simultaneously due to Poisson's ratio. However, the outer wall expands more than the inner wall of the DWNT because of a larger diameter. Therefore, the gap of the two walls in a DWNT increases during the compression. Because of the widened gap of the two walls, attraction will be initiated between the two walls due to the van der Waals effect. Such attraction force makes the outer wall to be prone to expand in the length direction, while it makes the inner wall to shrink further in the longitudinal direction. Because of the smaller diameter, the magnitude of the reduction of the inner wall in the longitudinal direction is higher than that of the extension of the outer walls, making the DWNT to be shortened further as a whole. Such a trend of further decrease in the length of the DWNT due to the van der Waals effect during the compression obviously weakens the resistance of the DWNT structure subjected to compression, and hence leads to lower in-plane stiffness of the DWNT. On the other hand, for longer DWNTs under same compression with shorter ones, the change of the gap of the two walls becomes less because of the smaller strain in the radial direction, which results in less attraction between the two walls due to van der Waals effect. Therefore, the difference of the material properties between DWNTs and SWNTs is negligible for larger sizes, which can be seen from the convergence of the two curves in Figures 22.1 and 22.2. Additional observation is that the convergence rate of the difference of the stiffness is different for zigzag and armchair DWNTs. In zigzag DWNTs, the difference of the stiffness diminishes for DWNTs larger than 12 nm. However, for armchair DWNTs larger than 5 nm, imperceptible difference is acknowledged. The lower stiffness of MWNTs was also reported from experimental observations. In these reports, in addition to the van der Waals effect on the stiffness, the occurrence of the wavelike ripples was also found to be attributed to the lower stiffness of MWNTs measured in experiments [5].

Figure 22.3 shows the variation of the in-plane shear modulus, Gt vs. the length of (8,8) @ (13,13) armchair DWNTs by the curve marked with triangle symbols. The prediction of the modulus varies from the value of $Gt = 115.11$ J/m^2 for the CNT with the length of 2.657 nm to the asymptotic value of $Gt = 123.83$ J/m^2 for the CNT with the length of 11.353 nm. The length-dependent shear modulus is also found in the simulations for (8,8) armchair SWNTs by the curve marked with square symbols. Our results are close to the prediction of the stiffness, $Gt = 150$ J/m^2, by Chang et al. [6] in which $t = 0.34$ nm was enforced, and also in good agreement with some existing predictions, such as those from lattice dynamics by Popov et al. [18]. Furthermore, it is again noted that van der Waals effect leads to the smaller shear modulus of DWNTs compared with SWNTs. The difference of the shear modulus diminishes for DWNTs at larger sizes. For the armchair DWNTs, the size effect on the difference of the shear modulus between DWNTs and SWNTs is negligible when the length of CNTs is greater than 6 nm.

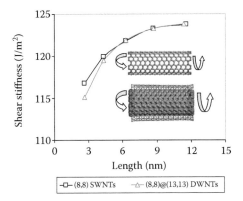

FIGURE 22.3 **(See color insert.)** Comparison of shear modulus between armchair DWNTs and SWNTs.

22.4 CONCLUSIONS

In-plane stiffness and shear modulus of zigzag and armchair DWNTs are calculated through molecular simulation via Materials Studio developed by Accelrys. Elastic rod theory is used to link the calculations of the strain energy and the corresponding second derivative with respect to the deformation from the molecular mechanics simulations to the properties to be estimated. The simulations show that the in-plane stiffness of zigzag (8,0) @ (17,0) and (8,8) @ (13,13) DWNTs are length-dependent, and the asymptotic values are about 375 J/m^2 and 377 J/m^2, respectively. In addition, the shear modulus of the armchair (8,8) @ (13,13) DWNTs are length-dependent as well, and the asymptotic value is about 124 J/m^2. These length-dependent material properties approach their asymptotic values for tubes longer than 12 nm. Molecular simulations also identify the lower stiffness of DWNTs compared with SWNTs. van der Waals effect between the two walls of DWNTs is discussed and found to be attributed to the weakened properties of DWNTs. The difference of material properties between DWNTs and SWNTs diminishes at larger sizes, that is, armchair DWNTs longer than 6 nm. It is expected that the application of molecular simulations, especially via Materials Studio, will uncover more accurate predictions of the material properties, especially the chirality-dependent properties, of CNTs, as well as their mechanical behaviors.

ACKNOWLEDGMENTS

This research was undertaken, in part, thanks to funding from the Canada Research Chairs Program, the National Science and Engineering Research Council, and the Canada Foundation for Innovation. The support and kind service from Accelrys for the applications of the powerful Materials Studio are highly appreciated by the author for all the research findings on nanoscience in his group.

REFERENCES

1. S. Iijima. Helical microtubules of graphitic carbon. *Nature (London)* **354**, 1991, 56.
2. E. W. Wong, P. E. Sheehan and C. M. Lieber. Nanobeam mechanics: Elasticity, strength, and toughness of nanorods and nanotubes. *Science* **277**, 1977, 1971.
3. A. Krishnan, E. Dujardin, T. Ebbesen, P. N. Yianilos and M. M. J. Treacy. Young's modulus of single-walled nanotubes. *Phys. Rev. B* **58**, 1998, 14043.
4. J. P. Salvetat, G. A. Briggs, J. M. Bonard, R. R. Basca, A. J. Kulik, T. Stockli, N. A. Burnham and L. Forro. Elastic and shear moduli of single-walled carbon nanotube ropes. *Phys. Rev. Lett.* **82**, 1999, 944.
5. P. Poncharal, Z. L. Wang, D. Ugarte and W. A. de Heerm. Electrostatic deflections and electromechanical resonances of carbon nanotubes. *Science* **283**, 1999, 1513.
6. T. C. Chang, J. Y. Geng and X. M. Guo. Chirality- and size-dependent elastic properties of single-walled carbon nanotubes. *Appl. Phys. Lett.* **87**, 2005, 251929.
7. G. M. Odegard, T. S. Gates, L. M. Nicholson and K. E. Wise. *NASA/TM*-2002-211454, 2002.
8. Q. Wang. Effective in-plane stiffness and bending rigidity of armchair and zigzag carbon nanotubes. *Int. J. Solids Struct.* **41**, 2004, 5451.
9. Q. Wang, Q. K. Han and B. C. Wen. Length dependent mechanical properties of carbon nanotubes. *J. Comput. Theor. Nanosci.*, **5** (7), 2008, 1454–1457.
10. J. M. Gere. *Mechanics of Materials*. Belmont, CA: Brooks/Cole Thomson Learning, 2001.
11. B. I. Yakobson, C. J. Brabec and J. Bernholc. Nanomechanics of carbon tubes: Instabilities beyond linear response. *Phys. Rev. Lett.* **76**, 1996, 2511.
12. K. N. Kudin, G. E. Scuseria and B. I. Yakobson. C_2F, BN and C nano-shell elasticity by *ab initio* computations. *Phys. Rev. B* **64**, 2001, 235406.
13. Accelrys. Materials Studio. Available online at http://www.accelrys.com/products/mstudio/.
14. D. Rigby, H. Sun and B. E. Eichinger. Computer simulations of poly(ethylene oxide): Force field, PVT diagram and cyclization behaviour. *Polym. Int.* **44**, 1997, 311.
15. Q. Wang, W. H. Duan, K. M. Liew and X. Q. He. Inelastic buckling of carbon nanotubes. *Appl. Phys. Lett.* **90**, 2007, 033110.
16. D. Robertson, D. Brenner and J. Mintmire. Energetics of nanoscale graphitic tubules. *Phys. Rev. B* **45**, 1992, 12592.
17. S. Gupta, K. Dharamvir and V. K. Jindal. Elastic moduli of single-walled carbon nanotubes and their ropes. *Phys. Rev. B* **72**, 2005, 165428.
18. V. N. Popov and V. E. Van Doren. Elastic properties of single-walled carbon nanotubes. *Phys. Rev. B* **61**, 2000, 3078.

Index

Page numbers followed by f and t indicate figures and tables, respectively.

T - #0186 - 221019 - C15 - 234/156/19 - PB - 9780367382117